Mosquito Eradication

The Story of Killing *Campto*

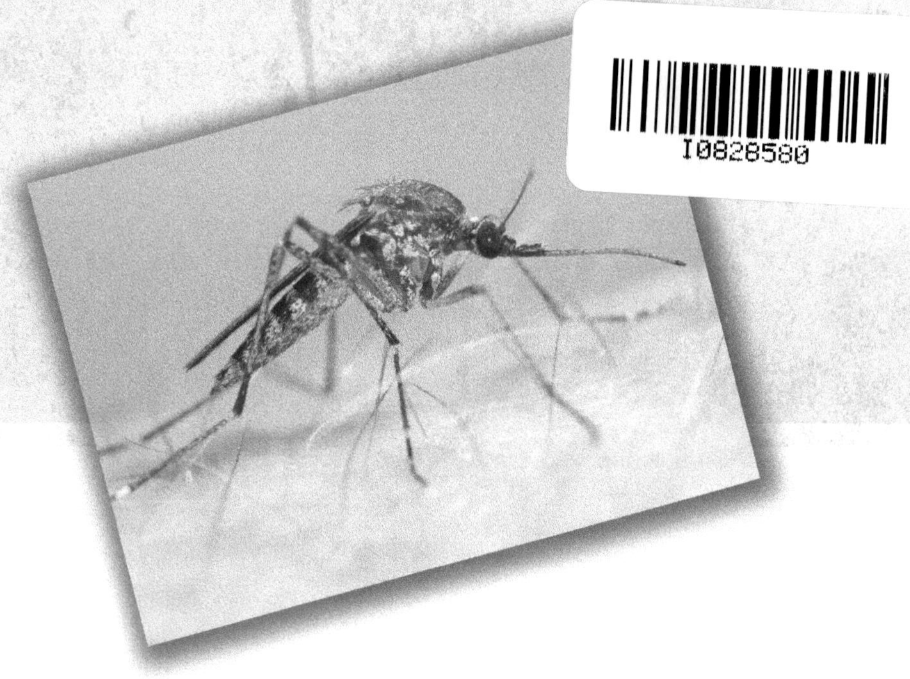

Editors: Brian H. Kay and Richard C. Russell

National Library of Australia Cataloguing-in-Publication entry

Mosquito eradication: the story of killing "Campto" /
edited by Brian H Kay and Richard C Russell.

9781486300570 (paperback)
9781486300587 (epdf)
9781486300594 (epub)

Includes bibliographical references.

Mosquitoes - Control - New Zealand.
Mosquitoes as carriers of disease - New Zealand.
Ross River virus.
Communicable diseases - New Zealand - Prevention.

Kay, Brian H. (Brian Herbert), editor.
Russell, Richard Charles, editor.

614.43230993

Published by
CSIRO PUBLISHING
36 Gardiner Road, Clayton VIC 3168
Private Bag 10, Clayton South VIC 3169
Australia

Telephone: [+613] 9545 8555
Local call: 1300 788 000 (Australia only)
Fax: +61 3 9662 7555
Email: csiropublishing@csiro.au
Website: www.publishing.csiro.au

Front cover: photo by Stephen L. Doggett
Back cover: map by Mark Disbury

Set in 10.5/12 Minion
Edited by Peter Storer Editorial Services
Cover and text design by James Kelly
Typeset by Thomson Digital
Printed by Ingram Lightning Source

Feb26_RP_ILS

Contents

Preface

> *'There is no such thing as partial success. It is either glorious success or dismal failure.'*
> *(Dr Fred Soper, Pan-American Health Organization and Master Mosquito Eradicator)*

In December 1998, residents of Westshore in the rural hinterland of Hawke's Bay on the east coast of the North Island of New Zealand (NZ) contacted government health authorities in Napier, an important regional centre with 54 000 people, to report an unusual abundance of day-biting mosquitoes at Westshore near the airport. Hawke's Bay is noted for its agriculture and particularly for its red wines.

The Australian southern saltmarsh mosquito, *Aedes camptorhynchus* (which we will call *Campto*), has never been previously recorded in NZ, but is known from its homeland as a significant pest species and vector of the arboviruses Ross River and Barmah Forest virus, which cause debilitating illness in humans.

NZ had no equivalent mosquito species and if *Campto* became further established and spread within NZ, it would present a serious health risk as well as becoming a nuisance pest affecting the quality of life of many communities. A case was put to demonstrate that a 'virgin soil' outbreak of Ross River virus was likely and such an occurrence, left uncontrolled, would necessitate the setting up of clinical and pathology services for arboviruses and generate the need for routine mosquito control.

The finding of *Campto* at this site galvanised the Ministry of Health (MoH) and was to be the genesis of the NZ Mosquito Industry. With no practical experience of saltmarsh species to call on, the MoH brought in Australian expertise to contribute to the Southern Saltmarsh Mosquito Technical Advisory Group (TAG). A short-term response programme using *Bacillus thuringiensis israelensis* (*Bti*), a benign microbe originally discovered in Israeli soil, to suppress breeding in some 650 or so hectares of habitat on the north-west side of Napier, as well as at a smaller (40 ha) site south of Napier, by aerial and ground delivery was instituted. A more substantial plan was developed based on using a slow release formulation of ProLink® XR-G granules, containing S-methoprene (a new age insect growth regulator) as the treatment agent. Take a deep breath and consider this:

- a country with no experience of mosquito control
- a target species that was an exotic import with no track record in NZ
- a species that had desiccation-resistant eggs that may lay around for months awaiting inundation by rain or tides, and did not all hatch at the same time
- an eradication plan that was reliant on unpredictable wet events to flush out the target species
- a treatment formulation that was untried in NZ and Australia, although it had been through one summer of trials in the USA
- considerable gaps in knowledge of the life history of the target mosquito.

Some would say that it was with a 'leap of faith' that the MoH endorsed the eradication plan and, with government backing, the eradication programme began in 1999. A government-funded programme to attack and eliminate the mosquito from the Hawke's Bay area was quickly instituted, but, over the next decade, the mosquito was found to have invaded and become established in other areas of NZ. These new infestations were dealt with in a systematic programme of delimitation surveys, intensive consultation with local stakeholders and then aerial and ground applications of S-methoprene. Again, the results were similar to that of Napier: that is, a collapse of the mosquito population once the treatments had begun.

All seemed rosy until 'Murphy' struck back and this time the infested site was one of the southern hemisphere's biggest natural harbours – the Kaipara, a remote windswept waterway with some 800 km of coastline. Identified habitat was first assessed at a heart stopping 22 000 ha, soon to be reassessed to around 2728 ha of positive habitat. This was a potential 'deal-breaker'; the successes of Napier, Porangahau, Mahia and Gisborne were gained over smaller and more accessible sites. The TAG determined that it was feasible to embark on an eradication programme based on the Napier model. The political will was found and the Kaipara eradication programme implemented. The budget was estimated to be around NZ$30 million, but in the end it cost around NZ$70 million, because extra infestations were detected.

The national eradication programme was to be further challenged by finding *Campto* in the Wairau wetlands and at Lake Grassmere in the north-east corner of the South Island in May 2004. This Aussie invader was now threatening livelihoods of vintners and purveyors of fine sauvignon blancs around Marlborough! Again, the TAG convened and they recommended that eradication was feasible, and once more the NZ Government 'bit the bullet' and endorsed an eradication plan for the new found *Campto* sites.

After 12 years of dedicated activity on the part of many, the NZ Government was able to proclaim final success on 1 July 2010 in eradicating this saltmarsh pest and vector mosquito – a world-first accomplishment. This book presents the story of the programme, with accounts of the different aspects and phases of the programme as related by those directly involved in the strategic, political, administrative, operational and scientific roles. Various technical problems and nasty surprises had to be coped with, and in an environmentally friendly manner in keeping with the NZ image! As technical advisors for different phases of the programme, and now as joint editors for this publication, we are proud to present what we hope you will find to be an interesting and enjoyable read about a signal success in mosquito eradication history, perhaps providing 'lessons' for others who may face similar challenges.

Brian Kay and Richard Russell (with thanks to JR Gardner)

THIS BOOK IS DEDICATED TO THE MEMORY OF

Professor Marshall Laird, BSc, MSc (Hons), PhD, DSc

An internationally recognised mosquito biologist from New Zealand

Marshall Laird was born in Wellington, in January 1923. During World War II he served in the Royal New Zealand Air Force as entomologist on the island of New Britain (Papua New Guinea), sparking his interest in mosquito and other aquatic fauna. He received his PhD from the Department of Zoology, Victoria University, Wellington in 1948, which started his life long career in biological sciences. Once he had completed his doctorate, he started to be recognised for the quality of his science, winning the 1951 Hamilton Memorial Prize for scientific or technical research carried out in New Zealand or the South Pacific Islands.

His early involvement with mosquitoes and their ecology persisted throughout his career. He had many years' field experience with mosquito habitats in Fiji and other parts of the Pacific, and later also in Asia, Africa and the Americas, culminating in his widely lauded book *The Natural History of Larval Mosquito Habitats*, published by Academic Press in 1988. Further, an enduring interest of his was the transport of mosquitoes and other insects of concern by aircraft. He was a vigorous advocate of aircraft disinsection to deter introductions of exotic pest and vector species, and was an adviser on such matters to the World Health Organization. In 1984, he edited a book published by Praeger, entitled *Commerce and the Spread of Pests and Disease Vectors*, on which Richard worked with him. He had earlier befriended Brian and Richard at the World Entomology Congress in Kyoto in 1980.

Unfortunately, although internationally prominent in his disciplines, he was not well known in New Zealand, because he spent most of his professional career overseas. His scientific career saw him working in such different places as the islands of the South Pacific, the (then) University of Malaya in Singapore, the World Health Organization in Geneva, Switzerland, and McGill University, Montreal and the Memorial University of Newfoundland, St Johns, both in Canada. Upon retiring from formal academic life in 1983, he returned to New Zealand and moved to Pahi in Northland, where he lived with his wife Elizabeth on a lifestyle block with an expanse of native bush. He maintained his scientific interests in mosquitoes as an Honorary Research Fellow with the Department of Zoology, University of Auckland, initiating surveys and publishing the results, and advising/consulting with interested parties on local and exotic species, which involved both of us.

In 1996, for his long and distinguished career, Marshall received the Medal of Honour from the American Mosquito Control Association – its highest award – and he greatly prized this as a prestigious accolade from his peers. He had promoted his long-term interest in sustainable mosquito control with a two volume book he edited with JW Miles, *Integrated Mosquito Control Methodologies* (Academic Press, 1983, 1985), and he certainly had an interest in the Southern Saltmarsh mosquito eradication programme that is the subject of this present publication. On hearing we had discovered an exotic species established in NZ, he was congratulatory, with a ringing: 'Well done!' Presciently, just a few years earlier, Marshall had organised for Gene Browne to study the use of *Bacillus thuringiensis israelensis* and methoprene products, believing that methoprene was a powerful tool for mosquito control; moreover, he strongly supported the push for its use in our eradication programme, and would have felt pleased when methoprene turned out to be the key to successful eradication. Sadly, Marshall, a remarkable New Zealand scientist, passed away on 30 September 2007 before the eradication of *Aedes camptorhynchus* from New Zealand was completed.

Brian Kay and Richard Russell (with thanks to Gene Browne)

Abbreviations

Ae.	*Aedes*
Bti	*Bacillus thuringiensis israelensis*
Campto	*Aedes camptorhynchus,* Southern Saltmarsh mosquito
Cx.	*Culex*
GIS	Geographic Information System
GPS	Global positioning System
ha	hectare
kg	kilogram
L	litre
MAF	Ministry of Agriculture and Forestry (now the Ministry of Primary Industries)
MoH	Ministry of Health
m	metre
NZ	New Zealand
NZ BioSecure	New Zealand BioSecure, a division of Southern Monitoring Services Ltd
TAG	Southern Saltmarsh Mosquito Technical Advisory Group

List of contributors

Ms Val Aldridge
15 St Johns Tce, Tawa, 5028, Wellington, NZ. Email: vlaldrid@anet.co.nz

Dr Nigel Beebe
Senior Lecturer, School of Biological Sciences, University of Queensland, St Lucia, Qld 4067, Australia. Email: n.beebe@uq.edu.au

Dr Michael Brown
Managing Director, Australian Wholesale Chemical Technologies, PO Box 984, North Lakes, Qld 4509, Australia. Email: drmichaelbrown@netspace.net.au

Dr Gene Browne
Bioscientific Solutions Ltd, 53 Wingate Street, Avondale 0600, Auckland, NZ. Email: gene@biosci.co.nz

Mr Mark Bullians
Manager Surveillance and Incursion Investigation Plants and Environment Compliance and Response, Ministry for Primary Industries, PO Box 2095, Auckland, NZ. Email: Mark.Bullians@mpi.govt.nz

Ms Rachel P Cane
Mosquito Consulting Services (NZ), 48 Bethels Road, Springston, NZ. Email: rachel.cane@mcspty.com

Mr Steve Crarer
17 Gladstone Road, Bluff Hill, Napier 4110, NZ. Email: steveandhopec@gmail.com

Mr Mark Disbury
National Manager, Mosquito Consulting Services (NZ), 2-4 Bell Road South, Lower Hutt, PO Box 38-328, Wellington Mail Centre, NZ. Email: Mark.Disbury@MCSPty.com

Mr Henry Dowler
Principal Consultant, HankStar Consulting Ltd., PO Box 5141, Springlands, Blenheim 7241, NZ. Email: Henry.Dowler@hankstar.co.nz

Andrew Forsyth
Team Leader, Public Health Legislation, Public Health Group, Ministry of Health, PO Box 5013, Wellington, NZ. Email: Andrew_forsyth@moh.govt.nz

Dr Ruth Frampton
Director, Critique Ltd, C/- 587 Springston-Rolleston Road RD 5, Christchurch, NZ. Email: framptonr@critiquelimited.co.nz

Mr John Gardner
Senior Advisor, Environmental and Border Health, Public Health Group, Ministry of Health, PO Box 5013, Wellington, NZ. Email: john_gardner@moh.govt.nz

Mr Steve Garner
Principal, Monarch Consulting, PO Box 25 042, Panama St, Wellington 6146, NZ. Email: garner@monarch-consulting.co.nz

Ms Helen Gear
Consultant, In Gear Global Ltd, Glen Garioch, 53a Coroglen Rise, Pukerua Bay 5026, Wellington, NZ. Email: helen@ingearglobal.com

Mr Ian Gear
Director, In Gear Global Ltd, Glen Garioch, 53a Coroglen Rise, Pukerua Bay 5026, Wellington, NZ. Email: ian@ingearglobal.com

Ms Sally Gilbert
Manager, Environmental and Border Health, Public Health Group, Ministry of Health, PO Box 5013, Wellington, NZ. Email: sally_gilbert@moh.govt.nz

Prof Travis R. Glare
Professor of Applied Entomology, Bioprotection Research Centre, PO Box 85084, Lincoln University, Lincoln 7647, NZ. Email: Travis.Glare@lincoln.ac.nz

Mr Bryn Gradwell
12 Tara Place, Snells Beach 0920, Rodney, Auckland, NZ. Email: bryngradwell@xtra.co.nz

Prof. Brian Kay
Group Leader, Mosquito Control Laboratory, Queensland Institute of Medical Research, Private Bag 2000, Post Office Royal Brisbane Hospital, Brisbane, Qld 4029, Australia. Email: brian.kay@qimr.edu.au

Mr Ruud Kleinpaste
Bugman, 86 Milns Road, Halswell, Christchurch, NZ. Email: Ruud@Bugman.co.nz

Dr Graham Mackereth
Coordinator, Health Intelligence Team, ESR NCBID-Wallaceville, PO Box 40158, Upper Hutt 5140, NZ. Email: Graham.Mackereth@esr.cri.nz

Mr Shaun Maclaren
Environmental Manager (SMS), 1135 Scenic Drive (N), Swanson, Auckland 0614, NZ. Email: maclarenshaun@gmail.com

Dr Andrew McFadden
Veterinary Epidemiologist/Principal Adviser, Investigation and Diagnostic Centre, Ministry for Primary Industries, PO Box 40 742, Upper Hutt, NZ. Email: Andrew.McFadden@mpi.govt.nz

Mr Darryl McGinn
Director and Consultant, Mosquito Consulting Services Pty Ltd, PO Box 339, Mt Ommaney, Brisbane, Qld 4074, Australia. Email: darryl.mcginn@mcspty.com

Dr Maureen O'Callaghan
Principal Scientist, AgResearch, Private Bag 4749, Christchurch 8140, NZ. Email: Maureen.O'Callaghan@agresearch.co.nz

Prof. Scott Ritchie
Faculty of Medicine, Health and Molecular Sciences, James Cook University, PO Box 6811, Cairns, Qld 4870, Australia. Email: scott.ritchie@jcu.edu.au

Prof. Richard Russell
Professor of Medical Entomology, Sydney Medical School, University of Sydney, Sydney, NSW 2006, Australia. Email: richard.russell@sydney.edu.au

Dr Robert Sanson
AsureQuality, PO Box 585, Palmerston North 4440, NZ. Email: Robert.Sanson@Asure-Quality.com

Ms Monica Singe
Technical Manager, Southern Monitoring Services Ltd – NZ BioSecure, 515 Whitehills Road, RD1 Silverdale 0994, NZ. Email: monica.singe@smsl.co.nz

Mr Matthew Stone
Director Animal and Animal Products, Standards Branch Ministry for Primary Industries, PO Box 2526, Wellington, NZ. Email: matthew.stone@mpi.govt.nz

Mr David Sullivan
Zocor Inc., PO Box 169, 98 N Kennedy, Belgrade MT 59714, USA. Email: zanusco1@msn.com

Miss Jessica Taylor
Mosquito Consulting Services, 2-4 Bell Road South, Gracefield, Lower Hutt 5010, PO Box 38-328, Wellington Mail Centre 5045, NZ. Email: jess.taylor@mcspty.com

Dr Barbara Thomson
Senior Scientist, Food Safety Programme, ESR (Institute of Environmental Science and Research Ltd), Christchurch Science Centre, PO Box 29-181, Christchurch 8540, NZ. Email: Barbara.Thomson@esr.cri.nz

Mr Noel Watson
Health Protection Officer, Public Health Unit, Planning & Performance, Hawke's Bay District Health Board, PO Box 447, Napier, 4110 NZ. Email: Noel.Watson@hawkesbaydhb.govt.nz

Dr Craig Williams
School of Pharmacy & Medical Sciences, University of South Australia, P6-13 City East Campus, GPO Box 2471, Adelaide SA 5001, Australia. Email: craig.williams@unisa.edu.au

Mr David Yard
Response Manager, Compliance and Response Branch, Ministry for Primary Industries, PO Box 2526, Wellington, NZ. Email: David.Yard@mpi.govt.nz

Acknowledgements

The editors are grateful to Pauline Fraley for administrative help and to Stephen Doggett for the fine image of *Aedes camptorhynchus*. The book could not have been compiled without the enthusiasm and diligence of all co-authors, but the extra assistance provided by Sally Gilbert and Mark Disbury deserves special mention.

Locations in New Zealand mentioned in this book

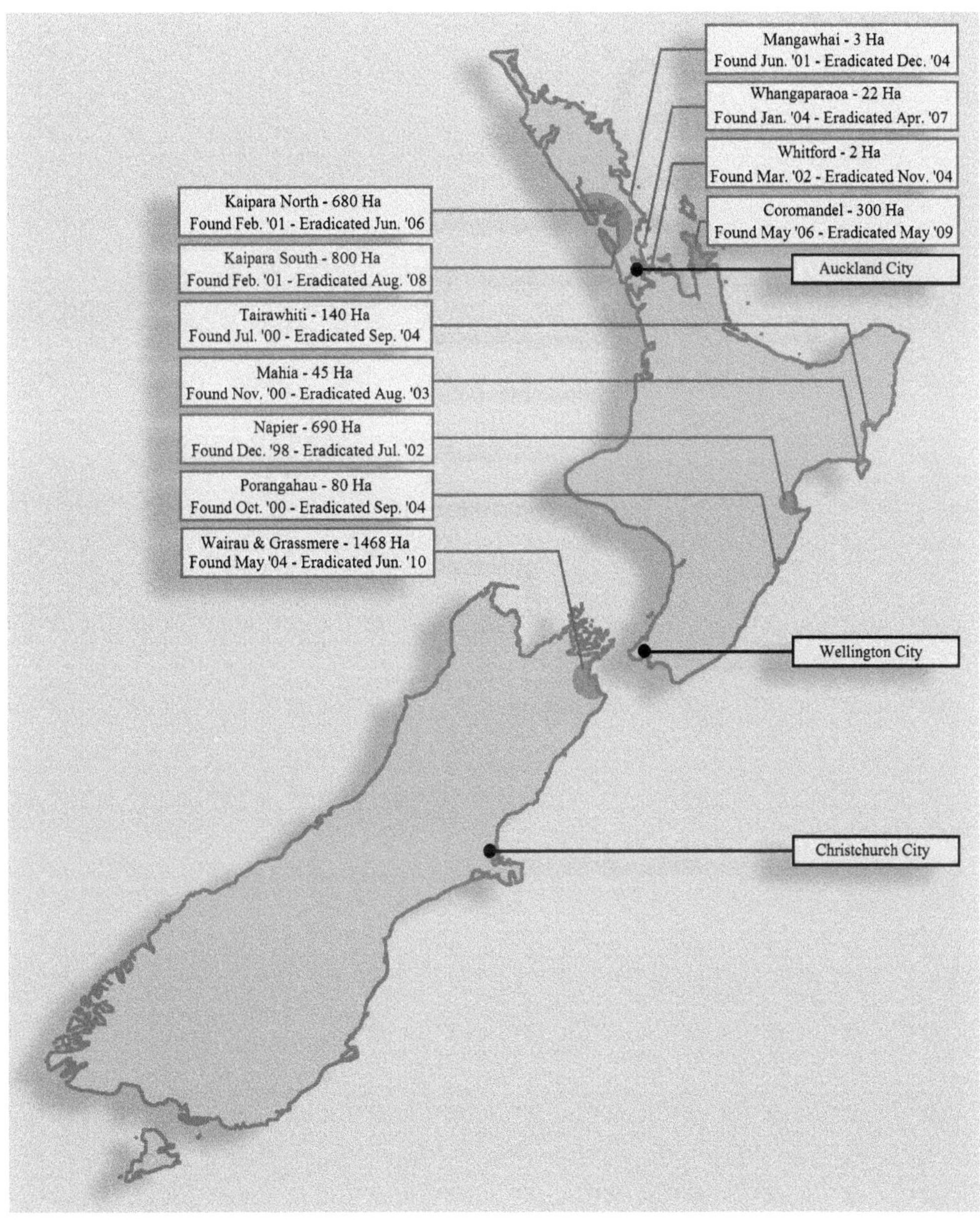

1

How it was before: historical perspectives on receptivity and risk

Brian Kay, JR Gardner and Gene Browne

NZ has a wealth of natural beauty, wonderful natural tourist attractions and relatively few dangerous or annoying insect pests. You can breathe in without fear of inhaling bush flies, as can happen in Western Australia, for example. However, a notable exception is in Milford Sound in the Fiordland National Park at the bottom of the South Island, where black flies can be so numerous and their biting so intense it seems that they could carry you away. Therefore, Milford Sound aside, it was a real shock when, in December 1998, hordes of vicious day-biting mosquitoes threatened to exsanguinate the residents of Hawke's Bay on the east coast of the North Island. Holidays in this popular destination were cancelled and some workers refused to enter forests for fear of attack. There was even a media story about a 'Pincushion Boy'.

Truly, Aotearoa, the Land of the Long White Cloud had potential to morph into the Land of the Long Black Cloud because the Southern Saltmarsh mosquito *Aedes camptorhynchus* (*Campto*) had arrived from Australia. How did it get to and establish in Hawke's Bay, and to other localities? We speculate, but don't really know.

The local public health unit responded to the residents' loud complaints and specimens of the pest were collected and sent to Gene Browne at the University of Auckland for identification. For some leading biosecurity practitioners, Christmas holidays were also cancelled as fears of a new invader crossing the border were confirmed. By 12 January 1999, Richard Russell in Sydney had confirmed the identification.

Why were we concerned?

The mosquito and its compatibility

First, *Campto* is known mostly from just south of Sydney, through Victorian intertidal areas, into Tasmania, South Australia and south-western Western Australia. Ballard and Marshall (1986) demonstrated that *Campto* was capable of transmitting Ross River virus in laboratory studies and it has been connected with numerous epidemics in all of these areas (Kelly-Hope *et al.* 2004).

This mosquito thrives in relatively cool climates and can become predominant in the southern part of its Australian range. It is particularly numerous in the coastal woodland belts for up to 49 km inland through eastern Victoria. Although it primarily breeds in brackish swamps, the fact that it also can colonise fresh water, and has even been located from the bottom of rabbit holes (Lee *et al.* 1984), has left us with some challenging questions. Would we really have to survey the entire coastline as well as every farm dam and waterlogged rabbit burrow in NZ?

In Victoria, it breeds all year round. In the Murray River delta, South Australia, Howard (1973) noted the apparent erratic distribution of larvae in pools and related this to changing pool levels and hatching requirements of *Campto* eggs. This could result in continuous production of *Campto* through late autumn, winter and spring. To add to this nightmare scenario, we had to consider the viability of desiccation-resistant eggs that may lie dormant in mud and vegetation for months. But for how many months? How long would treatment be necessary to get rid of the invaders? The major clue we had was from Mani Pillai's 1962 PhD thesis (Pillai 1962) in which eggs of different species, including *Campto,* were kept at different humidities for months to determine their viability. It seemed that ~9 months and three floodings would account for most eggs.

The next problem to consider was that its sister species, *Aedes vigilax,* is known to disperse over large distances. *Campto* is permanently established in the Mildura area along the Murray River, some 363–400 km from the Victorian coastline. Where would it go in NZ? We were about to find out.

Second, the climate of NZ (Table 1.1) is dominated by westerly air flows, its oceanic environment and its central mountainous areas, which modify weather systems as they pass eastwards. Northerly intrusions of warm, moist air move southwards into NZ

Table 1.1. Demographic data and daily average temperatures for 17 major population centres in NZ (modified from Kay 1997)

City	Latitude (°S)	Population (1991)	Growth projection*	Daily average temperature (°C)	
				Mid-Summer (max)	Mid-Winter (min)
Auckland	36.55	885 571	+2.2 to 2.8	22.8	8.2
Wellington	41.17	325 682	+1.1	20.3	5.9
Christchurch	43.33	307 179	+2.0	22.6	1.6
Hamilton	37.46	148 625	+1.5	23.0	0.0–3.9
Dunedin	45.52	109 503	+1.1	18.9	3.1
Palmerston North	40.20	70 951	+1.5	20.1–22.9	0.0–3.9
Tauranga	37.42	70 803	+3.3	23.7	4.9
Hastings	39.39	57 748	+0.5	24.0	4.0
Rotorua	38.07	53 702	+0.9	23.0	0.0–3.9
Napier	39.29	52 468	+0.6	24.5	4.2
Invercargill	46.26	51 984	–0.2	18.5	1.1
New Plymouth	39.03	48 519	+0.6	21.7	5.4
Nelson	41.18	47 391	+2.0	22.2	1.3
Whangarei	35.43	44 183	+1.1	23.9	4.0–7.9
Wanganui	39.56	41 213	+0.7	20.1–22.9	5.4–5.9
Gisborne	38.41	31 484	+0.2	24.8	4.5
Timaru	44.23	27 637	+0.2	20.1–22.9	< 0.0
Total (NZ)		3 592 400	+1.4		

*Based on estimated population change 1994–1995.

latitudes during summer. Temperature extremes are mainly confined to places east of the main ranges (Statistics NZ 1996), but some are equivalent to localities in Australia where Ross River virus occurs.

It is noted that Whangarei (35.43°S) is at approximately the same latitude as Batemans Bay, New South Wales, Australia (35.45°S), where Ross River virus has been shown to be periodically active in *Aedes vigilax* and *Aedes camptorhynchus* (Russell *et al.* 1991). In this regard, Kay (1997) concluded that

> 'Aedes camptorhynchus *could be considered as a more likely introduction from Australia than Aedes vigilax because it predominantly inhabits southern Australia.*'

Auckland (36.55°S) could be compared with Bega, New South Wales (36.41°S; average temperature, mid-summer 20.8°C, mid-winter 9.3°C), whereas Wellington (41.17°S) equates to the latitude of Launceston, Tasmania (41.25°S; 18.2°C, 7.2°C). Christchurch (43.33°S) is further south than Hobart (42.52°S). Isolated cases of Ross River virus disease (epidemic polyarthritis) have been notified on occasion in Tasmania as far south as Hobart, where *Campto* is common, and a major vector, *Culex annulirostris*, was recorded from collections made at Coles Bay on the east coast of Tasmania during 1985 (McManus and Marshall 1986) but not subsequently (McManus *et al.* 1992). These suggest that the southern NZ localities are under reduced threat, but climatic conditions may change from year to year even to facilitate annual arbovirus (**ar**thropod-**bo**rne viruses; i.e. carried by insects such as mosquitoes) activity.

Possible virus impact

The risk of introducing an arbovirus to the NZ population depends on the inbound traffic (e.g. people returning from overseas travel) and, of course, the presence in NZ of sufficient numbers of a competent mosquito vector. Because the human residents are not generally exposed to arbovirus infection, they lack protective antibodies. Therefore, the scene was set for what is known as a 'virgin soil' epidemic with widespread infection, perhaps as high as 1600 cases per 100 000 people as occurred in south Western Australia in the 1995–96 summer. The dramatic nature of what the late Professor Neville Stanley called 'Australia's gift to the Pacific' was evidenced in 1979–80 where many of ~50 000 Ross River victims in Fiji, Samoa, Cook Islands and elsewhere required emergency care in hospitals and clinics, thus overloading the health systems. In this case, the major vertebrate hosts in the transmission cycle were humans. This was the risk for NZ if *Campto* established, aided by infection in other animal hosts. Furthermore, the lack of awareness of the symptoms of epidemic polyarthritis (and arbovirus illness generally), combined with the lack of rapid diagnostic services for these illnesses, suggested that an arboviral outbreak in humans could proceed unchecked for quite some time, enabling the infection to spread widely. Ross River virus most commonly causes a debilitating polyarthritis, rash and fever, but is not fatal.

NZ presently has four known arboviruses: Whataroa (Maguire *et al.* 1967) isolated from the indigenous mosquitoes *Culiseta tonnoiri* and *Culex pervigilans*; Johnston Atoll (Austin 1978) from *Ornithodoros capensis* ticks off seabirds; Saumarez Reef virus from *Ixodes eudyptidis* ticks off seabirds; and an unidentified Hughes group virus from *Ornithodoros capensis* from seabirds (Austin 1984). Based on serological evidence in fowl and human sera from Westland and human sera from Tauranga (Maguire and Miles 1960; Hogg *et al.* 1963), an unidentified flavivirus that causes human disease would also seem to be present.

Whataroa virus is named after a location on the west coast of the South Island (Maguire *et al.* 1967; Miles 1973), where it survives in cycles between birds and cold-hardy mosquitoes. Mean winter and summer temperatures are 5°C and 16°C, respectively. Average temperatures during mid-summer greater than 20°C are conducive for transmission within an extrinsic incubation period that is not likely to be greater than the life of the mosquito itself. In laboratory experiments, *Culex tonnoiri* and *Aedes australis* transmitted Whataroa virus to suckling mice after 10 and 17 days, respectively (Miles 1973). In general, warmer average temperatures in the top half of the North Island would be more likely to facilitate faster transmission. In Tasmania, serological evidence of Ross River virus infection was common in Bennett's wallaby, brush-tailed possums and a range of other marsupials (McManus and Marshall 1986). It is considered that Ross River virus is probably enzootic in marsupials in Tasmania.

From this, we concluded that the presence of *Campto* increased the risk of outbreaks in NZ and most likely in the North Island. Subsequently, Kay and Jennings (2002) discovered that Ross River virus favoured more moderate temperatures for its replication in mosquitoes, with more efficient transmission at 18°C and 25°C than at 32°C. The extrinsic incubation periods, or time for the virus to mature in the mosquito and be ready for transmission, were 5, 4, and 3 days, respectively. This is unexpectedly quick, and well within the lifespan of your average mosquito. This study was done with *Aedes vigilax*, the northern saltmarsh counterpart of *Campto*, and so it supported the risk assessment based on known epidemiological data.

Vertebrate hosts

Apart from temperatures, this conclusion was supported by volume of traffic, cargo and passengers, by air and by sea, and by established pathways to and from Australia and into the Pacific. During 1995, 389 581 Australians visited NZ and this comprised 29% of total visitations. This indicated that a large pool of people visiting NZ were potentially or actually viraemic for Ross River, Barmah Forest and other arboviruses. Of course, it is also possible that NZ thoroughbred racehorses (notorious raiders of Australia's holy grail, the Melbourne Cup), could bring Ross River virus home from Victoria after the early November race meeting when summer had arrived (Kay *et al.* 1987).

Kelly-Hope *et al.* (2002) estimated that, of short-term Kiwi visitors to Queensland and Queensland residents to NZ departing from tropical Cairns or subtropical Gold Coast and Brisbane, ~100 could be expected to have subclinical or clinical infections each year and that more than 60% entered through Auckland. Her studies quantified the risk and demonstrated that, with *Campto* throughout NZ, it really was a numbers game. Of course, this had not been defined so elegantly when this all began.

Third, unlike dengue viruses, which mainly rely on a human–mosquito transmission cycle, the main vertebrate hosts of Ross River virus in Australia are considered to be kangaroos, wallabies and other marsupials. The prevalence and distribution of marsupials (primarily brush-tailed possums) in NZ provided further concern. Later, Ann Marie Boyd *et al.* (2001) demonstrated that brush-tailed possums were in fact very efficient hosts of Ross River virus and could spread the virus to feeding mosquitoes, but this was not the case for Barmah Forest virus, also known to be connected with *Campto*.

The brush-tailed possum *Trichosurus vulpecula,* originally liberated near the southernmost city of Invercargill around 1837–1840 in a bid to set up a fur industry, is now widely distributed throughout NZ (Johns and MacGibbon 1986). Because possums are a vector for bovine tuberculosis and destroyers of native ecosystems, they are the subject of intensive national control programmes in agricultural and high-value nature conservation areas. However, despite annual control expenditure approaching NZ$120 million (a February 2000 Parliamentary briefing stated that possum control for conservation and bovine tuberculosis purposes and related research currently requires an ongoing annual expenditure by the Crown of NZ$42.7 million, and by the private sector of NZ$74.8 million), it has been reported that the total possum population may be 60 million. Less agile possums are also subject to road 'accidents', analogous to what sometimes happens to the introduced and hated Cane toad, *Bufo marinus*, in Australia.

Six species of wallabies occur in NZ as discrete populations: Bennett's wallaby, *Macropus rufogriseus* (Hunter Hills behind Waimate to Oamaru, South Island); swamp wallaby *Wallabia bicolor,* Tammar wallaby *Macropus eugenii,* black-striped wallaby *Macropus dorsalis,* brush-tailed rock wallaby *Petrogale penicillata* and the Parma wallaby *Macropus parma* (Kawau Island); brush-tailed rock wallaby (Raugitoto and Motutapu Islands) and Tammar wallaby (Rotorua area). Control operations have been in place since 1940.

Rabbits, sheep, pigs, horses, cattle, chickens and other birds are capable of low to moderate viraemias after inoculation with Ross River virus (Kay *et al.* 1986). Dogs remain 'man's best friend' because they don't spread the virus and actually may act to divert mosquito feeding and transmission (Boyd and Kay 2002). Cats also do not contribute to the transmission cycle.

Some of the animal species named above were found infected in the Cook Islands and American Samoa (Rosen *et al.* 1981; Tesh *et al.* 1981) although humans were the primary hosts (and victims). Stock densities, particularly near urban areas (or international airports), could be considered to increase risk of Ross River virus. For example, following inoculation, 100% of sheep developed viraemias of $\log_{10}$ 3.8 ± 1.1 and suckling mouse LD_{50}/mL for an average of 2.4 ± 1.1 days (Kay *et al.* 1986). However, Ross River virus failed to establish in the Pacific Islands, notably Fiji and American Samoa, when it was introduced via viraemic travellers in 1979–80 (Aaskov *et al.* 1981). On the basis of genetic analysis, Sammels *et al.* (1995) demonstrated that all seven Pacific isolates belonged to the south-eastern Australian genotype and therefore there was nothing different about them. Perhaps, it could be hypothesised that failure to establish was because these Pacific Island countries lacked marsupial populations, although it was also established that humans develop viraemias high enough to infect blood-sucking mosquitoes (Rosen *et al.* 1981).

Invasive species in NZ

Humans have played a key role in transporting invasive species into NZ, and also propagating and dispersing them, thereby providing repeated introductions and establishing multiple populations. Most accidental introductions can be connected with agricultural and industrial activities, such as the importation, sale and distribution of seeds, soil (usually as a contaminant) and machinery. Surprisingly, most of NZ's animal pests are intentional introductions, most notably possums, rats, rabbits, stoats, and feral goats, deer and pigs.

Over 19 000 plants have also been introduced, including gorse, which naturalised in 1867, broom in 1872, heather in 1910, old man's beard and Kahili ginger in 1940, *Hydrilla* in 1963 and *Spartina* in 1981 (Owen 1998). 1987 saw the marine invasion of the exotic Asian kelp (Sinner *et al.* 2000). The Department of Conservation recognises more than 240 naturalised plant species as ecological weeds, and it is estimated that two environmental weeds have naturalised each year since the 1860s.

Other established exotic animals include white-tail spiders, the red imported fire ant (eradicated), Argentine ant, Varroa bee mite, white spotted tussock moth (eradicated), Australian painted apple moth (eradicated), black widow spiders and the springbok mantid (Butcher 2002). The cost to the NZ economy of the impact of such animals is high. The Varroa bee mite has cost at least $10 million and government estimates the cost may rise to $1.5 billion over 35 years. This excludes the impact on the $1 billion per annum fruit industry. Over $12 million was spent on the white spotted tussock moth (Butcher 2002).

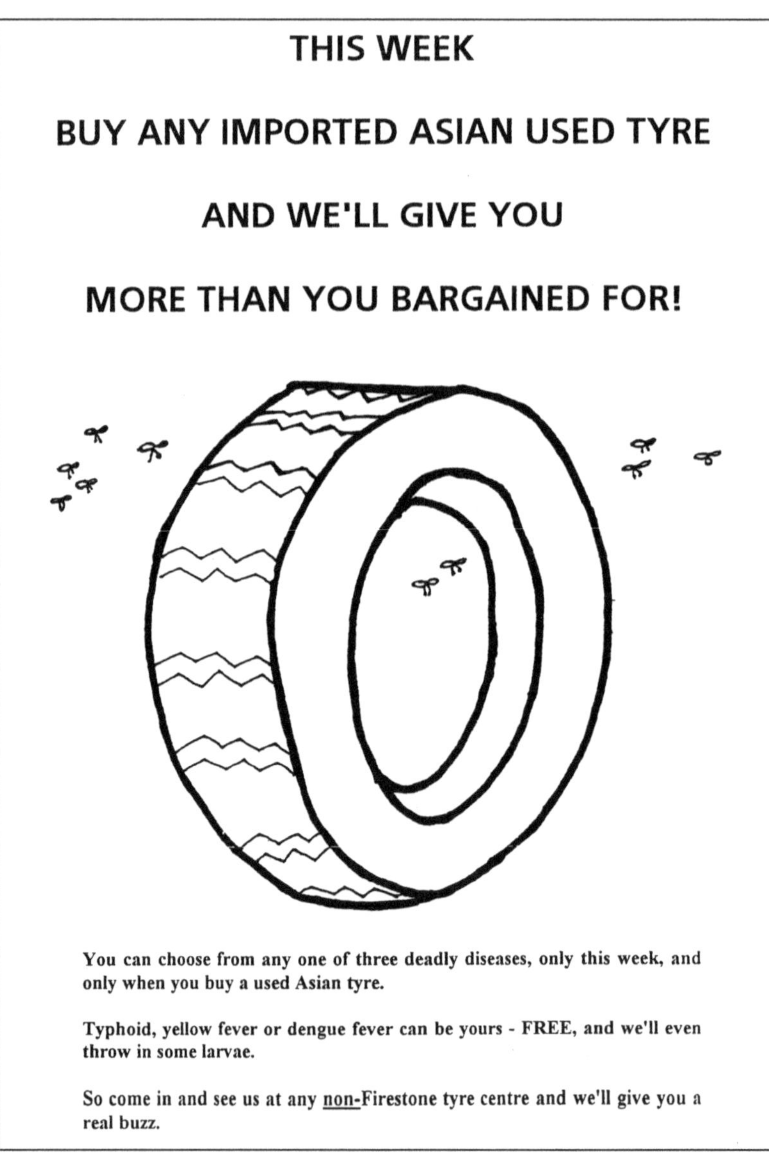

Figure 1.1 A flyer distributed by a tyre company recognised the risk of importing exotic mosquitoes into New Zealand in used tyres. (Reproduced from Laird *et al.* (1994) with permission from the *Journal of American Mosquito Control Association*)

No wonder, given these past experiences, NZ maintains one of the strictest biosecurity border control systems in the world. NZ gives a very high priority to protecting its natural ecosystems, people and economically important agricultural and horticultural industries.

Mosquitoes are frequent invaders and NZ has been no exception. To date, 28 exotic mosquito species have been confirmed as breaching the NZ border (Laird 1951; Graham 1939; Farr 1999; Mark Disbury *pers. comm.*), some on more than one occasion (see Chapter 13).

To date, only *Cx. quinquefasciatus, Ae. australis* and *Ae. notoscriptus* have been recorded as being established and, from 1998, *Campto. Culex quinquefasciatus* probably entered via the water casks of 1830s American whalers; *Aedes notoscriptus,* recognised by Miller in 1920 around Auckland and also intercepted by Graham in 1929 from two steamships from Sydney, has now established as far south as Christchurch on the South Island; and *Ae. australis,* found in oceanic rock pools at the southern end of the South Island, possibly hitched a ride via timber vessels from Tasmania during the 1960s (Laird 1990, 1995). At least six species – *Ae. aegypti, Cx. annulirostris, Ae. albopictus, Ae. japonicus* and *Tripteroides* spp. – have been intercepted either on aircraft or within imported used vehicle tyres (Laird *et al.* 1994) (Figure 1.1).

An example of an actual incursion response involving three exotic mosquito species over a 3-day period from 15–19 March, 1999 is described in Box 1.1.

Ships are capable of transporting large numbers of mosquito larvae long distances. Even though they may emerge into adults when still at sea, containers, holds, vehicle bodies, machinery and tyres may offer some degree of protection and the mosquitoes may not be blown away into the ocean. Air traffic is fast, but mosquito hitchhikers need to negotiate the rigours of aircraft disinsection. Some may do so in cargo such as 'lucky bamboo' plants and even in aircraft wheel housings; however, fewer usually arrive by air than by sea. Farr (1999) listed 682 dead mosquito specimens collected from disinsected aircraft originating from the Pacific Rim and arriving in NZ over a 1-year period.

Box 1.1

The scenario was that *Ae. japonicus* was found at the Ports of Auckland after a container was unloaded from a ship on 15 March 1999. The ship was further inspected and subsequently cleared by local health protection officers, as were surrounding ships. One of these surrounding ships then continued on its route to Tauranga where a half open container of wet tyres was unloaded, revealing a second exotic species, *Ae. albopictus*. Although the ship was requested to remain in port, it had already left, making its way through to Napier. Upon docking, the ship was boarded immediately and searched thoroughly by a select group of investigators. Deep within the bow of the ship, a third exotic mosquito, an adult *Cx. annulirostris,* was found.

During the discovery of *Cx. annulirostris* at the Napier port, the endemic *Cx. pervigilans* was also discovered on board the ship. Because this species is only found in NZ, and the ship completes a strict Australia, Papua New Guinea, NZ route, the mosquito could only have gained entry at another NZ port (either Auckland or Tauranga). Fortunately, all the mosquitoes were eliminated.

Early surveillance and international obligations

Dr Marshall Laird (1995) nominated 1902 to 1920 as the first phase when Dr Ernest Robinson reasoned that, since the NZ Government had set up a 'properly organised Department of Public Health, control of malaria will be easily stopped'. As it was now realised that malaria was carried by *Anopheles* mosquitoes rather than swamp miasmas (bad air as in **mal**aria), concern related to their introduction on board ships from tropical Australia or Melanesia. In 1916, a Scottish entomologist David Miller, was seconded from the then Department of Agriculture to the Department of Health as the 'Government Entomologist'. His remit was whether the Yellow Fever mosquito or any malaria carrying anophelines had established in the warmer parts of the North Island. Miller completed his survey by horseback at the end of summer 1918–1919 and his negative findings provided some reassurance as large numbers of malarious troops were returning home after World War I.

The second phase up until World War II featured the formation of the Auckland Mosquito Control Committee (later the Auckland Mosquito Research Committee) for which Dr David Graham was the research officer. He discovered the arrival of *Ae. notoscriptus* on board two steam ships from Australia in 1929, and documented the finding of several thousand *Cx. annulirostris* larvae and pupae in barrels in a hold of a vessel from Fiji. His activities in identifying the mosquito fauna of NZ, including the presence of the cosmopolitan *Cx. quinquefasciatus*, had emphasised the danger of foreign invaders.

Phase 3 saw Japanese entry into World War II and the despatch of armed service personnel into tropical islands in the Pacific. From this arose problems with dengue, filariasis and malaria. Growing arrivals of military flights were usually subjected to disinsection with insecticides upon arrival. The Royal NZ Air Force retained responsibility for disinsection of inbound aircraft until 1950 when the numbers of military arrivals had dwindled and the Plant Quarantine Service of the Department of Agriculture assumed control. In 1960, the World Health Organization in Geneva considered disinsection procedures but by 1980, this was less well practiced due to growing passenger resentment. In 1987, the NZ Government considered this and in 1988, concluded 'that aircraft disinsection should remain but be designed to minimise inconvenience to passengers'. The MoH also specified that 'clean and dry tyres' be accepted as imports, but, although it is easy to spot residual water, it is not so easy to see little poppy seed-like eggs sticking to the insides of dark tyre casings.

The 1988–1989 survey, in phase 4, organised by Grant Martindale from the MoH, was more comprehensive in examining 2304 mosquito samples from 11 different categories of habitats, including swamps. It covered the North Island from North Cape (34.23°S) to the southerly point in Auckland province (38.40°S) but did not detect any new invaders (Laird 1989, 1990). The emphasis remained with international cargo, especially tyres. However, in 1992–1993, *Ae. albopictus* and *Ae. japonicus* were detected from used tyre importations off ships from Japan (Laird *et al.* 1994).

During 1994, the Institute of Environmental Science and Research Ltd (ESR) subcontracted Dr Philip Weinstein to produce a report for the MoH entitled *The Real and Potential Risks of Arboviral Disease in New Zealand* (Weinstein 1994). The report considered that maintenance of regular surveys was essential to exclude introduced mosquito species and that contingency plans should be developed in the likelihood that such an event would happen. Weinstein *et al.* (1995) subsequently considered that there was a real risk of the introduction of Australian arboviruses into NZ and Maguire (1994) provided serological

Box 1.2: *International Health Regulations 1969 – Article 19*

'1. Every port and the area within the perimeter of every airport shall be kept free from Aedes aegypti in its immature and adult stages and the mosquito vectors of malaria and other diseases of epidemiological significance in international traffic. For this purpose active anti-mosquito measures shall be maintained within a protective area extending for a distance of at least 400 metres around the perimeter.
2. Within a direct transit area provided at any airport situated in or adjacent to an area where the vectors referred to in paragraph I of this Article exist, any building used as accommodation for persons or animals shall be kept mosquito-proof.
3. For the purposes of this Article, the perimeter of an airport means a line enclosing the area containing the airport buildings and any land or water used or intended to be used for the parking of aircraft.
4. Each health administration shall furnish data to the Organization once a year on the extent to which of its ports and airports are kept free from vectors of epidemiological significance in international traffic.'

International Health Regulations 1969 – Articles 70 and 71

The regulations stipulate that a ship be regarded as 'suspected' if it has left an infected area less than 6 days before arrival or, if arriving within 30 days of leaving such an area, *Aedes aegypti* or other vectors of yellow fever are found on board.

In the NZ context, it is considered that 'suspected' should be those vessels which have arrived from ports known to have *Aedes aegypti* or *Aedes albopictus* at least (extension of this could be considered). Thus the following could be applied:

'71 (b) inspection of the ship or aircraft and destruction of any Aedes aegypti or other vectors of yellow fever on board; in an area where the vector of yellow fever is present, the ship may. Until such measures have been carried out, be required to keep at least 400 metres from land.'

evidence to demonstrate that both Ross River virus (causing polyarthritis, fever and rash) and dengue viruses (body aches, fever, rash and as dengue haemorrhagic fever/dengue shock syndrome, some mortality) had probably been introduced into NZ from various sources on numerous occasions.

Mosquito surveillance was modelled around a rotational 10-year plan. The MoH's 10-year plan (1995–2005) scheduled two surveys (1999–2000 and 2004–05) and eight spot checks. The 1995–1996 spot checks covered the Waikato and Taranaki districts first with a 2-day training course, which resulted in only 106 positive samples, of which 77% were from tyres (Browne 1996). Only one positive sample was collected from a swamp! It was anybody's guess as to when the rotational system would come back to a previously surveyed district. Furthermore, the irregularity of the mosquito surveillance programme meant that it could be years before an incursion was detected. Much of the surveying concentrated on artificial habitats, such as tyres, bottles, and other man-made containers, with an accent on dengue and importation of major vectors *Ae. aegypti* and *Ae. albopictus* or even *Ae. polynesiensis.*

What does it take for a pest to establish? A lot of luck helps!

Pests are not pests in all circumstances and it is worthwhile examining how their abundance is regulated in nature. Van Emden (1974) provided an example of how natural processes can have an impact on pest populations. For a species to establish, it must overcome climatic variables in the environment, as well as outcompete other species already existing within the same area (see Chapter 7) or fill a niche that nobody else wants.

Worryingly, Laird (1990, 1995, 1996) believed that mosquito larval habitats in NZ were largely under-used and therefore amenable to new colonisations. He was correct! Another risky foible of NZ farmers is their use of old tyres to keep waterproof covers on their silage (bulk stock feed) pits. This practice simply increases the numbers of potential mosquito habitats (i.e. water-filled tyres, available for colonisation).

Mosquitoes are covered by the *Biosecurity Act 1993*: an unwanted organism is: 'any organism that a Chief Technical Officer believes is capable or potentially capable of causing unwanted harm to any natural and physical resources or human health' (see Chapter 3).

New Zealanders only began to take mosquitoes seriously in the 1990s, when it became clear that the country's 'sea gap' no longer provided a barrier to the introduction of unwanted species. Measures were put in place to reduce risk, such as mandatory fumigation of used tyre imports, residual treatment of aircraft hulls (work that had commenced before 1990 in association with Australia) and detailed inspections of imported risk cargo. However, the work was of varying quality. It focused on haphazard surveys that tried to deal with the obvious risk associated with mainstream aircraft and ships, but did not really consider the potential problems of leisure yachts or corporate jets. In addition, quarantine officers often could not survey the soft-top container tops on the decks of vessels because they did not have ladders!

The principle of border control as laid out in the *International Health Regulations 1969* (WHO 1983) is to ensure that vessels, passengers and vectors are separated, thus preventing importation or exportation of either pathogen or vector (Box 1.2). NZ is signatory to this agreement (in a modified form) and therefore exclusion of mosquitoes for a 400 m zone around international sea- and airport buildings, runways and aprons, and wharves was mandatory.

1996–1998: Thank goodness someone did their homework!

Even though the *Biosecurity Act* became law in 1993, for the next 3 years the NZ public health sector remained grossly under-resourced and relatively ineffective when it came to mosquito surveillance and control. Border health services were largely delivered by around 14 independent public health units. The national approach was loosely directed and coordinated by the MoH.

From late 1995, however, things began to change very quickly for the better. This rapid period of improvement was due, in no small part, to the leadership of Henry Dowler and the support he received from the Director-General of Health, Dr Karen Poutasi, and Director of Public Health, Dr Gillian Durham.

After joining the MoH as Senior Advisor (Health Protection) in August 1995, Henry Dowler recognised the public health significance of exotic mosquito surveillance and control. He also determined that action was needed to tackle the many shortcomings of NZ's surveillance approach, limited knowledge and resources, and almost complete absence of health sector planning for a mosquito eradication response programme.

In early 1996, Henry Dowler joined the Mosquito Control Association of Australia (MCAA). In July that same year, he attended the MCAA 'Mos101' mosquito control training course on the Gold Coast, Queensland Australia. After the discovery of *Campto* in Hawke's Bay more than 2 years later, it became apparent just how important Henry's newly acquired knowledge of Australian saltmarsh mosquitoes and ongoing professional relationships with several Australian mosquito experts during that week-long course would be! He also ran into Darryl McGinn (who spearheaded operational design) later at a course at Mount Macedon in Victoria.

Later on, as Deputy Chief Technical Officer (Health) under the *Biosecurity Act 1993*, Henry Dowler commissioned or led several important pieces of work. Henry also led the operational response to the *Campto* discovery in Hawke's Bay before leaving to join the then Ministry for Emergency Management.

By coincidence, Steve Garner, the Hawke's Bay Senior Health Protection Officer, who led the public health elements of setting up the initial 1998–1999 response, was a very close friend of Henry Dowler. Steve had actually been Henry's best man 16 years earlier at his Invercargill wedding. Their ability to have some very challenging discussions without coming to blows might help to explain how establishing the operational response went so smoothly!

However, a great deal more work was carried out before December 1998, which contributed to the ultimate success of eradication responses in Hawke's Bay and in many other locations throughout NZ. Some particularly noteworthy examples of work by the MoH were:

- the seminal 'Kay Review' in 1997. This review, discussed in more detail below, resulted in a November 1997 report to Hon. Simon Upton, Minister for Biosecurity entitled *Exclusion and Control of Exotic Mosquitoes of Public Health Significance*.
- a draft (February 1998) eradication contingency plan for exotic mosquitoes of public health significance – which substantially informed the early phases of the response to the Hawke's Bay *Campto* discovery
- a very successful first NZ national mosquito control conference in Auckland (February 1998) – despite the conference venue being significantly affected at the time by the disastrous 5-week-long power outage across Auckland
- AgResearch (Travis R. Glare and Maureen O'Callaghan) report (July 1998) on the environmental and health impacts of *Bacillus thuringiensis israelensis* (*Bti*) (i.e. in anticipation, correctly as it turned out, of the importance of this mosquito control product in the early 'containment and control' phases of an eradication response programme)
- new training programmes for public health personnel on mosquito exclusion and surveillance in NZ (including, somewhat fortuitously, a national training seminar in Wellington on 3–4 December 1998, just 3 weeks before the Hawke's Bay *Campto* discovery)
- technical guidance notes (February 1999) on the biological profile and surveillance for '*Campto*'
- AgResearch (Travis R. Glare and Maureen O'Callaghan) report (March 1999) on the environmental and health impacts of the insect juvenile hormone analogue, S-methoprene.

The Kay Review 1997

As noted above, in 1996, Brian Kay was commissioned to review the NZ programme for the exclusion and surveillance of mosquitoes of public health significance. This report can be regarded as a seminal piece of work that laid the foundations for mosquito surveillance in NZ. There were some 14 recommendations made in the Kay report and they have, over time, all been implemented.

The Kay report identified a 'hit' list of potential risks and included the saltmarsh species. Establishment of the following mosquito species was noted with concern:

- Importation of *Ae. albopictus*, a vector of both dengue and Ross River viruses, which spread widely through both North and South Islands. *Aedes albopictus* has been previously intercepted at Auckland and Lyttelton in 1993 (Laird *et al.* 1994).
- Importation of *Ae. aegypti*, the primary vector of dengue (dengue haemorrhagic fever) and a capable vector of Ross River virus possibly into Northland. This species has previously been intercepted in NZ (see Laird 1995).
- Establishment of *Cx. annulirostris* through Northland freshwater pondages, seepages and artificial wetlands. In Australia, this species is recognised as a major vector of Ross River, Barmah Forest, Murray Valley and Japanese encephalitis, Kunjin, Kokobera and Alfuy viruses, all linked with human disease. *Culex annulirostris* has previously been intercepted in NZ (see Laird 1995) and it is abundant in many parts of the Pacific, including Fiji.
- Colonisation of Northland-Coromandel intertidal zones with either Australian or Pacific Island (particularly Fiji) *Ae. vigilax*, a primary vector of Ross River and Barmah Forest viruses (or *Campto*, which also is a major Ross River virus vector in southern Australia). Note: we did not get this quite right because Hawke's Bay is too far south for mangroves and part of the index site was in inundated pasture, which was slightly brackish due to the farm being on an uplifted coral base from the 1931 earthquake.
- Introduction and generalised spread of Japanese *Ae. japonicus*, which has been intercepted on three occasions in NZ ports (Laird *et al.* 1994). *Aedes japonicus* has been shown to transmit Japanese encephalitis virus via infected eggs and also to laboratory animals (Takashima and Rosen 1989). Its role in natural cycles, including virus overwintering, has yet to be studied. The major cycle is between mosquitoes, particularly banded-proboscis *Culex*, and pigs. According to Statistics NZ (1996), there were 422 766 pigs in NZ in 1996.
- Introduction of *Ae. polynesiensis*, a vector of dengue, Ross River virus and filariasis, in the South Pacific may be limited by cool temperatures.
- That other NZ mosquito species, both exotic (*Ae. australis, Cx. quinquefasciatus*) or indigenous (e.g. *Cx. pervigilans*), could act as vectors of introduced viral pathogens.

It was recognised that little was known about the vector competence, host feeding patterns and general bionomics of NZ mosquitoes. With regard to Ross River virus, however, eight populations of *Cx. quinquefasciatus* from Tonga, Florida and Australia have been found to either be refractory or poorly susceptible to infection (Kay *et al.* 1982).

To some extent, the lack of definitive information about NZ mosquitoes made risk assessment more tenuous. A review of risk of exotic mosquitoes and particularly dengue, Ross River and Barmah Forest viruses, was carried out in three ways:

- first principles assessment based on demographics, climate, international traffic, trade and the likely animal hosts of such arboviruses
- review of the scientific literature and discussion with recognised NZ experts
- on-site inspection and discussion with personnel at seven localities.

The broad service objective was to review NZ's current arrangement and capabilities with respect to border inspection and surveillance of exotic mosquitoes.

The general finding was that incursion of exotic mosquitoes or vector-borne disease would seem to be only a matter of time, given the rudimentary and non-audited programme currently being undertaken, and a general lack of expertise. At the time of review, NZ authorities did not seem to be fully complying with the International Health Regulations to which they were a signatory.

Border inspection and surveillance activities must be efficient and expedient and supported by an emergency response capability. At the time, this was lacking. Kay recommended and justified annual expenditure of between NZ$1–2 million per annum to carry out several recommendations (in lieu of NZ$140 000 for the 10-year plan):

- Extension of MAF and MoH memoranda to include mosquito inspection of first port of call yachts and for formalising arrangements for disinsection procedures.
- Methyl bromide fumigation of containers with tyres should be subject to audit by public health authorities.
- Quality assurance and audit procedures should be upgraded generally.
- MoH should upgrade training of all stakeholders and commission national guidelines for these activities.
- Stakeholders should prepare specifically designed protocols to adequately address potential risks in their localities.
- A national mosquito surveillance and control laboratory should be designated to guide, review and report on enhanced activities by accredited health protection officers within public health and within other authorities.
- Health protection officers within public health units should be adequately provisioned with equipment in order to foster self-sufficiency and a more timely response capability.
- An exotic mosquito (and arbovirus) contingency plan should be developed.
- Arrangements should be made for availability of *Bacillus* and S-methoprene mosquitocides for routine control purposes and appropriate connections established before the event of emergency. (Who would have thought it was to be needed for emergency control of *Campto* ~1 year later?)
- MoH should enhance community awareness in order to gain greater participation in mosquito surveillance and control activities.
- To improve surveillance, MoH should improve awareness of medical practitioners with respect to the symptoms of vector-borne diseases, particularly dengue, Ross River and Barmah Forest viruses, and encourage greater use of diagnostic testing facilities.
- Research should be commissioned to define the potential role of common NZ mosquito species and brush-tailed possums with respect to arbovirus transmission, particularly with respect to Ross River and Barmah Forest viruses.
- Border inspection and mosquito surveillance activities should be continuous and funds should be allocated on the basis of perceived risk at each locality.

THE DOMINION

THURSDAY APRIL 17 1997

Heed alarm on mosquito threat

'MOST New Zealanders regard the "mozzie" as an unlikely harbinger of calamity, and see it as a nuisance rather than a menace. But the exotic mosquito is a hazard that, unless recognised and countered, has the potential to spread devastating diseases and damage the economy.

Australian mosquito expert Brian Kay has given authorities a timely warning .over the inadequate state of NZ border barriers against exotic mosquitoes. His background as head of the Queensland Institute of Medical Research mosquito control laboratory has made him well aware of the consequences of the introduction to NZ of exotic mosquitoes capable of carrying diseases such as dengue fever, its more fearsome relative, dengue haemorrhagic fever, Ross River fever and the Barmah Forest virus.

Even more alarming is the suspicion that possums could become hosts to Ross River fever and the Barmah Forest virus. That would be the ultimate irony - an Australian pest, destroying our native forests and birds and spreading bovine TB, then assisting in the spread of particularly virulent mosquito-borne diseases.

The personal consequences are fearsome. Dengue fever has symptoms so painful it is known as break-bone fever, while dengue haemorrhagic fever causes abnormal blood clotting, bleeding noses and gums and sometimes liver enlargement. Ross River fever and Barmah Forest virus both have symptoms which range from the flu-like to arthritis.

The financial consequences are equally worrying. The Health Ministry has estimated the economic impact of an outbreak of dengue fever in NZ would be $250 million. An outbreak of Ross River fever in Australia in the mid-1980s was costed at A$10 million. It is still spreading into southern parts of Australia from the north.

Dr Kay's assessment is that NZ is unlikely to be able to keep mosquitos out. The fruit fly made it in. So did the forest-threatening tussock moth. Officials are optimistic that today will see the last aerial and ground spray to kill the tussock moth, but it has taken a year and $8.7 million to contain the problem.

The present allocation for mosquito detection and control is a mere $140,000, for 10 years. Dr Kay notes that a mosquito invasion "would seem to be only a matter of time, given the rudimentary and non-audited programme currently undertaken, a general lack of expertise and a glaring deficiency with respect to inspection of yachts".

He advises yacht inspections, stockpiles of mosquitocides in case of outbreaks, a contingency plan, education for doctors on the symptoms of mosquito-borne diseases, and fumigation of imported tyres, which are a notorious breeding site for mosquitoes. Increased Government spending should never be lightly entered into, but to ignore the mosquito threat would be to inflict on future New Zealanders a blight almost as great as that of the possum.

Figure 1.2 *Dominion*, 17 April 1997.

THE DOMINION

WEDNESDAY JANUARY 20 1999

Move now on killer bugs

ERADICATE, whatever the cost - and do it soon. That must be the message the Government sends to health, conservation and biosecurity officers who have spent the past fortnight strategising over what they should do about the noxious new mosquito found lurking in Napier's Ahuriri Estuary.

Any delay will be a bonus for the *Aedes camptorhynchus*, the nasty Australian invaders which somehow sneaked into the estuary, found the brackish water to their liking; and settled in to breed.

They must be stopped before they have a chance to spread through Hawke's Bay and 'the rest of the North Island, just as the white-spotted tussock moth was eliminated from Auckland in 1997.

So far there have been no confirmed cases of Ross River virus caused by the mosquitoes, but the wider they roam, the greater the chance that one will alight on someone who has picked up the disease overseas, and in fect a new victim.

In the same way it would be only a question of time before the virus spread to possums, which alarming Australian research has just confirmed can be carriers of the virus. The 70 million possums in NZ would then pose a potential risk to human health on top of the bovine tuberculosis they already visit on cattle.

This threat puts the cost of immediate, action to eradicate the mosquito into its proper perspective. At worst, the bill for attacking the mosquitoes could be "millions of dollars", though some estimates fall well short of that, especially since their Ahuriri bridgehead is not a populated area.

Contrast that with the certainty of having to spend hundreds of millions over the years in health costs and blitzes on mosquitoes and possums to keep the disease at bay, and there is no contest: the money has to be found, and quickly.

In addition to such quantifiable costs of the virus and its carrier becoming established in NZ, there would be an intangible cost in the diminished quality of life, not only for people who get the virus·(which can affect them for years), but also those who live in fear of it.

Having sprayed till there is no sign of the mosquitoes or their larvae, health and biosecurity agencies must then sharpen their alertness at ports, airports and wherever else this and other pests, such as *Aedes albopictus, Aedes aegypti*, dengue fever and Barmah Forest virus, could come in.

In a 1997 report, Queensland mosquito expert Professor Brian Kay was appalled at the "rudimentary and non-audited" mosquito programme then in force, and the ludicrously inadequate allocation for detection and control - $140,000 over 10 years.

The division of responsibility among the Health Ministry, Health Funding Authority and hospital management around the country, not to mention MAF and local authorities, must not be allowed to hinder swift preventive action.

These matters need to be sorted out promptly, so that the emphasis remains on keeping the mosquitoes and diseases out rather than fighting them once they are here.

Figure 1.3 *Dominion*, 20 January 1999.

- The risk-based analysis suggested that at least 60% of funds should be allocated for upgrading activities in Auckland and Whangarei –Bay of Islands, 20% for North Island ports (including Gisborne and Napier) and Christchurch, with the remaining 20% to other localities considered in this report. Note: this risk-based approach was unique at the time and pre-dated the more universal application of risk-based approaches to border and public health described, such as those in the *International Health Regulations 2005*.

When this report became public, it was clear from media editorials that the public did not want to face life with new 'killer bugs' and urged positive action (Figs 1.2 and 1.3). The first article was written before *Campto* had actually arrived! The second advocated an unequivocal response to the Australian invader.

The Kay Review of the NZ Programme for the Exclusion and Surveillance of Exotic Mosquitoes of Public Health Significance was expanded by the MoH with material from the ensuing discussion document and presented in June 1997 (revised in November 1997) to Hon. Simon Upton, Minister for Biosecurity as *Exclusion and Control of Exotic Mosquitoes of Public Health Significance*. Henry Dowler, who commissioned the review for the MoH and who partnered Brian Kay in review-related engagements throughout NZ, said:

> *'Without the tremendous collegial support, expert advice and guidance from Brian and his Australian and international mosquito control networks, we would have been well and truly caught with our pants down. The NZ public should be very grateful that the ANZAC spirit [Australian–NZ army mateship developed during World War I] is alive and well as we work side-by-side again to battle a common foe – albeit an Australian insect!'*

References

Aaskov JG, Mataika JU, Lawrence GW, Rabukawaqa V, Tucker MM, Miles JAR, *et al.* (1981) An epidemic of Ross River virus infection in Fiji, 1979. *The American Journal of Tropical Medicine and Hygiene* **30**, 1053–1059.

Austin FJ (1978) Johnston Atoll virus (Quaranfil group) from *Ornithodoros capensis* (Ixodoidea:Argasidae) infesting a gannet colony in New Zealand. *The American Journal of Tropical Medicine and Hygiene* **27**, 1045–1048.

Austin FJ (1984) Ticks as arbovirus vectors in New Zealand. *New Zealand Entomologist* **8**, 105–106. doi:10.1080/00779962.1984.9722481

Ballard W, Marshall ID (1986) An investigation of the potential of *Aedes camptorhynchus* (Thom.) as a vector of Ross River virus. *The Australian Journal of Experimental Biology and Medical Science* **64**, 197–200. doi:10.1038/icb.1986.21

Boyd AM, Kay BH (2002) Assessment of the potential of dogs and cats urban reservoirs of Ross River and Barmah Forest viruses *Australian Veterinary Journal* **80**, 83–86.

Boyd AM, Hall RA, Gemmell RT, Kay BH (2001) Experimental infections of Australian brushtail possums, *Trichosurus vulpecula* (Phalangeridae: Marsupialia) with Ross River and Barmah Forest viruses, using a natural mosquito vector system. *The American Journal of Tropical Medicine and Hygiene* **65**, 777–782.

Browne G (1996) 'Report on the New Zealand 1995–1996 mosquito survey in Waikato and Taranaki'. Ministry of Health (Health Policy and Regulation Section), Wellington, NZ.

Butcher M (2002) A bug's life. *North and South* **May**, 65–74.

Farr D (1999) 'Aircraft interceptions. Summary report for the Aircraft disinsection working party'. Report for Auckland International Airport, Auckland, NZ.

Graham DW (1939) Mosquito life in the Auckland district. *Transactions and Proceedings of the Royal Society of New Zealand* **69**, 210–254.

Hogg D, Ross RW, Miles JAR, Austin FJ, Maguire T (1963) Evidence of human arbovirus infection in New Zealand. *The New Zealand Medical Journal* **62**, 519–521.

Howard GW (1973) Aspects of the epidemiology of *Eperythrozoon ovis* in South Australia. PhD thesis. Department of Entomology, Waite Agricultural Research Institute, Adelaide.

Johns JH, MacGibbon RJ (1986) *Wild Animals in NZ*. Reed Methuen, Auckland.

Kay BH (1997) *Review of the New Zealand Programme for the Exclusion and Surveillance of Mosquitoes of Public Health Significance*. Ministry of Health,Wellington, NZ.

Kay BH, Jennings CD (2002) Enhancement or modulation of the vector competence of *Ochlerotatus vigilax*, (Diptera: Culicidae) for Ross River virus by temperature. *Journal of Medical Entomology* **39**, 99–105. doi:10.1603/0022-2585-39.1.99

Kay BH, Miles JAR, Gubler DJ, Mitchell CJ (1982) Vectors of Ross River virus: an overview. In *Viral Diseases in South-east Asia and the Western Pacific* (Ed. JS Mackenzie) pp. 532–536. Academic Press, Sydney.

Kay BH, Hall RA, Fanning ID, Mottram P, Young PL, Pollitt CC (1986) Experimental infection of vertebrates with Murray Valley encephalitis and Ross River viruses. *Arbovirus Research in Australia* **4**, 71–75.

Kay BH, Pollitt CC, Fanning ID, Hall RA (1987) The experimental infection of horses with Murray Valley encephalitis and Ross River viruses. *Australian Veterinary Journal* **64**, 52–55. doi:10.1111/j.1751-0813.1987.tb16129.x

Kelly-Hope LA, Kay BH, Purdie DM, Williams GM (2002) The risk of Ross River and Barmah Forest disease in Queensland. Implications for New Zealand. *Australian and New Zealand Journal of Public Health* **26**, 69–77. doi:10.1111/j.1467-842X.2002.tb00274.x

Kelly-Hope LA, Purdie DM, Kay BH (2004) Ross River virus disease in Australia, 1886–1998, with analysis of risk factors associated with outbreaks. *Journal of Medical Entomology* **41**, 133–150. doi:10.1603/0022-2585-41.2.133

Laird M (1951) *Insects collected from aircraft arriving in New Zealand from abroad*. Publication no. 11. Victoria University College Zoology, Wellington, NZ.

Laird M (1989) Aircraft disinsection review in New Zealand. *Travel Medicine International (London)* **1989**(2), 73–75.

Laird M (1990) New Zealand's northern mosquito survey, 1988–89. *Journal of the American Mosquito Control Association* **6**, 287–299.

Laird M (1995) Background and findings of the 1993–94 New Zealand mosquito survey. *New Zealand Entomologist* **18**, 77–90. doi:10.1080/00779962.1995.9722010

Laird M (1996) *New Zealand's Mosquito Fauna in 1995: History And Status*. Ministry of Health, Wellington, NZ

Laird M, Calder L, Thornton RC, Syme R, Holder PW, Mogi M (1994) Japanese *Aedes albopictus* among four mosquito species reaching New Zealand in used tyres. *Journal of the American Mosquito Control Association* **10**, 14–23.

Lee DJ, Hicks MM, Griffiths M, Russell RC, Marks EN (1984) *The Culicidae of the Australian Region. Vol. 3*. Monograph Series No.2. Commonwealth Department of Health, Commonwealth Institute of Health incorporating the School of Tropical Medicine and Public Health, University of Sydney.

Maguire T (1994) Do Ross River and dengue viruses pose a threat to New Zealand? *The New Zealand Medical Journal* **107**, 448–450.

Maguire T, Miles JAR (1960) Evidence of infection with arthropod-borne viruses in New Zealand. *Proceedings of the Otago Medical School* **38**, 25–27.

Maguire T, Miles JAR, Casals J (1967) Whataroa virus, a group A arbovirus, isolated in South Westland, New Zealand. *The American Journal of Tropical Medicine and Hygiene* **16**, 371–373.

McManus TJ, Marshall ID (1986) The epidemiology of Ross River virus in Tasmania. *Arbovirus Research in Australia* **4**, 127–131.

McManus TJ, Russell RC, Wells PJ, Clancy JG, Fennell M, Cloonan MJ (1992) Further studies on the epidemiology and effects of Ross River virus in Tasmania. *Arbovirus Research in Australia* **6**, 68–72.

Miles JAR (1973) The ecology of Whataroa virus, an alphavirus, in South Westland, New Zealand. *Journal of Hygiene (Cambridge)* **71**, 701–713. doi:10.1017/S0022172400022968

Owen SJ (1998) *Department of Conservation Strategic Plan for Managing Invasive Weeds.* Department of Conservation, Wellington, NZ.

Pillai JS (1962) Factors influencing egg survival in *Aedes eggs* with special reference to some Victorian species. PhD thesis. University of Melbourne.

Rosen L, Gubler DJ, Bennett PH (1981) Epidemic polyarthritis (Ross River) virus infection in the Cook Islands. *The American Journal of Tropical Medicine and Hygiene* **30**, 1294–1302.

Russell RC, Cloonan MJ, Wells PJ, Vale TG (1991) Mosquito (Diptera: Culicidae) and arbovirus activity on the south coast of New South Wales, Australia, in 1985–1988. *Journal of Medical Entomology* **28**, 796–804.

Sammels LM, Coelon RJ, Lindsay MD, Mackenzie JS (1995) Geographic distribution and evolution of Ross River virus in Australia and the Pacific Islands. *Virology* **212**, 20–29. doi:10.1006/viro.1995.1449

Sinner J, Forrest B, Taylor M, Dodgshun T, Brown S, Gibbs W (2000) 'Options for a national pest management strategy for the Asian kelp *Undaria*'. Technical Report 63. Ministry of Fisheries, Wellington, NZ.

Statistics NZ (1996) *NZ Official Year Book 1996*. Statistics NZ, Auckland.

Takashima I, Rosen L (1989) Horizontal and vertical transmission of Japanese encephalitis virus by *Aedes japonicus* (Diptera:Culicidae). *Journal of Medical Entomology* **26**, 454–458.

Tesh RB, McLean RG, Shroyer DA, Calisher CH, Rosen L (1981) Ross River virus (Togaviridae: Alphavirus) infection (Epidemic polyarthritis) in American Samoa. *Transactions of the Royal Society of Tropical Medicine and Hygiene* **75**, 426–431. doi:10.1016/0035-9203(81)90112-7

Van Emden HF (1974) *Pest Control and its Ecology*. Studies in biology No. 50. Edward Arnold Publishing, London.

Weinstein P (1994) 'The real and potential risks of human arboviral disease in New Zealand'. Report to the Communicable Disease Centre, Environmental Science and Research Ltd, Porirua, NZ.

Weinstein P, Laird M, Calder L (1995) Australian arboviruses: At what risk New Zealand? *Australian and New Zealand Journal of Medicine* **25**, 666–669. doi:10.1111/j.1445-5994.1995.tb02850.x

World Health Organization (1983) *International Health Regulations (1969)*. WHO, Geneva.

2

Strategy development and refinement at Hawke's Bay

Steve Garner, Noel Watson, Henry Dowler, Darryl McGinn, David Sullivan and Brian Kay

Boxing Day December 1998, Steve Garner's pager chirped. Damn Boxing Day and 10 km back to cell phone coverage. 'Steve Garner, Public Health', he said as he got through to his operator. 'You need to call Henry Dowler at the MoH – something about a Biosecurity incursion'. Henry was the Deputy Chief Technical Officer Biosecurity (Health) at the time among his other duties. His curiosity was piqued.

'Hi Henry, Steve here what have you got?'

'Our mosquito surveillance has identified the presence of Aedes camptorhynchus, *the southern saltmarsh mosquito, from Australia. I need you to organise a delimiting survey. Can you get a team together for the 28th? I am heading through tonight can we meet at your office to plan the attack?'*

'Right, I'll get right on it Henry. I'm in Wanganui at the moment with my folks so will need to pack the car up. I should be back by 5 pm.'

And so it began. It's hard to believe looking back that a public health generalist, Steve Garner, would lead a major mosquito eradication programme starting in Napier, then extending south to Haumoana and Porangahau, and then north to the top end of Hawke's Bay at Mahia, Gisborne, the Kaipara Harbour and the Wairau area near Blenheim. This chapter sets out the early beginnings of the programme; establishing an operations base; delimiting teams; the choice of a control product and trials and tribulations associated with it; and the initial surveillance programme, which led to refined models of the eradication programme successfully used in all of the other incursion centres.

There were significant obstacles to overcome in establishing an effective response against the arrival of *Campto* in Hawke's Bay. When the programme began, there were few trained staff, almost no larval sampling equipment and no adult mosquito traps, a single mosquito entomologist based in Auckland, no larvicidal experience or infrastructure and equipment and no digital aerial colour photography of the area. Much of the habitat was around the Napier airport, which was a secure area, or on Landcorp farm or Department of Conservation wetlands. Access to these areas was by negotiation. All sampling information was collected on paper and subsequently had to be entered into a database, meaning it wasn't immediately available for analysis. Access to GIS mapping software was limited and hand-held GPS devices were expensive and inaccurate. To further complicate things, the initial response was funded through the Health Funding Authority who had to reprioritise resources to make funding available.

In the beginning

Following public complaints to the Napier City Council on 20 December 1998, strange mosquitoes were recognised by Noel Watson, another Health Protection Officer and colleague of Steve Garner's. On 24 December, Gene Browne, the national mosquito entomologist retained by the MoH, identified the invaders as *Campto,* and this was confirmed by Mark Bullians of Agriquality, also in Auckland on Christmas Day 1998. Noel remembered the events as follows:

> *'I was fortunate to get a week long training trip to Brisbane in the mid 1990s to do a basic mosquito course with the Brisbane City Council. While there, we visited coastal swamps to view the type of habitat that the Northern Saltmarsh mosquito* Aedes vigilax *breeds in. I can recall thinking as coming into Napier airport on return that we had some of the same type of habitat about us as I saw over there. The MoH's focus though at that point, understandably, was on intercepting container breeding exotic mosquitoes such as* Aedes aegypti, *the dengue and yellow fever mosquito.*
>
> *On 21 December, I was visited by Steve Turpin of the Napier City Council's Environmental Health team who had been given some adult mosquitoes by the caretaker of the local Westshore Primary School. The caretaker had told him that he had caught them at the school and that there were a lot of them around, even during the day, biting the children/staff at the school. At that time, I had had no training in the morphological identification of different mosquitoes and neither did we have any microscope of any sort at the public health unit. What I did know from previous summers in Hawke's Bay, and from being involved in a nationwide MoH mosquito sampling programme, was that* Aedes notoscriptus *was normally in high numbers at this time and I'd had several complaints about it. So I packaged the samples up and sent them off by courier to the then MoH contracted entomologist, Gene Browne, for identification. I then got back to my piles of food complaints and controlled pesticide permit applications for possum-control operations (incidentally originating from Australia). The exotic mosquito work was around 10% of my total work load at the time.*
>
> *When the samples should have arrived with Gene Browne in Auckland, they hadn't, and enquiries with the courier company concerned were not encouraging: they didn't seem to know where the samples were or when they were going to get them to Gene. They seemed to be 'on the wrong side of Christmas' and struggling to cope with the demand for their service at that time of the year. Therefore it was eventually decided that some more samples were needed to be sent, so on the afternoon of 23 December I went out to a drain running between the highway and the Napier Aero club buildings to look for adult mosquitoes. It did not take me long – the mosquitoes came to me! I soon had a sample of them, and was running for cover back to the car. I do remember a lot of mosquitoes all over the front windscreen of the car and thought temporarily*

that this was like the scene out of the movie Jumanji. *I had recently watched this film with my children where the giant mosquitoes attacked the car. I decided that, although I could not identify these mosquitoes by eye, they were like nothing in my brief experience of mosquitoes because of their aggressiveness. I decided then that I would try and get them sent by air freight that afternoon. I had the mosquitoes alive in a plastic sampling tube with the lid securely on, but had not taken any other packaging material with me. I used the insurance envelope in the glove box of the work car to be used in case of an accident, and then put that into the air courier envelope supplied by the airport. Obviously, this was all before NZ started getting serious about administering the* Biosecurity Act 1993. *I do know that the envelope was empty. I rang Gene on return to the office and advised him when the mosquitoes would be coming to him and that they would be alive'.*

MoH protocols required the identification to be confirmed by an Australian medical entomologist. On 12 January 1999, Professor Richard Russell from the University of Sydney at Westmead Hospital confirmed the identification.

Steve Garner's backyard cricket had ended in Wanganui, some two-and-a-half hours away from Mossie Central at Napier; Henry Dowler was packing his bags from Wellington while Gene Browne found himself deployed to Napier by 28 December. So much for a long summer holiday break, Steve thought.

Getting started

Overnight on the 28th, Henry Dowler, Gene Browne and Steve Garner began planning an initial delimiting survey to determine the extent of the infestation. Public health staff and council environmental health officers were enlisted to assist with field surveying, which commenced the next day after a training session led by Gene Browne and Noel Watson. Noel recounts:

'On Monday 28 December, I recall an initial delimiting survey occurred with a team for the first time. Persons involved were Steve Garner (Senior Health Protection Officer, Hawke's Bay Public Health Unit), Henry Dowler (Senior Advisor, MoH), Bill Littley (Manager Health Protection, Hawke's Bay Public Health Unit), Gene Browne (MoH contracted entomologist), and me. From this survey, and Gene's identification of the samples, we confirmed that Campto *was widespread, and in high numbers around the land that had been brought up above sea level by the 1931 Napier Earthquake. My notebook indicated the remaining days of 1998 was spent either doing delimiting, and finding out about Ross River fever and* Campto, *and also starting to look at what resources were available for things like the blood testing of people who thought they might have Ross River fever, what weather and tide data were available that we could utilise, seeking initial emergency funding through the MoH, what treatment products were available and what restrictions would apply under the* Resource Management Act *administered by the local Hawke's Bay Regional Council. Were there any provisions in the* Biosecurity Act 1993 *that could assist with the widespread application of a treatment product(s)?'*

By 5 January, most of the health protection officers were back from holiday. Noel took a previously planned 1 week holiday with his wife and four children at the beach in the Bay of Plenty. However, he knew that, at some stage soon, the presence of *Campto* in Napier would come to the inevitable attention of the media and that this would elevate things to another level.

The Watsons were packing up from the beach in the Bay of Plenty and coming home, and were listening to Rudd Kleinpaste's gardening and bug show. It was entirely about the discovery of *Campto* in Napier. People were ringing up from Napier and Bayview saying that they had all the symptoms of Ross River fever –Noel knew his brief respite was over! Noel recalls:

> *'The MoH and Hawke's Bay Public Health Unit media staff began to release statements which included that if anyone thought they had a* Campto, *bring them into us. Unfortunately, the major national and the local newspaper articles included an enlarged photo of a* Campto *mosquito fully engorged with blood. Consequently, we were inundated with a lot of red wasps and very few mosquitoes!'*

Work on developing and providing information for local residents and media releases for local and national media was done in association with the MoH. The surveying work was hampered by a lack of four-wheel-drive vehicles to access difficult-to-reach habitat, trained staff and the hot dry conditions that rapidly dried out habitat. In addition, no adult mosquito traps, an invaluable surveillance tool, were available in Napier. The response team was operating out of the Public Health Unit offices, which was less than ideal.

In the true spirit of Australia/NZ cooperation, Steve Garner made a hasty phone call to Dr Michael Lindsay in Western Australia. After introductions and an explanation of our predicament, Mike offered to provide 10 adult traps as soon as they could be despatched. Mike was also able to provide valuable information about the behaviour of the *Campto* in Western Australia.

Meanwhile, back in Napier reinforcements arrived – Henry Dowler had called in some help from 'across the ditch' (i.e. Australia). Expert help arrived in the form of Prof. Brian Kay from the Queensland Institute of Medical Research, Brisbane and Darryl McGinn from the Brisbane City Council. Brian and Darryl both had considerable experience in developing, evaluating and running programmes against saltmarsh *Aedes vigilax* in south-east Queensland: Brian from the research and Darryl from the operations perspective. Brian and Darryl were able to add immediate value to our sampling strategy, including identifying the need for colour aerial photography of the district to assist with identifying the extent of habitat, advice on sampling equipment, training staff and advising on the placement of adult traps.

Health protection officers from around NZ were seconded to Napier to assist local health protection officers with the delimiting and surveillance.

In addition (pending a decision by the NZ Government), preparations were made to undertake some treatment of habitat with a biological larvicide known as *Bacillus thuringiensis israelensis* (*Bti*) originally discovered in Israel (Margalit and Dean 1985). Thankfully *Bti* was registered for use in NZ, but our initial operation would use up the entire year's supply for NZ! Planning involved identifying and contracting a suitable helicopter firm for aerial treatments, identifying a suitable helicopter loading zone, procuring mechanical and hand-pumped knapsack sprayers and sourcing the product from Auckland. Equipment had to be calibrated and a quality management system established to monitor the applications. Darryl McGinn was asked to organise this (see Chapter 5).

Public information was also prepared and letter drops made to nearby residents providing information about the mosquitoes and the planned larvicide applications. At the time the operation commenced, only a two-lane road separated houses from the nearest habitat, so a lot of community liaison was undertaken to minimise the impact on those residents and to assure them of the safety of the product being used.

The Mosquito Response Centre fights back – the *Bti* chronicles

After heavy rainfall in mid January, a major hatch of mosquito larvae was identified by field staff and the response was upgraded to a national biosecurity response. Henry Dowler requested the Hawke's Bay District Health Board establish a staffed and operational Mosquito Response Centre within a week. Again lady luck was smiling on us. The Hawke's Bay District Health Board had recently begun to relocate hospital services from a site on Napier Hill to Hastings. Part of the recently abandoned maternity unit was hastily taken over and Steve, Henry and Noel Watson, with the full support of Hawke's Bay District Health Board, established a fully operational response centre with an on-site mosquito identification laboratory, offices, telephones, a large briefing area and meeting rooms. The office had a view over some of the major habitat around the Ahuriri estuary and the airport, and so we were well placed to keep an eye on tidal and rainfall flooding of habitat.

Back in Wellington, the MoH had established a Southern Saltmarsh Mosquito Technical Advisory Group (TAG). At their first meeting on 19 January 1999, they recommended the MoH commence treatment of the infested habitat with *Bti*. Authorisation to undertake aerial treatment was gained from the Minister for Biosecurity the following day using emergency powers under the *Biosecurity Act 1993*. The first aerial application of *Bti* in NZ commenced on 21 January 1999. This was widely reported in local and national media. One intrepid TV cameraman secreted himself in one of the wetlands among the tall reeds. As the helicopter lined up for a treatment run, flying just above fence height, the cameraman stood up causing the run to be aborted. A shaky cameraman was removed.

Around the same time, pre-embarkation disinsection of aircraft departing Napier was initiated. This was because the airport was surrounded by prime habitat and anecdotally was inundated with adult mosquitoes, which proved to be vicious day biters (see Chapter 5 for Darryl McGinn's comments about the pre-dawn scene).

Aerial and ground-based applications of *Bti* were scheduled after each habitat flooding event. Applications were monitored by using dye cards that changed colour when hit by drops of liquid, and larval and adult mosquito sampling before and after applications. Monitoring results showed the applications were very effective at controlling the mosquito population with adult numbers reducing significantly within the first 100 days of treatment.

By mid February, the NZ Government had agreed to fund continued operations in Napier and Haumoana, covering some 680 ha of habitat, continuing until April when the feasibility of eradication would be reviewed.

The Napier Mosquito Response Centre (MRC) was able to gear up and begin employing local staff to supplement the health protection officers being brought in each week from around NZ to undertake larval surveillance. Brisbane City Council provided a contract staff member to assist with organising treatments and undertaking quality assurance activities. The Brisbane staff trained local staff to undertake their roles.

A routine of continual surveillance of habitat through sampling for larvae and trapping adult mosquitoes helped to fully delimit the spread of the mosquito and range and size of habitat. Rainfall and tidal monitoring was undertaken to assist with determining treatment frequency. Over time, a high level of local expertise developed.

Eradication planning – the evolution of a plan

With *Bti* applications underway, our minds turned to the future. The day after Richard Russell had confirmed the invader as *Aedes camptorhynchus* (Thomson), Steve Garner, Brian Kay and Darryl McGinn each cradled a beer and considered the unfolding events. It was Saturday afternoon in the Napier MRC. Outside, the view was great over the Ahuriri estuary, the airport and the Landcorp farm where this had all ignited (Fig. 2.1). Inside, they stared at a blank whiteboard.

'Well what can we do about it?' Steve questioned.

In 1973, Brian had helped to develop the Abate (temephos) 5% sand granule to be used as the mainstay of broad-scale aerial control in Australia for 26 years and, with Michael Brown, pioneered the usage and evaluation of S-methoprene and *Bacillus* products mainly for saltmarsh mosquito control in south-east Queensland. Darryl was well recognised as the Head of Brisbane City Council operations against pests, especially mosquitoes, which involved aerial treatment of up to 25 000 ha of saltmarsh every year, with individual treatments of up to 3000 ha.

> *'Well, you need something like S-methoprene with residual action but because of the large size of the pellet formulation, treatment of every hoof print at Landcorp farm would not be possible'*, they said.
>
> *'You get great residual life with pellets in saltmarsh in Queensland, about 60–90 days judging from Scott Ritchie's studies at Pine Shire (Ritchie and Piggott 1994) but the cost would be prohibitive'* Brian added.
>
> 'Bti *has no residual action and therefore would be unsuitable as a long-term tool for eradication'* Darryl finalised the assessment. Steve and the whiteboard waited in anticipation.
>
> *'We need an S-methoprene product which has granules of intermediate size between Altosid pellets (3–120 mm extrusions), which may give up to 6 pellets per square metre, and Altosand (0.5–1 mm diameter), which can blanket a square metre of ground with up to 300-400 granules'*. Brian said. *'We need to contact David Sullivan in California because I think that there is an experimental S-methoprene product which may fit our specifications.'*

This was to be Altosid® X-RG (or later registered in NZ as ProLink® XR-G), which was 1.0–2.4 mm in diameter and theoretically should result in an expected 164 granules/m^2 when dispensed at 6 kg/ ha (Russell *et al.* 2009).

The three squeezed into Steve's office and listened to the ring tone: 'David Sullivan here'. After introductions, Steve said, 'We have a wee bit of a problem here in Hawke's Bay...' The 'silver fox' listened intently and replied 'I think you need some XR-G, but it has only been used in California for one summer'.

Figure 2.1 The Ahuriri lagoon from Napier Hill. (Image: G Mackereth)

The initial programme was born; to treat some 650 or so hectares around the airport and a further few hectares at a discrete site south of the city. We collectively outlined the following programme to eradicate *Campto*:

- Brian Kay and Michael Brown will do initial field research.
- Darryl McGinn and crew will do applications.
- The treatments must be comprehensive.
- The plan was to eradicate by April 2000 with 22 treatments of Altosid XR-G starting at the end of April 1999.
- Treat at a rate of 3 kg/ha according to a recommended 3–4 kg/ha Australian label rate.
- Order enough product for at least two treatments.

Characteristics of methoprene products

In 1973, David Sullivan started working at Zoecon as Division Manager of the Pheromone Supply Division. Part of David's responsibility was to package the Altosid products including the 'new' Altosid briquettes. In 1979, he became Product Manager for all insect growth regulators, including methoprene products used in the control of mosquitoes, fleas, stored-product pests, flies, and so on. In 1985, he left Zoecon and formed his own distribution company, Zanus Corporation, and specialised in selling mosquito-control products in the western United States. By 1996, Zanus was the largest Altosid distributor and had worked with almost all of the Mosquito Abatement Districts from Canada to Mexico, west of the Dakotas and Texas. During this period, Altosid products containing S-methoprene were developed, including a 5% and 20% liquid, pellets and granules, and Zanus was involved in many research projects to establish the efficacy of these products under all environmental conditions. Methoprene has two chemical isomers, the S-form being the cleaner environmentally preferable one.

In late 1998, David Sullivan received a phone call from Henry Dowler who had heard his Altosid talk at the conference at Noosa Heads and inquired to see if he was interested in reviewing a draft report entitled *Environmental and Health Impacts of the Insect Juvenile Hormone Analogue S-methoprene*. The MoH wanted to validate the report so that S-methoprene could be registered in NZ for mosquito control as a consequence of the Kay review in 1997.

The draft was well done but it included data on formulations that would not be used for mosquito control and there was little information that could be used for registration purposes. David outlined some of the difficulties with using S-methoprene, which, to the uninitiated, did not kill things within a reasonable timeframe:

- Many early researchers did not understand the mode of action of S-methoprene whereby a dose of this 'insect growth regulator' or 'synthetic juvenile hormone' kept most individuals in their non-adult state. Without an instant 'kill', they believed the product didn't work.
- Much of the literature used for evaluation relied on emulsifiable concentrates used indoors to control fleas and stored product pests, and not mosquitoes.
- S-methoprene is effective in controlling mosquitoes below the level of detection by analytical methods (Ross *et al.* 1994). It is difficult to detect S-methoprene in water, as we found out later (see Chapter 6). Trials have shown that S-methoprene concentrations are less than 0.2 parts per billion after 24 h but larvae are still controlled for 21–30 days for XR-G and pellets, respectively. Yes folks, can you imagine what two-tenths of one part of a billion might look like? Try mixing the contents of a tea or coffee cup into a billion litres.
- If Altosid is used as directed on the label, it will not cause frog deformities or harm fish or other non-target organisms. Solvents were used to increase solubility of S-methoprene in water when doing fish studies, resulting in the apparent fish toxicity. All in all, it seemed to be an environmentally suitable product to build our strategy around.
- All Altosid formulas have a density greater than 1 and sink towards the bottom of water. When released, S-methoprene has a specific gravity less than water and will rise to the surface.
- Many asynchronous populations of mosquitoes mature in more than 7 days and Altosid liquid formulations will not control them without an additional treatment. All solid formulations of Altosid have 21 days or more activity.

S-methoprene registration in NZ

In December 1998, Henry Dowler had informed David Sullivan of the *Campto* incursion at Hawke's Bay, so it was no surprise when Steve Garner, Brian Kay and Darryl McGinn phoned on 13 March. The only Altosid product available was Biorational Resources' Altosand, which, at the time, was not working satisfactorily in Brisbane. Novartis had never registered granules or pellets in Australia.

The MoH had contacted three NZ companies, including Novartis NZ, but none of these companies expressed an interest in supporting the registration of Altosid XR-G or pellets in NZ. A couple of days after the discussion on 13 March, the MoH contacted David Sullivan for assistance in getting a registration for Altosid in NZ, and in supplying product for the programme. David was advised that Novartis Australia would support the MoH if they would be responsible for all fees associated with the registration of Altosid products in NZ. They also noted '… Novartis will not write any part of the application for NZ registration' (source: Mr Neutze, Novartis email, 15 March 1999.) The MoH applied for an experimental permit to apply Altosid IGR granules and pellets to obtain data for registration.

David Sullivan contacted Greg Braithwaite and John Neberz at Zoecon to see if they would help him in getting a new registration. They could not do it directly because it would violate their sales agreement with Novartis. (Note: a merger of Ciba-Geigy and Sandoz formed Novartis. When they merged, they sold off the Zoecon Division and their insect growth regulators in the United States, but retained the products for the rest of the world.) Greg and John were supportive of Zoecon assisting Zanus in getting the registration, but others at Zoecon thought it would not be financially feasible. After a few days of discussion, and with Greg's and John's support, Zoecon agreed to support the registration for ProLink IGR granules and pellets.

In order to register products in NZ, a company needed a local contact and Zoecon did not have, nor would they have, a NZ agent. Zanus, with the help of Michael Brown, arranged for an individual with dual Australian and NZ citizenship to be a member of two new companies being formed to register ProLink: Zancor Ind. NZ Ltd and Zancor Ind. Australia Ltd.

The initial order for 9000 kg Altosid XR-G and 1000 kg pellets was placed at the end of March 1999 and the initial shipment was received in May. The shipments of product were under an experimental use permit, using the Altosid label, and could not be resold by MoH. The remaining orders were under the ProLink IGR label.

Bruce Evans of EvaTech, who was going through the registration process, noticed that the USA label for XR-G had application rates from 5.6 to 22.4 kg/ha, while the experimental label copied from the Australian label had rates from 3 to 4 kg/ha. Zoecon and Zanus recommended that the application rate range from 5.5 to 11.0 kg/ha, and also recommended a minimum of 6 kg/ha to ensure appropriate coverage required by an eradication project.

Doug Van Gundy (Zoecon/Wellmark) and David Sullivan were invited to participate in the TAG meeting in Wellington to review the progress of the *Campto* eradication programme. There was concern that it was difficult to detect S-methoprene in water at low levels (< 0.2 ppb), although it still controls mosquitoes at that level. Secondly, there were no scientific data available on the effects of S-methoprene on *Campto*. Doug gave a summary of the 25-year history of the use of S-methoprene, based on known effective rates on various *Aedes* spp., application rates, water conditions and temperatures. David also shared his experiences with many field people who were uneasy when using S-methoprene because there was no immediate effect. In evangelistic fashion, he told the group that 'You will not see any immediate effect; you just have to have faith'. David's other favourite was 'The mosquitoes know it's there'.

There was an initial moment of disquiet – the MoH's boffins couldn't find any trace of the treatment product in the water column and yet they had spent thousands for product and application. JR Gardner from the MoH colourfully enquired if 'the Kiwis been conned by a bunch of Yankee snake oil merchants and Sydney Harbour Bridge retailers?' There were serious technical questions to sort out.

ProLink XR-G and pellets were registered in NZ in September 1999 and the registration numbers are P05537 and P05536, respectively. Based on the presence of registered product, a plan was hatched (Box 2.1).

Egg survival in relation to treatment regimen

The other issue related to the capacity of eggs to withstand desiccation. Little was known of the capacity of *Campto* eggs to lie dormant in damp substrates such as mud or vegetation awaiting inundation by rain or tide. The main decision about how long eggs could withstand desiccation came from Pillai (1962) who compared the attributes of eggs of various species of *Aedes* and from that it was concluded that just about all of dormant eggs would hatch after three inundations within 9 months. Out in the field, eggs also are eaten by

Box 2.1: Eradication of the Southern Saltmarsh Mosquito from the Hawke's Bay: Eradication Plan, Revised 17 July 1999

Reassessment rules

- Should any of these assumptions prove to be incorrect, the response should be immediately reviewed.
- The southern saltmarsh mosquito is not established at any other site in New Zealand beyond those already identified in the Hawke's Bay.
- The southern saltmarsh mosquito will not spread beyond the known habitats in the Hawke's Bay (650 ha).
- Information on egg desiccation rates, instalment hatching, and the life cycle of the southern saltmarsh mosquito are correct.
- That the full range of breeding sites are treated to maintain a continuous lethal concentration throughout the water body for 10 months.
- That disinsection is ongoing for aircraft leaving Napier airport for the duration of the treatment programme.
- Risks associated with further incursions and/or interceptions will be cost-effectively managed.
- That road traffic is not a significant risk for spreading mosquitoes.
- Import health standards will be reviewed from the perspective of mosquito control and the standards are effectively enforced.
- Efficacy tests for S-methoprene will demonstrate that it is effective and environmentally sustainable against the southern saltmarsh mosquito.
- Zero larvae, pupae and adults will be achieved after 13 applications of the control agent, as described in the operational plan.

Uncertainties

- Longitudinal environmental monitoring shows no significant adverse environmental impacts.

Proposal for eradication

- Initial application of appropriate S-methoprene products over the known habitat (~650 ha) including areas currently dry or with falling water levels (mixture of pellets applied at 4 kg/ha and granules applied at 6 kg/ha). It was noted that S-methoprene is more cost effective than *Bti* because it has:
- a lower requirement for personnel to apply and monitor
- a lower risk of missed sites and hang-up in vegetation
- less need for over-treatment to ensure eradication.
- Repeated blanket application of S-methoprene products (mixture of pellets applied at 4 kg/ha and granules applied at 6 kg/ha) over the known habitat (approx. 650 ha) after 21 and 30 days for the granules and pellets respectively, as specified by the manufacturer to maintain an effective residual.
- The eradication plan allows for S-methoprene applications every 21 days over the whole habitat (approx. 650 ha). However, S-methoprene applications may only be required over restricted areas of the habitat because of limited water events (probably more likely to be ~500 ha). This would reduce costs but cannot be guaranteed.
- Reapply S-methoprene products 11 further times, after 21 and 30 days for the granules and pellets, respectively, over areas which are inundated. Dry areas will

have S-methoprene products re-applied every second application run. Sites subject to variable water levels should be identified. GPS and sentinel trays would be used to check application rates.

- *Bti* will be used for spot treatment on detection of untreated breeding areas.
- Ongoing monitoring of larvae, pupae and adults according to the operational plan (see below).
- Longitudinal eco-monitoring of sentinel species in the water column will be required to meet the requirements of the resource consent, to provide data for the application for full registration of S-methoprene and to provide information for risk communication with environmental and community groups.
- Habitat modification and elimination (filling depressions, clearing drainage channels, maintaining water levels in other channels and impounding) will be undertaken.
- Continue aircraft disinsection throughout the treatment period.
- Maintain national surveillance to the national standard and review import health standards.
- Check the sensitivity of the adult traps (e.g. by using crushes to provide animal-bait traps), particularly when numbers are low. Adult trapping will continue for at least 6 months after no adults have been found to increase the detection capacity for 'strays' but thereafter trapping would be instituted after inundation events only. The interval between monitoring may be increased as well.

Definition of eradication

- Adult and larval numbers decrease to zero following 13 treatments with control agents and/or habitat modification.
- Minimum water event to hatch eggs (habitats submerged to an identified level) and no larvae or adults found and
- Two years of surveillance with no evidence of eggs, larvae and adults, according to WHO.

predators such as ants and cockroaches, and may also be subject to fungal attack (Russell *et al.* 2001). In South Australia, Howard (1973) noted that a female after taking a blood meal may lay as many as 138 eggs per batch and observed that 41% would hatch on first flooding after 24 weeks, but this again was laboratory data.

Subsequently, the survey team noticed that on occasions there seemed to survival of some eggs longer than 9 months, but this was not the norm. Bader and Williams (2011) also produced laboratory data on egg survival and instalment hatching and found that no more than 56% of eggs would hatch on the first inundation and that by applying average hatching rates to a mathematical model, they concluded that there should be zero eggs present after three inundations within 11 months. When stored dry for 15 months, they also demonstrated that 13% of eggs remained viable, so it is possible that our working hypothesis of 9 months was optimistic.

Choice of S-methoprene

The review on environmental and health impacts of S-methoprene (Glare and O'Callaghan 1999) provides useful data but, not unexpectedly, much of the data related to the old formulations that contained both S- and R-isomers, which are not as relevant or as clean as new S-isomer-based products.

Nevertheless, this formed the bulk of our knowledge to work from. Isomers of methoprene hydrolyse in water by day 7 (Hangertner *et al.* 1976), although Wright (1976) listed a half-life as short as 2 days, but this may not be applicable to the new XR-G product. The stability of the XR-G product was a key issue in the eradication plan but the answer was an unknown. Thus, with the decision to base the eradication campaign on the new XR-G product, it was imperative that parameters were estimated in local context where possible.

In the reported absence of decay curve data (D. Sullivan, Zanus Corporation), the MoH strategy was based on the following assumptions:

- 100% of *Campto* will die at concentrations of S-methoprene from 0.1 to 0.5 ppb and above (Kay 1999). On the basis of known data on the effects of sub-lethal dosages, it is assumed that any survivors at 1 ppb would not have sufficient fitness to feed and reproduce normally.
- Applications at 4 kg/ha will deliver 40 ppb total/L in pools of mean 15 cm depth but it was surmised that the S-methoprene release profile would not be equal over the 21-day period of mosquito control – the period specified by the manufacturers. On recalculated figures, which reduce the half-life of S-methoprene to 3 days, and to some extent allowing for the impact of organics and microorganisms, we believed that in summer, the peak concentrations could reach 5–10 ppb, but stabilise at a level of around 5 ppb. In the absence of hard data, it was difficult to state any figure with authority, and the level may well be less.
- Nevertheless, the south-east Queensland aerial control programme against the saltmarsh *Ae. vigilax* ran successfully on 8 ppb total over 5 days, and the susceptibility of these two species seemed to be similar.
- The original conceptualisation of the eradication plan, therefore, was done on an understanding that sufficient flexibility would be built into the working protocol to facilitate adjustment and on an understanding of the decay of XR-G in Hawke's Bay.

Thus, for field monitoring, we believed that it may be theoretically possible to set cut-off points to initiate retreatment for *Campto* of 10–20 times above its LC_{100}. Given that microscale mapping of pools was now occurring, we also planned supplementary treatments of deeper breeding sites, or those with fluctuating levels, by suspending 'black socks' of pellets at defined heights on stakes. These socks would release product on inundation.

In Chapter 6, we detail what we learnt about using S-methoprene against *Campto* larvae, which were found to be susceptible to XR-G at levels below laboratory detection levels. We were horrified to learn that the entire active ingredient did not release from the granule, as we naively expected. Bioassay had to be used, rather than chemical analysis, to monitor treatments. After a couple of tense months, the evidence of the S-methoprene efficacy came through as the mosquito biomass collapsed and the programme tracked towards success.

Eradication begins

In June 1999, the NZ government agreed to fund a world-first eradication attempt. Applying the liquid *Bti* was relatively straightforward and both ground application equipment and aerial application was well developed in NZ at the time, but possibly not at the miniscule dosage rates applied for mosquito control. The application of ProLink XR-G granules posed several new challenges. Drawing on experience from Australia and the USA, local application equipment was retrofitted to deliver the granular- and pellet-based material and this went through several iterations (see Chapter 5). Many days were spent calibrating the equipment to ensure S-methoprene could be spread at the required application rates. This involved walking or riding quad-based equipment past sheets or greased trays to collect the pellets or granules where they fell and then counting the individual pellets or granules to work out swath widths and application rates. The same system was used for calibrating aerial applications. A technique of recording the flight path of the aircraft when applying product, using the helicopter Differential GPS system, and then printing the results onto an aerial photograph enabled us to ensure all of the identified habitat received an appropriate treatment.

Now all that was required was to:

- get the product registered in NZ
- apply for and gain permission for aerial and ground-based applications
- inform the public, community groups and environmental groups of the proposed campaign, application methods and product safety
- gain government support and funding
- demonstrate the efficacy of the new XR-G methoprene product
- import sufficient product from the USA to start and maintain the programme.

After much preparation and action of our 'to do list', S-methoprene applications commenced in August.

However, nature is a contrary thing and, while we were celebrating the good progress in Napier and Haumoana, news of infestations of *Campto* were reported at three additional sites on NZ's East Coast, at Gisborne, Mahia and Porongahau. Gisborne (Tairawhiti) and Mahia were ~95 km north of Napier in Hawke's Bay, whereas Porongahau was about the same distance south.

Although Mahia and Porongahau were in the same health district as Napier, and were therefore managed by the same public health unit, the Gisborne sites were in a separate public health unit district. Alongside assistance from the Napier MRC, the Gisborne Public Health Unit established their response team. Bryn Gradwell from NZ BioSecure was contracted to establish a local team of surveillance staff. He moved into the local hotel and made friends. Hiring local staff became an important feature of the evolving national programme. Local staff could ensure the local community understood the programme goals and ensure interest groups received the information they needed to support to the programme. Local Maori consider themselves custodians of the land, and *Campto* was an uninvited pest. A MoH review of the programme saw a further commitment of funding and resources and a reorganisation of the work. Changes included the establishment of a stand-alone mosquito laboratory in Napier with two full-time staff, and contracting the eradication programme to a commercial entity, NZ BioSecure.

By the time these new infestations were identified, we had overcome many of the initial obstacles. We had a full-time team of trained mosquito hunters, dedicated operation centres, four-wheel-drive vehicles and quad bikes, and sampling and treatment equipment and systems.

A further advantage was that our helicopter provider Helicopters Hawkes Bay Ltd worked in close partnership with us and was able to assist with investing in treatment equipment and technology to support our programme. In addition, their experienced pilots were also adept at identifying potential habitat from the air. It was relatively straightforward to scale up the operation to deal with the new areas. The new areas were dealt with in a systematic programme of delimitation surveys, intensive consultation with local stakeholders and then aerial and ground applications of S-methoprene. Again, the results were similar to that of Napier (i.e. a collapse of the mosquito population once the treatments had begun).

Again, the new areas were systematically brought under control using the same methodology successfully employed in Napier. However, once again as the mosquito population was nearing zero, *Campto* was found in the Kaipara Harbour, north-west of Auckland. This was a deal-breaker because Kaipara Harbour was said to be the largest harbour in NZ. After some deliberations, the NZ Government agreed to support a control operation in the Kaipara using *Bti* while they decided whether or not eradication was feasible (see Chapter 8).

The same issues arose when *Campto* was found in the Wairau estuary near Blenheim. Again, well-trained teams were deployed by helicopter and four-wheel-drive vehicles to systematically survey the extent of the habitat and to undertake initial control activities with *Bti*.

The advantages of building a core team of expertise meant that this could be deployed nationally. Mark Disbury had been employed to identify *Campto* but was soon able to expand his role into managing the GIS system and developing systems to map treatments and surveillance. Mark became an invaluable member of the MRC management team and went on to be involved in all of the subsequent finds of mosquitoes around NZ. Mark is now Manager of the National Mosquito Surveillance Programme (see Chapter 10).

Hope and Steve Crarer also began work as field staff at the start of the programme. Steve became operations manager at the Napier MRC and, like Mark, was involved in eradication activities around NZ managing the initial phases of the Wairau response. Hope became a surveillance team leader and was involved in training numerous staff around the country in surveillance techniques. Alongside their Gisborne counterparts, staff from the MRC were able to rapidly establish delimiting surveys in the Kaipara Harbour and Blenheim and support the new operations established in those areas.

Another key success factor in the eradication programme was the establishment of a comprehensive stakeholder engagement programme and communication strategy. The MoH liaised closely with the Hawke's Bay Public Health Unit to develop a communications strategy. A community liaison group in Hawke's Bay was established by 19 January 1999. Local networks were used to identify key stakeholders and affected landowners were identified through council records. Information collected was used to support resource consent applications for treatments in all areas. Feedback received was used to shape the programme. Media coverage moved quickly from the sensational 'Anti-spray group issues warning' in January 2000 to an objective 'Spray may have ended pest' by February 2000. Even in the early days of Thursday 28 January 1999, the *Dominion* newspaper (Fig. 2.2) through its local correspondent was spreading considered support and putting an identity to the planned action against *Campto*.

Although consultation took considerable resource to implement, (e.g. over 3 months of visiting landowners, conducting community meetings and visiting community groups including indigenous groups to enrol their support for eradication), the results meant that operations in all areas enjoyed widespread community support (see Chapter 12). No official complaints to consent authorities were received, despite highly visible aerial treatment of areas that lasted more than 12 months.

The importance of achieving eradication was brought home to Steve Garner during a trip to the Mildura area of Victoria, Australia. One school in the area had cancelled all outdoor activity and covered all open walkways between school buildings and all windows with mosquito netting due to presence of swarms of local *Campto*, with their predilection to feed during the day.

BILL KEARNS

Mr Dowler at his headquarters in Napier. In the background is where the battle to control the mosquitoes is being fought

Untouched, but he's out for blood

By PHILIP KITCHIN

NEW ZEALAND'S top Aussie mozzie fighter, Henry Dowler of Wellington, has never even been bitten by one of the exotic bloodsuckers.

In the weeks since Mr Dowler, 35, hung up his home telephone on Christmas Eve and told his wife and two children he would have to go back to his offices at the Health Ministry he has been exposed to possibly hundreds of the aerial invaders.

But aedes camptorynchus — the Aussie mozzie that can carry the debilitating Ross River virus — doesn't seem to like his blood. But the feeling's not mutual.

Since being whisked from Wellington to Hawke's Bay to lead an Anzac attack on the mozzies, he has worked 16-hour days as resources and people were scrambled in an effort to contain the invaders.

As a 20-year-old, Mr Dowler cut his teeth in his field of environmental health when he was working for a local authority doing what it could to contain the disastrous 1984 Invercargill floods.

Three years ago he shifted to the Health Ministry as deputy chief technical adviser and since then he has been thrown from one potential environmental crisis to the next.

First there was the Ruapehu eruption and the dangers from ash and water contamination, then there was the tussock moth infestation that threatened forestry — one of the biggest export earners.

Then last year, again during his Christmas break, Mr Dowler got the call to arms for the fight against the vicious Asian tiger mosquito which had landed in Auckland and Wellington.

The latest infiltration by an unwanted insect could prove the most difficult battle Mr Dowler has been involved in, but the jury on that score is out, at least till the Government decides whether it wants to raise the stakes to a war aimed at eradication.

When the spray programme began, Mr Dowler's gut feeling as to whether elimination of the pests could be achieved was "optimistic".

After yesterday's discovery of two more breeding colonies which came to life after widespread rain in Hawke's Bay, that gut feeling has deflated to a diplomatically phrased "cautious optimism".

"It really is too early to say whether it can be eradicated ... it depends on just how widespread it has become," he says. "At the moment, what we find one day is determining what we do the next."

But if eradication proves impossible because the mosquitoes have spread too far to make regular spraying practical or worthwhile cost-wise, Mr Dowler agrees that it will be inevitable that the Ross River virus will arrive in New Zealand.

The worst cases of the virus are crippling for victims who are stricken with several nasty severe flu-like symptoms, but the average person gets away with an attack not unlike a bad dose of flu. Lucky people are virtually unaffected.

Mr Dowler says that if the worst happens and we have to live with the virus, decisions on long-term control and containment could be made by local authorities in whatever areas the mosquitoes had spread to.

In the meantime, the war continues with seemingly ever-increasing amounts of marshland in Hawke's Bay being added to the target sights of helicopter pilots spraying breeding grounds.

Mr Dowler insists the spray is harmless to people and, to back his faith in it, he swears he would be prepared to allow the helicopters to douse him with it.

Asked about the likelihood of a blanket spraying operation where all gullies, dams, waterways and likely breeding grounds would be sprayed, Mr Dowler nods his head.

That broad-brush approach has been considered but "it's awfully expensive" and if, after blanket spraying a 10-kilometre radius of Napier, the mozzies are found a few kilometres further out, the exercise could be said to have been a waste of resources.

Mr Dowler says finding how far the mozzies have spread and then hitting the breeding grounds with spray is a more cost effective way of containment till a decision is made on elimination.

Figure 2.2. *Dominion*, 28 January 1999

References

Bader CA, Williams CR (2011) Eggs of the Australian saltmarsh mosquito, *Aedes camptorhynchus*, survive for long periods and hatch in instalments: implications for New Zealand. *Medical and Veterinary Entomology* **25**, 70–76. doi:10.1111/j.1365-2915.2010.00908.x

Glare TR, O'Callaghan M (1999) 'Environmental and health impacts of the insect juvenile hormone analogue, S-methoprene'. Report for the Ministry of Health, Wellington, NZ.

Hangertner WW, Suchy M, Wipf HK, Zurfluch RC (1976) Synthesis and laboratory and field evaluation of a new, highly active and stable insect growth regulator. *Journal of Agricultural Food Chemistry* **24**, 169–175.

Howard GW (1973) Aspects of the epidemiology of *Eperythrozoon ovis* in South Australia. PhD thesis. Department of Entomology, Waite Agricultural Research Institute, Adelaide.

Kay BH (1999) 'Laboratory evaluation of the efficacy of S-methoprene and three 1200 ITU/mg *Bti* products for control of *Aedes camptorhynchus,* compared to other Australian vectors'. Report to Ministry of Health, Wellington, NZ.

Margalit J, Dean D (1985) The story of *Bacillus thuringiensis israelensis* for the control of mosquitoes. *Journal of the American Mosquito Control Association* **1**, 1–7.

Pillai JS (1962) Factors influencing egg survival in *Aedes* with special reference to some Victorian species. PhD thesis. University of Melbourne.

Ritchie SA, Piggott D (1994) Efficacy of Altosid pellets against saltmarsh mosquitoes in SE Queensland. *Annual Report of the Local Authorities Research Committee* **1993–1994**, 70–74.

Ross DH, Judy D, Jacobson B, Howell JR (1994) Methoprene concentrations in freshwater microcosms treated with sustained-release Altosid® formulations. *Journal of the American Mosquito Control Association* **10**, 202–210.

Russell BM, Kay BH, Shipton W (2001) The survival of *Aedes aegypti* (Diptera: Culicidae) eggs in surface and subterranean breeding sites over the northern Queensland dry season. *Journal of Medical Entomology* **38**, 441–445. doi:10.1603/0022-2585-38.3.441

Russell TL, Gatton M, Ryan PA, Kay BH (2009) Quality assurance of aerial applications of larvicides for mosquito control: effects of granule and catch tray size on field monitoring programs. *Journal of Economic Entomology* **102**, 507–514. doi:10.1603/029.102.0207

Wright J (1976) Environmental and toxicological aspects of insect growth regulators. Environmental Health Perspectives **14**, 127–132.

3

Solid government legislation and support as a key to success

Sally Gilbert, Andrew Forsyth, Ian Gear, David Sullivan and David Yard

In early June 2001, the NZ Government agreed to fund the attempted eradication of *Campto* (the Southern Saltmarsh mosquito, *Aedes camptorhynchus)* in Napier, Gisborne (Tairawhiti), Mahia and Porangahau. In June 2002, the government approved an eradication programme for the Kaipara (including Mangawhai and Whitford), and in November 2004, it approved an eradication programme for the Wairau region (including Grassmere). The re-emergence of *Campto* in the southern Kaipara in 2005 resulted in the extension of the programme for that area. In May 2006, the Coromandel Peninsula was included in the programme.

In this chapter, we examine the national management and coordination of the eradication programme, including the roles of the government and government agencies, technical advice, structural arrangements and funding.

The role of government

The NZ Government enters into agreements with its Directors-General or Chief Executives of all its Ministries and Departments for the provision of specific services or outputs. During the *Campto* eradication programme, the Director-General of Health signed an agreement with the Minister for Biosecurity each year that the MoH managed the programme (i.e. from 1999 to 2006).

The objective of the annual agreements was to provide the Minister for Biosecurity with information to assess the strategic importance and value of the departmental outputs and to make comparisons with similar outputs across both the public and private sectors. Outputs and key result areas were detailed in the agreement. The Director-General of Health was accountable to the Minister for Biosecurity for the delivery of the eradication programme to the quality, quantity and cost specified, including the delivery of any outputs subcontracted to a third party (i.e. contracts for the delivery of the eradication programme at all the sites around NZ).

During the MoH's management of the programme, quarterly progress reports were provided to the Ministers for Biosecurity, Health and Finance that included progress with/success of eradication to date, implementation of the reassessment rules (see Chapter 4 for a discussion of the reassessment rules), national surveillance, expenditure against budget to date, advice from the TAG and any emerging issues or other relevant information.

Table 3.1. Dates of briefings and papers provided to government

	Jan 1999	Feb 1999	March 1999	April 1999	May 1999
Briefings to Ministers	6, 15, 20, 21	3, 10, 11	5, 12	8, 20, 22, 23	18, 24, 26, 31
Reports to Government	22	8, 15 (2 papers)	12, 22, 26	–	2 June 1999

These regular reports were supported by *ad hoc* briefings to inform joint Ministers of any new findings, unusual events and noteworthy achievements, and by informal fortnightly meetings with the Minister for Biosecurity to discuss the programme.

During the eradication programme, several Ministers were given responsibility to oversee the response to the mosquito incursion. There were also several Parliamentary elections and a change of government during the time. The Ministers and Associate Ministers with responsibility for the eradication programme at various times during its implementation were:

- Ministers and Associate Ministers for Biosecurity (Hon. Simon Upton, Hon. John Luxton, Hon. Jim Sutton, Hon. Marian Hobbs, Hon. Jim Anderton, Hon. David Carter and Hon. Nathan Guy)
- Ministers and Associate Ministers of Health (Hon. Wyatt Creech, Hon. Annette King, Hon. Damien O'Connor and Hon. Pete Hodgson),
- Ministers of Finance (Hon. Bill English and Hon. Michael Cullen)
- The Treasurer (Hon. Bill Birch).

Ministers received numerous papers (Table 3.1) and were required to consider funding, legislative, *Official Information Act* releases, and administrative decisions throughout the programme. The volume of briefings, reports to Ministers and the government during the initiation of the response programme, until the end of the 1998–99 financial year on 30 June 1999, was substantial.

Hon. Marian Hobbs, Minister for Biosecurity during much of the eradication programme, recalls the response:

'Unlike other pests, such as the blessed insect that could have wiped out pine forests, that we had to spray in West and South Auckland, the southern saltmarsh [mosquito] has the potential to be a carrier of Ross River virus, and also to really affect the outdoor/indoor lifestyles of Kiwis in many months of the year. The barbecue culture would have been hit hard.

'It was an unusual one to get through Cabinet, because there was no immediate economic threat to any sector, but there was no opposing argument from Treasury that I could recall ... The [Minister] of Finance did live in the area, although I do not think he was aware of the threat.

'I think beginning in Napier, the team managed to eradicate the invader there. I can remember a helicopter ride to look at the terrain that was being sprayed and treated. There was strong public acceptance of the spraying. It was not heavily populated and the work with those living in the area explained the threat to lifestyle and health that this mosquito presented. I can remember being thanked by locals in the following years.

'While we were in a self-congratulatory mode, we learnt that the mossie had got into the water area around Dargaville and south of it. This was much more extensive breeding area, and the fight here was longer and more intensive. Yet again unlike West Auckland, there was an acceptance of the need to do this. So the communications team really worked successfully as well as the clever and dedicated teams working on foot, as I remember, as well as using spray from above.

'As I left office in 2005, I do not think it had been totally wiped out, but it had been contained.

'It was costly, but no-one ever argued against that. Maybe we were lucky that it was not on the North Shore, but was in inhospitable rural land.

'I have good memories of a clever team working well with communities and using a range of methods to hunt the mossies down and more particularly the 'eggs' for want of a clearer scientific term!'

By 12 March 1999, the most urgent decisions had been made. Issues around *Resource Management Act 1991* requirements had been resolved. Immediate funding decisions had been made and ongoing funding issues were now being considered as part of the routine government Budget processes. Notwithstanding this, nearly all reports to government, and briefings to Ministers, during April, May and June 1999 related to funding.

By the start of the 1999–2000 financial year (i.e. July 1999), the government's involvement with the eradication programme had moved to a more routine status. On 29 July 1999, the MoH submitted its first quarterly progress report to Ministers overseeing progress with eradication, results of S-methoprene field efficacy testing, reassessment rules, the financial situation, local government assistance, consultation and next steps.

The third routine progress report was submitted to joint Ministers on 31 January 2000. There had been a change of government between the second and third progress reports, so this was the first report to the incoming Labour-led government. The report continued to cover progress with eradication, financial situation, local government assistance and consultation, but for the first time also included an assessment of the programme against the reassessment rules.

Administrative arrangements

The *Campto* eradication programme was managed by the NZ Government, through its public service. The government sets the policies and priorities for the public service to deliver, makes the laws, establishes the administrative structures (Ministries), provides the funding, and holds its Ministries accountable for the delivery of programmes and stewardship of Crown funding.

When the *Campto* incursion was detected, biosecurity responsibilities (in particular legislative powers, functions and duties) were undertaken by four government departments: the MAF, MoH, Ministry of Fisheries and Department of Conservation. The MAF's Biosecurity Authority provided a leadership and coordination role in biosecurity activities, which included administering the *Biosecurity Act 1993*. A series of memoranda of understanding documented an overarching framework for the agencies to work together on biosecurity matters. Under these arrangements, the MoH was acknowledged as the government's principal advisor on all matters related to human health. At the time of the incursion, the MoH had responsibility for responding to exotic organisms of public health significance.

In 2003, the government gave the Chief Executive of the MAF accountability for end-to-end management of the biosecurity system, and therefore for the Chief Executive of the MAF to be accountable to the Minister for Biosecurity for health, environment, economic and social/cultural outcomes associated with pests and unwanted organisms under the *Biosecurity Act*. Following a transition period, the *Campto* eradication programme was transferred to the MAF's Eradication Programmes section on 1 July 2006.

The role of the Ministry of Health

Campto is a vicious biter and a competent vector of human arboviral diseases, including Ross River virus disease and Barmah Forest virus disease. Because the harm associated with this mosquito is primarily to people, the MoH accepted responsibility for the incursion response.

The MoH provided the overall leadership for the response. It was responsible, and accountable, for ensuring the programme had appropriate strategic and operational eradication plans in place, scientific and technical advice and support was available, funding was secured, statutory and other obligations were complied with, sufficient human and other resources were available, and information was provided to Ministers, other agencies, relevant statutory bodies (such as the Conservation Authority and the Pesticides Board), public health units, territorial authorities and regional councils, community groups and the public. This included providing information to the media, affected communities and the wider public, and ensuring any complaints or queries were appropriately recorded and followed up.

When the incursion response was first identified in December 1998, the MoH had significantly reviewed and revised its mosquito surveillance and response programmes in the previous 2 years. The MoH had commissioned Brian Kay in 1997 to review and advise on the exclusion and surveillance of exotic mosquitoes of public health significance in NZ (Kay 1997).

As a result, the MoH had developed standard operating procedures for interception and incursion responses, and had commissioned several reports including health and environmental impact assessments of two potential mosquito control agents: *Bacillus thuringiensis israelensis* (*Bti*) and S-methoprene.

The MoH (and health protection officers in public health units) had acquired some (limited) experience in the surveillance of container-breeding mosquitoes and in responding to interceptions of exotic mosquitoes of public health significance found in risk goods (such as used tyres and machinery) being imported into NZ. However, these responses were at the border, in defined goods (or on specific vessels) and limited to container-breeding species.

In addition, the MoH (and government sector generally) had not yet adopted the Coordinated Incident Management System (CIMS) that was such a successful feature of later emergency responses, including responding to further *Campto* incursions and to other pests and diseases of human health significance such as pandemic influenza (H1N1 2009). On the other hand, MoH officials and health protection staff were generally experienced in emergency responses to events such as natural disasters and chemical spills, and had a strong culture of sharing information and resources, and of mutual support and assistance.

Notification of the incursion

On Christmas Eve 1998, when Gene Browne (the MoH's contracted entomologist) phoned to report the positive identification of *Campto* from Napier, the MoH had effectively already closed for Christmas. Sally Gilbert recalls receiving the phone call:

> *'I was the duty on call officer for the Christmas period that year, and was the only person remaining in the office when the call came. I had had little involvement with the mosquito programme at that point, as Henry Dowler was the MoH's expert and had been leading the development of a comprehensive mosquito surveillance and response programme.*
>
> *'I recognised Gene's name, but I did not know the significance of Gene's advice, other than it was very important and extremely urgent. I had no inkling that this was the start of a more than 10-year challenge!*
>
> *'After taking the details from Gene, I phoned Henry who, of course, immediately recognised the significance of the identification. He changed his family's holiday plans, so instead of Rotorua, the Dowler family went to Napier. Henry was able to develop and lead the local response with the Hawke's Bay public health staff.*
>
> *'We had a very lucky break that, after the heavy rain that had triggered the hatching, the Hawke's Bay then enjoyed several weeks of its usual hot and dry summer weather. This gave us time to work out what we needed to do in the immediate term, to give us time to assess the long-term response.'*

The MoH's immediate response

The MoH's standard operating procedures outlined seven actions for an immediate response (MoH 1999). This period is from the initial notification (24 December 1998) until the end of the containment phase (30 April 1999).

Action 1 – Confirming the identification of the mosquito

The specimens identified by Gene Browne on 24 December 1998, were confirmed by Mark Bullians (Agriquality) on 25 December 1998. Specimens were also sent to Richard Russell (University of Sydney at Westmead Hospital, Australia) for international confirmation, and this was received on 12 January 1999.

Action 2 – Undertaking a risk assessment

By 30 December 1998, Hawke's Bay health protection staff had identified *Campto* as a competent vector of human arboviral diseases. A draft report *A Health Risk Assessment Relating to the Establishment of the Exotic Mosquitoes* Aedes camptorhynchus *and* Culex australicus *in Napier, New Zealand* (Hearnden 1999) was submitted at the first TAG meeting on 19 January 1999 (for further information see 'Health Risk Assessment' in Chapter 4).

Action 3 – Informing public health units

The Hawke's Bay Public Health Unit was immediately informed by Gene Browne of his identification of *Campto* (i.e. before his notifying the MoH). Chapter 2 describes the response of the health protection staff who received the notification. Information was provided to all public health units on 8 January 1999, in the first of a series of circular letters:

- 14 January 1999: public health units were directed to undertake enhanced surveillance for saltmarsh mosquito species in their regions.
- 21 January: this advice was reinforced.
- 26 January: copies of the *Australian Mosquito Control Manual* were sent to assist with the enhanced national surveillance.
- 12 February: copies of a report on the biological profile and surveillance for *Campto* were sent.
- 26 February: copies of the health risk assessment were sent.
- 14 April 1999: further information and an update on the response were sent.

Action 4 – Defining policy and strategy

NZ traditionally experiences a shut-down period over Christmas–New Year, when many organisations close for staff to enjoy the summer break that coincides with the beginning of the school summer holidays. At best, only skeleton staff will be on call in most government departments and council offices. With the incursion being notified late in the day on Christmas Eve, contacting officials on other agencies as well as key communications, legal, and financial staff within the MoH was difficult.

However, hot dry weather in the Hawke's Bay was very helpful. On 30 December 1998, Henry Dowler noted that:

> *'Current hot, windy conditions in Napier will keep the adult mosquito numbers down as well as quickly reducing the available habitat for larvae/pupae to complete their life cycle ... However, we can expect another mosquito population explosion if there is a further significant rain event in the area this summer. As the immediate problem has passed, it is this possibility that we now are working through ...'*

During the weekend of 16–17 January 1999, however, there was a significant rainfall event in the Hawke's Bay. It was assumed that this would trigger egg hatching and larval development. On 17 January, Sally Gilbert warned staff at the MoH that there was a 'need to make some firm decisions and get things moving...'

The incursion was upgraded to a national biosecurity issue. By 18 January 1998, MoH staff had prepared an Eradication Contingency Plan and the first TAG meeting was held on 19 January. The TAG provided expert advice to the MoH, recommending immediate containment. (For further information about the TAG see Chapter 4). An interagency meeting (MAF, Department of Conservation, Ministry of Research, Science and Technology, and Treasury officials) was also held on 19 January 1999 to provide advice to MoH officials, and to discuss assistance and support for the response.

On 20 January 1999, the Minister for Biosecurity exercised statutory powers under the *Biosecurity Act* to enable the application of *Bti* over the infested habitat (see later in this chapter for further information).

On 10 February 1999, the NZ Government agreed to Phase One until 30 April 1999. This would enable applications of *Bti* over the infested habitat, as well as ongoing surveillance and the development of a long-term response plan.

Considerable activity focused on a cost–benefit analysis of response options (see Chapter 4 for more detail). It required the development of detailed and robust costings for the proposed eradication programme and the epidemiological advice relating to disease rates if the mosquito established in NZ. The establishment of the TAG (see Chapter 4) was crucial in developing the eradication programme policy and strategy, and identifying the key assumptions that underpinned the cost–benefit analysis and health risk assessment.

Action 5 – Establishing the Napier Mosquito Response Centre (MRC)

By 18 January 1999, the Napier MRC was sufficiently resourced and effectively operating independently of the host public health unit, and moved out of the public health unit offices to a separate office (coincidentally overlooking the saltmarsh habitat). Chapter 2 describes the establishment and operation of the Napier MRC in detail.

Action 6 – Liaison

Ministers were first briefed on the incursion on 6 January 1999; the Health Funding Authority (which contracted and funded public health units) was informed on 30 December 1998. Other government agencies, including the MAF, Department of Conservation, Treasury, Ministry of Research, Science and Technology, and Department of Prime Minister and Cabinet were notified of the incursion during the first week of January 1999. (The support of other agencies is described in more detail later in this chapter.)

Action 7 – Field actions

Field operations are detailed elsewhere in this book (see Chapters 5, 8 and 9) and included a delimiting survey, habitat survey and mapping, surveillance (adult and larval trapping), treatment and mitigation of habitat. It also included media communications, information for key stakeholders including local government and medical practitioners. Although aircraft disinsection was implemented, other movement controls were not applied. Trace-back was not applicable, although work on pathways of entry was undertaken.

Moving to eradication

On 26 April 1999, the NZ Government decided to proceed with Phase Two, the eradication of *Campto*. The eradication plan involved treating the habitat for 13 months with S-methoprene, and at least 2 years of surveillance with no evidence of the presence of the mosquito. Eradication activities in the Hawke's Bay were supported by national saltmarsh surveillance. The eradication plan included reassessment rules and criteria for demonstrating eradication.

Over the next 6 years, *Campto* was found in several other sites, as described elsewhere in this book. However, the strategic eradication plan, developed for the MoH by the TAG, remained applicable, with minor revisions as the programme matured. The eradication plan described the key components of the programme:

- history of eradication programme describing sites under treatment or surveillance, and identifying sites where eradication had been declared and that were now part of the National Saltmarsh Surveillance Programme
- criteria for completing the treatment phase of an eradication programme:
  - treatment with S-methoprene over two summers
  - no adults for at least 9 months (thus allowing for egg desiccation)
  - three water events (thus allowing for 99.9% instalment hatching)
  - no *Campto* larvae detected.

- criteria for reducing the frequency or intensity of surveillance after treatments have been completed:
  - time period after last light trap find
  - sufficient light trapping and dipping
  - water event occurred
  - reliable local advice on water events and inundation
  - time period since last treatment
  - definition of water event or inundation that triggers surveillance
  - seasonal (i.e. approaching summer or winter)
  - extent of proposed surveillance (e.g. post-inundation or sentinel sites).
- process of eradication:
  - adult and larval numbers decrease to zero as a result of repeated treatments with control agents and/or habitat management
  - a minimum of three water events to hatch eggs (habitats submerged to an identified level) and no larvae or adults found.
- declaration of eradication:
  - follows 2 years of active surveillance with no evidence of eggs, larvae, pupae or adults detected (as recommended by the World Health Organization).

The operational plan that then applied to each specific operational site included:

- the site history (detection, delimiting and progress)
- a treatment plan: maintaining an effective lethal concentration of S-methoprene in all identified positive habitats, for at least two treatments; and then an effective lethal concentration of S-methoprene in all inundated habitats for two summers and for three water events. The treatment plans used an ongoing, rolling, zone-based approach. S-methoprene was applied as pellets to drains and deep habitat on a 30-day cycle or granules for other habitat on a 21-day cycle. Delivery platforms included aerial (helicopter) and ground-based equipment.
- accommodating the needs of the organic farming and other operations without compromising the eradication programme, through habitat mitigation and use of organic-certified *Bti*
- assessing treatment efficacy through the surveillance programme, and through the field collection of pupae and monitoring successful adult emergence
- a communication strategy, which included identification/confirmation of target audiences, roles and responsibilities, key messages, methods (e.g. hui/meetings, newsletters, email networks, public notices, advertising, media releases, etc.), stakeholder and community liaison, media strategy and health education materials
- habitat mitigation: clearing drains, artificial flooding, filling depressions to eliminate ponding, removal of dense/thatched vegetation and identification of any land modification that potentially produces new habitat
- surveillance of adults, larvae and pupae: the entire wet habitat is surveyed after each water event within the time appropriate to maximise finding larvae, and any live larvae or pupae collected and returned to the laboratory for screening identification. In addition, the area-wide adult light-trapping programme will continue with a mix of sentinel and random sites. Larval sampling is also used to establish before and after densities for trend analysis over time.
- monitoring of potential adverse effects associated with the application of treatment agent(s) to the environment: active monitoring was not undertaken but assessment

of bird survey records and any other harvesting/monitoring programmes identified in the area were undertaken if available.

Coordinated Incident Management System (CIMS)

The other significant change to the MoH's national leadership of the eradication programme was the concurrent development and adoption of CIMS to respond to all emergencies in the health sector.

CIMS is the model adopted by central and local government agencies and emergency services in NZ for the command, control and coordination of emergency response. It is intended to provide a flexible structure, allowing the multiple agencies or units involved in an emergency to work together as a team, and can be scaled up or down as the response requires. It is also consistent with the World Health Organization's guide for public health emergency contingency planning at points of entry, required for compliance with the new *International Health Regulations 2005.*

Because the CIMS framework has been widely adopted, it means that incursion responses can be conducted inter-operably with other agencies and local government. Roles and responsibilities in emergency responses are clearly understood, and the framework ensures all aspects of the response are considered and delivered. The key features of CIMS, when applied to an incursion response, include:

- national (or regional or local) coordinator: responsible for the overall direction of the response
- key decision maker (statutory or technical): Chief Technical Officer or equivalent
- planning and intelligence: collation and analysis of surveillance data, scientific research, technical advice, strategic and operational planning, forecasting, health risk assessment, cost–benefit analysis, operation of the TAG and preparation of situation reports (sitreps)
- operations: field response, operational and tactical planning and response
- liaison: interactions with government and government agencies and international liaison
- communications: media strategy and statements, public information and advice, health education, community and special interest groups
- logistics: internal (e.g. finance, legal, administration, duty rosters, record keeping, food and IT support) and external (e.g. stores and supplies, transport, staff secondments and accommodation).

The structure would be replicated at regional and local levels and in the MRCs as applicable, with the local (or regional) coordinator reporting to the coordinator at the next level. The incursion response initiated significant action from different sections of the MoH:

- significant roles:
  - strategic advice and statutory decisions: Chief Technical Officer (Health), Deputy Chief Technical Officers (Health) and Chief Advisor
  - management and resourcing: Deputy Director-General, Manager and Team Leader
  - policy advice, technical advice: senior analysts (3), analysts (3), senior advisors in environmental health (2), health protection (2), public health engineering, toxicology, epidemiology, communicable diseases, epidemiology, biostatistics, data analysis and biosecurity.

- Supporting roles:
  - governance and oversight: Director-General of Health and Chief Internal Auditor
  - purchasing: Director (Finance and Commercial) and Director (Purchase Interest)
  - communications: Manager and Senior Advisor
  - legal: Manager and Senior Solicitor
  - finance: Manager, financial analysts and vote analysts

The detail of much of this activity is recorded elsewhere but Box 3.1 provides an example of the task list on one day taken at random.

Box 3.1: Activities being undertaken and status on 10 October 2000

Items in italics are required to be completed urgently.

TAG

- *draft national surveillance report for the TAG 13 October – under action*
- *quarterly surveillance report due 10 October 2000 – contractor to advise*
- provide literature review to the MAF, and invite them to participate in the next TAG meeting if there are any outstanding issues they wish to raise – on hold
- request update reports for 1 week before next meeting – done 9.10.00
- write up meeting minutes and distribute as drafts – to be done

Communications

- media statement following TAG and officials meetings – to be done

Officials Group and meetings with officials

- officials meeting 16 October 2000 – to be done

Disease/spray drift complaints surveillance

- nil

Briefing to Ministers and reports to government

- *Napier eradication progress report to joint Ministers due 31 October 2000 – under action*
- progress briefing following delimitation survey and TAG and officials meeting – to be done

Authorised persons and other statutory appointments

- nil

Statutory actions

- nil

Treatment agents

- full registration of S-methoprene for sale and use in NZ – David Sullivan (manufacturer) – progressing

Finance and expenditure

- monitor and advise – ongoing
- update costings and potential expenditure – ongoing
- clarify HFA funding – meeting Thursday 12 October

Miscellaneous
- mosquito identification contract – with provider for signature
- check RRV incidence with ESR – to be done
- audit the July/Oct time delay in sample identification – under action
- ESR to investigate feasibility of DNA fingerprinting mosquitoes – to be done
- revise health impact assessment – to be done

Gisborne Mosquito Response Centre
- establish response centre – under action
- DCTO to visit Wednesday – under action
- delimiting survey – under action
- report to the TAG – to be done
- aircraft disinsection – under action

To think about later
- clarify whether resource consents needed and coordinate science support – to be done
- thank you letters to stakeholders, etc. who helped – to be done

Transfer from the MoH to the Ministry of Agriculture and Forestry

In 2003, the government agreed that the accountability arrangements for biosecurity be reorganised by giving the Chief Executive of the MAF accountability for end-to-end management of the biosecurity system and therefore accountable to the Minister for Biosecurity for biosecurity activities that contribute to health, environment, economic and social/cultural outcomes. In mid-2004, the MoH and MAF officials agreed that the MoH would retain responsibility for all saltmarsh mosquito eradication and surveillance programmes, at least until the end of the eradication programme, because the eradication programme was, by definition, finite. However, it was noted that there may be reasons to shift accountability for either activity earlier – if the MAF has developed the required capability, was ready, and both agencies agreed.

In July 2005, the MAF indicated it wished the transfer of the eradication programme to the MAF to occur by 31 December 2005. MoH officials requested assurance that 'MAF has developed the required capability and is ready' to take accountability for the eradication programme and could show that the continuity of the programme was guaranteed and not jeopardised. Suitable reassurances were given and, in late 2005, it was agreed that the *Campto* eradication programme would transfer to the MAF with effect from 1 July 2006. Sally Gilbert was seconded to work in the MAF for half a day each week to brief MAF staff and ensure the appropriate systems, structures and arrangements were in place before the transfer.

On 1 July 2006, the eradication programme transferred to the MAF. At that time, the programme covered Whangaparaoa, Wairau (including Grassmere), northern Kaipara, southern Kaipara and the Coromandel Peninsula. Sally Gilbert's secondment was extended into October 2006, because of the departure of the two MAF officials who had been the key contacts with MoH officials during the transition period. Early in 2007, David Yard, a permanent MAF staff member, was assigned responsibility for the eradication programme, and he was able to provide leadership of the programme until its successful completion.

As an indication of the scope of work required to successfully transfer such a complex programme, the following summary lists some of the key areas of activity undertaken during the preparation for the transfer. Note that this summary only refers to the national management of an ongoing and established programme: it does not include any operational activity or programme development.

Administration of transfer completed

- agreement on scope of transferring responsibility for the eradication programmes
- development and maintenance of a work programme and timeframes for the transfer
- development and implementation of a communications plan
- key stakeholders advised of the transfer of responsibility for the eradication programme
- appropriate government approvals and consequent processes to transfer funding for the eradication programme
- deeds of novation prepared to transfer eradication contracts and TAG letters of agreement
- advice provided on contract service specifications in MAF contracts with SMS
- analysis of terms of reference, payment arrangements and membership of the TAG
- development of work schedule for ongoing MAF work, with dates and tasks
- site visits for MAF officials and liaison with contractors.

Ongoing eradication programme implementation during the transfer period

- monitor eradication progress, including field visits and meetings with stakeholders
- convene TAG meetings including preparation of background papers, general administration, note taking and complete any action points resulting from meetings
- quarterly progress reports to Ministers
- quarterly progress reports to the TAG, public health units, officials group and other stakeholders
- maintain, revise and implement communications strategy
- contract management and negotiation of milestones for next financial year, contract report responses, invoice management, audits and foreign exchange management
- purchase of ProLink XR-G and foreign exchange management including audit of stocks in hand
- interim response to extended treatment in southern Kaipara and Coromandel incursion: funding, contracting, briefings, exemptions from *Resource Management Act* consents, eradication planning, stakeholder and media communications

Related work programme completed during the transfer period

- Agnique® environmental and health impact assessment report as a potential control agent: discussions with importer and report released publicly
- incursion response protocol redrafted to more explicitly fully adopt CIMS and liaise with the MAF's incursion response project team
- protocols reviewed
- peer reviews of reports, correspondence, Ministerial correspondence briefings and meetings, and media liaison
- ongoing scanning and commissioning of health and environmental impacts of potential control agents, including adulticides.

Some work intended to be undertaken by the MoH did not proceed.

- follow-up recommendations for mosquito-control agent and non-target species monitoring reports
- two recommendations from the *Pathways of Entry* report (two other recommendations were implemented): to consider laboratory studies examining the possibility of *Campto* being picked up on footwear and carried to another site and field, and laboratory research to critically examine the range of *Campto* breeding sites
- undertake ongoing scanning and commissioning of health and environmental impacts of potential control agents, including adulticides.

MAF stewardship

Following the transfer of the eradication programme to the MAF, ministry officials took the opportunity to review the programme and the principles that underpinned it. The MAF revised the terms of reference and membership of the TAG, moving it from a technical and operational role to a scientific role.

In August 2006, the MAF briefed the government seeking additional funding for 2006–07. In its advice, the MAF noted that, since the mosquito was first detected in NZ in 1998, eradication had been achieved at seven of 10 sites and was tracking well at Whangaparaoa. The MAF then noted the reappearance of the mosquito at southern Kaipara Harbour in November 2005, the continued presence of the mosquito in the Wairau and the detection of *Campto* on the Coromandel Peninsula in May 2006.

The MAF advised it had analysed possible response options and concluded that, at this stage, eradicating *Campto* from NZ remained the most desirable course of action. The MAF noted that although the TAG was confident that eradication of the mosquito from NZ was possible, 'there were a range of views on this'. The MAF was concerned that eradication could only be achieved if infested sites were detected and treated quickly enough to prevent further spread and that there may be infestations of the mosquito that had not yet been detected. The MAF recommended that the government should continue the *Campto* eradication programme until June 2007, by when the MAF would report back with advice on the long-term direction of the programme.

The MAF then commissioned scientific research on the key assumptions that underpinned the cost–benefit analysis and secondary disease response options. It also monitored and reviewed the results of national mosquito surveillance.

In 2007, MAF officials advised the government that the mosquito had been eradicated from eight of 12 sites. Of the four remaining sites, two were tracking well, while the MAF had extended treatment timeframes at two sites in response to further detections of the mosquito. National surveillance had found no new infested sites.

The MAF advised that technical experts' views on the feasibility of eradication varied. Its experts questioned whether all populations of *Campto* could be detected, but were more confident that, once detected, eradication could succeed. The MAF had identified three potential options:

- no further intervention
- a targeted control programme or
- continuing the eradication programme.

Taking no further intervention would have immediately halted the eradication programme and no further action would have been taken until the impacts of the mosquito generated 'sufficient pressure for a response', which would have been to nuisance biting and/or to cases of arboviral diseases.

The targeted control option would have reduced mosquito populations at sites where health and nuisance impacts would be most severe, and encouraged people to avoid exposure to mosquitoes via a social marketing campaign. Costs were estimated at NZ$11 million over 3 years in current sites near population centres. It was noted that, over time, mosquito control may have needed to be extended to additional communities and because eradication was not the primary aim, the costs would be incurred over a much longer timeframe.

Continuing eradication would involve at least 3 years of treatment and surveillance at the four sites currently within the programme, and would cost approximately NZ$11 million. An additional contingency of up to NZ$6 million was also made available, to be used if new sites were found or if it were necessary to extend treatment at existing sites.

The MAF's analysis suggested that continuing the eradication programme was the best response option available. The government agreed and directed the MAF to continue the programme for a minimum of 3 years, to eradicate *Campto* from NZ. The MAF would review the eradication programme if new populations of the mosquito were detected or if eradication was not achieved by 2009–10.

Although small-scale finds continued, particularly in the Wairau, there were no new areas of infestation identified and the programme continued to a successful conclusion in July 2010.

Structural Arrangements: Evolution from Local to Regional to National Eradication

Public health units throughout NZ are specialist arms of regional health service providers called district health boards. These boards are responsible for the health of populations in their regions and provide hospital-based services, community health services and public health services.

With the initial Hawke's Bay incursion falling within a single region, the Hawke's Bay District Health Board was responsible for delivering the eradication programme within its local area. Particularly in the initial response phase, it was supported by MoH officials, staff seconded from other public health units, from technical and scientific experts from throughout NZ and from overseas (principally Australia, but also the US), and from local community groups and council staff. As the programme matured, the public health unit established a stand-alone mosquito response centre, fully staffed with appropriate technical experts, field operators and administrative assistants.

On 4 October 2000, identification of specimens of *Campto* from the Tarawhiti region was made. As a result, Tairawhiti public health staff initiated a local response, with technical support and advice from the MoH and entomological specialists were contracted by the MoH. A mosquito response centre was established in Gisborne and experts seconded or contracted from throughout NZ to staff the response centre and provide advice and assistance. Southern Monitoring Services Ltd trading as NZ BioSecure became the primary contractor and effectively operated the Gisborne MRC. In addition, two staff from the Napier MRC travelled to Gisborne to assist with the response.

In December 2000, on the advice of its TAG to develop synergies between the operations in Tairawhiti and Hawke's Bay, the MoH assigned the overall planning and management of the regional response to the Napier MRC, using sub-contractual relationships with the Gisborne MRC. The operational aspects were also reviewed for optimal efficiency:

- screening laboratory maintained at the Napier MRC
- operational planning and technical audits managed by Napier MRC
- purchase/lease of equipment, hire of vehicles and employment of staff managed by Napier MRC, via the Gisborne MRC for Tairawhiti operational requirements
- treatment of all sites planned and managed by Napier MRC
- surveillance, monitoring, community liaison and liaison with the local public health unit managed by each MRC within the overall operational plan
- ongoing routine biosecurity responses, disease investigations and complaint investigations managed by the local public health unit
- Tairawhiti Public Health Unit provided GIS and administrative support to the Gisborne MRC.

On 12 June 2001, following the discovery of *Campto* in Kaipara and Mangawhai, the NZ Government announced a combination of options:

- continue the eradication activities in Napier
- commence eradication in Gisborne, Mahia and Porangahau
- contain/control in Kaipara and Mangawhai
- undertake secondary disease response in Auckland and Northland.

As a result of this decision, the MoH sought closed tenders for the provision of a national mosquito response service to:

- establish and maintain a national response centre with appropriately experienced and expert staff
- manage the response to incursions of exotic mosquitoes of public health significance, including consultation and liaison with the relevant public health unit(s)
- support public health units responding to interceptions of exotic mosquitoes of public health significance as required by the Chief Technical Officer (Health)
- support public health units with advice, information and guidance as required by the Chief Technical Officer (Health)
- provide an immediate and effective response to continue to completion the current eradication programmes in Hawke's Bay and Tairawhiti in coordination with the local public health units (the respective roles of the successful tenderer and the public health units in the eradication programmes will be defined in the contract for services and it is anticipated that these roles will be the subject of memoranda of understanding between the successful tenderer and the public health units)
- provide an immediate and effective response to develop and maintain a containment/control programme in Kaipara and Mangawhai in coordination with the local public health units (the respective roles of the successful tenderer and the public health units in the eradication programmes will be defined in the contract for services and it is anticipated that these roles will be the subject of memoranda of understanding between the successful tenderer and the public health units)
- develop an arboviral surveillance and response programme in conjunction with secondary disease response activities undertaken by the MoH and public health units
- monitor, and advise the Chief Technical Officer (Health) of the adequacy of, public health units' mosquito surveillance activities and response activities

- review and revise the nationwide surveillance programme for exotic mosquitoes of public health significance
- review and revise the national contingency planning guidelines
- review, and advise the Chief Technical Officer (Health) of the adequacy of, public health units' contingency plans
- develop a health promotion programme for travellers to areas with endemic arboviral diseases, including advice on implementation through public health units, travel agencies and medical practitioners
- review the existing resources and provide advice on appropriate health education resources to support public health units.

NZ BioSecure was the successful tenderer and was contracted to provide the national exotic mosquito response service. In establishing the service, NZ BioSecure secured and/or continued the employment of key staff from the Napier and Gisborne MRCs. This ensured continuity within the programme and that crucial experience and technical knowledge was not lost.

NZ Biosecure successfully operated the National MRC for the remainder of the period the eradication programme remained with the MoH. Following the transfer of the eradication programme to the MAF on 1 July 2006, the MAF contracted an operational manager to 'give comfort' to MAF officials on the effective operation of the eradication programme. The MAF has a well structured response model and the use of an operations manager to oversee all operational activities of a response is standard practice.

NZ Biosecure remained the main contractor until 2008, when a decision was made to put the contract out to tender to comply with government procurement processes. NZ Biosecure was unsuccessful in its tender application, and the contract was awarded to Flybusters Antiants Ltd, which remained the main contractor through to the declaration of eradication in July 2010.

The roles of other government agencies

The support of several government agencies was crucial to the success of the eradication programme. The *Mosquito Eradication Process Review* (Thomson and Cressey 2000) found that one of the key components of the Hawke's Bay response, which appeared to have been pivotal to its effectiveness, was the establishment of reference/advisory groups, including an interagency liaison group.

Prior to the establishment of the interagency liaison group, and throughout the entire eradication programme, officials at several agencies were informed of the incursion and kept updated with copies of the briefings prepared for Ministers and the circular letters sent to public health units. They were sent progress reports and updates on the programme. These agencies included the Department of Conservation, Department of Labour, Department of Prime Minister and Cabinet, Ministry of Foreign Affairs and Trade, MAF, Ministry of Research, Science and Technology, State Services Commission, and the Treasury.

A more formal interagency liaison group was formed as a forum for information sharing and cooperation between interested government agencies. The membership of the group was somewhat fluid, but included: Department of Conservation, Health Funding Authority, MAF, Ministry for the Environment, Treasury, and Ministry of Research, Science and Technology.

The first interagency meeting, held on 19 January 1999, was attended by officials from the Department of Conservation, MAF, Ministry of Research, Science and Technology, and the Treasury. The meeting followed the significant rainfall event in Hawke's Bay over the weekend of 16–17 January and discussed areas of assistance, support and advice on policy, strategy and operations. Further meetings of this group were held on 21 and 29 January 1999. Examples of the advice and support provided by these agencies included:

- Department of Conservation: asked staff to identify potential habitat in the conservation estate throughout NZ, obtained permits for the capture of wild animals for bleeding for the MAF serosurvey and provided letters of support for resource consent applications and/or exemptions from the need for consents
- Ministry of Research, Science and Technology: provided advice and checked Australasian insecticide stocks and advised on science support available to support the programme
- MAF: seconded staff to the MRC to assist with delimiting surveys, etc., agreed to two staff joining the TAG, undertook an animal serosurvey of farm animals and other potential hosts for Ross River virus disease, provided technical assistance to obtain an experimental use permit for the mosquitocides, assessed pathways of entry of *Campto* into NZ, added information to their 0800 number, advised on aircraft disinsection and made staff available to assist as required
- Treasury: provided ongoing input into preparation of reports to the government and progress reports and, in particular, the preparation of the cost–benefit information to inform decisions on eradication.

The interagency liaison group adopted a less formal involvement after their meeting of 29 January 1999. However, the lines of communication developed during the early stages of the response continued to generate value throughout the response, as representatives were provided with regular copies of quarterly progress reports.

In addition to the MoH and Health Funding Authority staff identified previously as involved in the programme, the following summarises the numbers and roles of key officials from other agencies involved in supporting the response:

- Department of Conservation: Chief Science Officer, Senior Advisor and policy analysts (2)
- MAF: Director-General, Biosecurity Authority Group Director, Biosecurity Policy Manager, Forest Biosecurity Director, Chief Technical Officer (Agriculture), Chief Technical Officer (Forestry), Agricultural Compounds and Veterinary Medicines Group National Manager (Toxicology and Residues), Animal Biosecurity Programme Coordinator (Surveillance and Disease Response), National Advisor (Risk Management), Policy Advisors (2), Pesticides Board Registrar, Technical Assessor (Toxicology and Residues), National Manager (Plant Residues) and MAF Legal
- Ministry of Research Science and Technology: Director, Senior Advisor
- Treasury: Social Policy Branch Health Director, Health Purchase and Regulation Section Manager, analysts (5), Regulatory Economics Directorate–Land and Natural Resources Manager

Liaison with local government

Although the key liaison with local government (district and regional councils) was undertaken by the operational staff in each eradication zone, MoH officials convened a meeting with local government officers from the Napier City Council, Hastings District Council and Hawke's Bay Regional Council on 7 April 1999.

This meeting was to provide an update on the response. It noted that national surveillance had been undertaken (and was continuing) but, to date, *Campto* appeared to be limited to the Hawke's Bay. The NZ Government had asked MoH officials to identify any potential contributions to assist with the response, and the meeting was part of that process. The local government representatives indicated support in principle, but were unable to provide specific commitments until there was further information about the government's decision on whether or not to eradicate the mosquito.

Although further detail is available in the chapters dealing with responses in each location, throughout the eradication programme local government was strongly supportive of the responses underway. Councils provided a range of support, including:

- assistance with resource management consents or exemptions to the requirement for a consent
- assistance with habitat identification
- secondment of staff and provision of vehicles
- assistance to identify residents, landowners, key stakeholders and individuals
- access to council land
- reporting nuisance biting complaints
- briefings for their councils.

In a specific example of council support, Napier City Council funded staff and contractors to complete significant drainage improvements and site filling around habitat zones. Local government officers were provided with regular updates on the programme from the local mosquito response centres but also the national progress reports and updates prepared by the MoH.

The *Biosecurity Act 1993*

The *Biosecurity Act 1993* provides the primary legislative framework to manage exotic and other organisms, other than diseases in humans. It was enacted to 'restate and reform the law relating to the exclusion, eradication, and effective management of pests and unwanted organisms'. It provides a wide range of powers to reduce risks from unwanted/exotic organisms and to deal with pests established in NZ.

In the context of the *Biosecurity Act 1993*, the MoH was particularly concerned with organisms capable of causing, or of contributing substantially to, the introduction or wider distribution of a serious human disease and/or the epidemic spread of a serious human disease.

Several actions were taken, using the powers in the Act, to respond to the *Campto* incursions.

Henry Dowler, Deputy Chief Technical Officer (Health), declared *Campto* to be an unwanted organism under section 102(1) of the Act on 8 January 1999:

> *'The Southern Saltmarsh mosquito,* Aedes camptorhynchus, *is an organism by virtue of the fact that it includes a genetic structure (i.e., DNA) that is capable of replicating itself (which structure comprises only part of the entity).*
>
> *'I believe that the Southern Saltmarsh mosquito,* Aedes camptorhynchus, *apparently established in Hawke's Bay (Napier) is capable of causing unwanted harm to human health. It is capable of transmitting Ross River Virus disease and is a possible vector of Murray Valley encephalitis.*

'I certify that I am a Deputy Chief Technical Officer appointed under section 102(1) of the Biosecurity Act 1993, *and that the* Aedes camptorhynchus *is an unwanted organism for the purposes of the Act.'*

The reason for determining *Campto* to be unwanted was so that certain powers under the *Biosecurity Act 1993* could be exercised against those organisms as and when necessary. An unwanted organism has that status throughout NZ. Where an organism has been declared an unwanted organism under the Act, biosecurity officers (such as inspectors and authorised persons) have a wide range of powers (such as powers to search premises) in relation to such organisms.

The *Biosecurity Act 1993* provides for the appointment of a range of statutory officers, with varying powers, to undertake the implementation and enforcement of the Act, including Chief Technical Officers, Deputy Chief Technical Officers, Inspectors, Authorised Persons and Accredited Persons. The Chief Technical Officer appoints non-MoH employees only as Authorised or Accredited Persons and delegates limited powers only to them. The Chief Technical Officer can appoint Authorised Persons under Section 103 of the Act if the person has 'appropriate experience, technical competence, and qualifications relevant to the area of responsibilities proposed to be allocated to that person'.

To ensure that these criteria were met, the Chief Technical Officer (Health) required Authorised Persons to have public health training, qualifications and experience, as follows:

- designation or appointment as a Health Protection Officer or Medical Officer of Health
- attendance at the MoH training courses
- appropriate experience and/or other relevant training and qualifications
- endorsement from the public health manager that the individual is competent and suitable for appointment.

Authorised Persons appointed by the Chief Technical Officer (Health) were given limited powers under the Act, rather than the full range of powers available. The powers available were considered sufficient for response activities. An Authorised Person may:

- require information to be provided (S.43)
- employ or request any person to assist them (S.106)
- enter and inspect any place other than a dwelling house, marae or building associated with a marae (S.109, 112)
- record information (S.113)
- use general powers to eradicate or manage a pest or unwanted organism (S.114)
- use devices (S.115)
- seize abandoned goods (S.119)
- examine organisms, organic material, goods or material (S.121)
- apply an article or substance to a place to ascertain the presence or absence of a pest or unwanted organism(S.121A)
- give directions to treat goods, water, places, equipment, etc.; to destroy any pest or unwanted organism or harbourage; or take steps to prevent the spread of any pest or unwanted organism (S.122).

The exercise of powers was subject to the Authorised Person:

- having appropriate experience, technical competence and qualifications to be appointed (S.103(4))

- complying with instructions from the Chief Technical Officer (Health) (S.104)
- acting in good faith, with reasonable cause, and taking reasonable care at all times (S.163, 164).

Authorised persons were also required to produce their letter of appointment if they exercised their power of entry on every occasion they exercised that power (i.e. if re-entering a property during a single visit and on any subsequent visits).

Conversely, the following powers were not considered to be likely to be required by an authorised person and the Chief Technical Officer (Health) did not authorise their use. If an Authorised Person expected to need to use any of these powers, they were required to contact the MoH urgently to discuss the matter. Those applicable to the *Campto* response were:

- applications for warrants to enter and inspect marae or dwelling houses etc. (S.110, 112)
- applications for warrants for the entry and search of any place where an offence punishable by imprisonment may have occurred or be occurring (S.111)
- seizing evidence (S.118)
- declaring a restricted place (S.130, 133, 134)
- investigating and/or instituting enforcement action for offences under the Act (S. 154, 155, 161).

The MRC and some public health units employed technicians doing surveillance and related activities. People could be accredited under section 103(7) of the *Biosecurity Act 1993* to 'perform… functions consequential upon the exercise of powers under the Act by an … Authorised Person'. This means that Accredited Persons could carry out functions required due to the exercise of powers by an Authorised Person such as administrative functions, analysing specimens and undertaking field tests. It did not require the Accredited Person to be under the supervision or direction of the Authorised Person.

Accrediting a person under section 103(7) recognised people had experience, technical competence and qualifications appropriate to the area of responsibilities allocated to them and ensured they had the full protection of the immunity provisions of the Act (sections 163 and 164).

The *Resource Management Act 1991*

Application of insecticidal sprays was not a permitted activity under the various regional air quality or water quality plans in place for the regions where incursions were detected. However, Section 7A of the *Biosecurity Act 1993* provided for the Minister to exempt actions taken in an attempt to eradicate an organism, from Part III of the *Resource Management Act 1991*. This meant that resource consents were not required unless the eradication programme lasted more than 20 working days.

In considering whether or not to exempt the actions, the *Biosecurity Act 1993* required the responsible Minister to be satisfied that it is likely that:

- the organism is not established in NZ, the organism is not known to be established in NZ, or the organism is established in NZ but is restricted to certain parts of NZ
- the organism has the potential to cause all or any of significant economic loss, significant adverse effects on human health, or significant environmental loss if it becomes established in NZ or if it becomes established throughout NZ

- it is in the public interest that action be taken immediately in an attempt to eradicate the organism.

MoH officials were required to brief the Minister with sufficient and adequate information for him to make this decision. The information provided by the Ministry covered:

- the establishment of *Campto* in NZ: status of current eradication programmes, areas where the mosquito had been successfully eradicated, the newly discovered site, assessment of potential habitat and likelihood of spread beyond its current distribution, results of national surveillance
- significant adverse effects on human health: summary of impacts of the epidemiology and symptoms of Ross River virus disease, likelihood of viraemic people coming to the area; advice that any long-term establishment of *Campto* represented an irreversible hazard with significant consequences – allowing circumstances for amplification of an introduced arbovirus (most probably RRV from viraemic tourists) and the occurrence of a large 'virgin soil' epidemics
- immediate action is in the public interest: essential to increase the probability of success of the application programme, because the control agent is only effective against mosquito larvae, not pupae or adults. Containment of the mosquito before it spread from the current limited habitat was in the public interest because it made the likelihood of a successful eradication programme higher. It also reduced the cost of eradication, limited the area being considered for larvicidal treatment, and reduced the area under consideration for potential habitat modification. Limiting the spread of the mosquito by immediate action for eradication was also in the public interest because it reduced the likelihood of the mosquitoes coming into contact with a viraemic individual and becoming a vector for Ross River Virus disease. It would also reduce the biting nuisance problem and public anxiety from the threat of Ross River Virus disease.

Section 7A(2) of the *Biosecurity Act 1993* also required the Minister to consult with the relevant consent authorities before making a decision. The Minister was not required to consult with anyone other than the regional council and even this is 'only to the extent possible in the circumstances'. However, the Act did enable the Minister to consult other people considered to be representative of those who may be affected.

Prior to seeking the Minister's exemption, the MRC contacted the consent authorities (usually the local regional council) and, if time permitted, other relevant stakeholders such as the district council, Department of Conservation (if Conservation reserve land was affected), other key landowners and the local *Iwi* (indigenous tribes). Their views were reported to the Minister but, in all cases, were supportive of the programme and of the exemption being issued.

After making the decision, the Act required the Minister to give public notice of the decision 'in such a manner as the Minister thinks fit'. The public notice had to specify the organism to be eradicated, the principal actions taken in the attempt to eradicate the organism, and the areas affected by the action.

In each case, therefore, MoH officials drafted a public notice that specified the matters required by the *Biosecurity Act 1993* and placed a copy of the notice in newspapers circulating in the area. A media statement was also released, along with background information.

Following the 20-day exemption, the eradication programme was required to have either obtained a resource consent or had regulations passed by the government to exempt the eradication programme from the requirement for a consent.

Box 3.2: Biosecurity (Resource Management Act Exemption) Regulations 2004 (No xx)

Governor-General
Order in Council
At Wellington this day of [Month Year]
Present:
His Excellency the Administrator of the Government in Council
Pursuant to section 7A(6) of the *Biosecurity Act 1993*, His Excellency the Administrator of the Government, acting on the recommendation of the Minister for Biosecurity, and on the advice and with the consent of the Executive Council, makes the following regulations.

Contents

Regulations
1 Title
These regulations are the Biosecurity (Resource Management Act Exemption) Regulations (No xx) 200x.

2 Commencement
These regulations come into force on [date].

3 Expiry
These regulations, unless sooner revoked, continue in force until the close of [date], and then expire and are deemed to be revoked.

4 Eradication actions and exemption defined
In these regulations,—
eradication actions means actions taken in accordance with Part 6 of the *Biosecurity Act 1993* in an attempt to eradicate *Aedes camptorhynchus* from the [location] area
exemption means the exemption that—
(a) was granted by the Minister for Biosecurity on [date]; and
(b) expired on [date]; and
(c) exempted eradication actions from the provisions of Part 3 of the *Resource Management Act 1991.*

5 Continuation of exemption
The exemption is continued.
Explanatory note
This note is not part of the regulations, but is intended to indicate their general effect.
These regulations, which come into force on [date], relate to actions taken under Part 6 of the *Biosecurity Act 1993* in an attempt to eradicate the Southern Saltmarsh mosquito (*Aedes camptorhynchus*) from the [location] area. Those actions were exempted from the provisions of Part 3 of the *Resource Management Act 1991* under an exemption granted by the Minister for Biosecurity on [date] under section 7A(1) of the *Biosecurity Act 1993*. The exemption expired on [date]. These regulations continue the exemption. These regulations will, unless sooner revoked, continue in force until [date].
Issued under the authority of the *Acts and Regulations Publication Act 1989*. The *Biosecurity (Resource Management Act Exemption) Regulations 2004 (No 2)* are administered in the Ministry of Health.

Promulgating regulations within 20 days is always challenging and MoH used the cooperation and support from several government agencies, including the Cabinet Office, to enable papers to be prepared and submitted urgently. (As a comparison, a routine Cabinet Paper for government consideration usually takes a month to draft and a further 2–3 months for consultation with other agencies and consideration by the government.)

In advice to the government, the Minister for Biosecurity advised that Ministerial powers had been exercised to authorise a 20 working-day exemption from Part III of the *Resource Management Act* under section 7A of the *Biosecurity Act 1993*. The exemption would expire after 20 days so regulations were required to continue the exemptions. The regulations would remain for 2 years unless sooner revoked.

The Regulations were generally very brief and straightforward and took the form shown in Box 3.2.

The *Health Act 1956*

The *Health Act 1956* includes provisions relating to statutory nuisances – circumstances that may be offensive or injurious to health, including conditions giving rise to the breeding of mosquitoes – and gives powers to councils to require such conditions to be abated. However, the provisions are most applicable to container-breeding species, because they can require residents to clean up their properties and were not helpful for widespread natural habitat. The *Health Act* also includes provisions for the detection and exclusion of disease vectors, such as rodents and mosquitoes, at ports and airports and on craft, as part of the requirements for border health protection. This framework comprises a multi-layered set of measures designed to protect human health and is generally effective. However, it does not guarantee inviolable public health security at the border.

The *Hazardous Substances and New Organisms Act 1996*

At the inception of the incursion response programme, approvals for the mosquitocides being used were obtained under the *Pesticides Act 1979*. When, the *Hazardous Substances and New Organisms Act 1996* came into force, the transition provisions transferred the approvals to the new legislative regime. These approvals covered the composition, labelling and controls on the use of *Bti* and S-methoprene.

The *International Health Regulations 2005*

The *International Health Regulations 2005* provide for the detection and mitigation of international threats to public health while avoiding unnecessary interference with travel and trade. The regulations take an all-risks approach (covering biological, chemical and radiological hazards) with responses being proportionate to the public health risk. The *International Health Regulations 2005* replaced the more limited 1969 Regulations and took effect in 2007, towards the end of the programme, so did not have a direct impact on it. The relevant sections of the *International Health Regulations 2005* at the time of the initial incursion are listed in Chapter 1.

The Regulations apply to all potential sources of harm and explicitly include vector and host species of diseases of human health significance. However, if a similar response was required in future, under the *International Health Regulations 2005*, NZ would likely need to inform the World Health Organization of the situation and the response measures being implemented. The 2005 Regulations include a decision tree and reporting framework for events of potential international public health significance. The MoH has established a National Focal Point and operationalised the decision tree to enable reporting if required.

Funding

All government agencies are funded each year through the annual Budget process. The funding is appropriated to specific output classes, to fund the work that has been agreed with the Minister for that output. New funding may be applied for as part of the Budget process and the government holds limited contingency funding for unforeseen developments and emergencies.

When considering response options over the years, Ministers viewed eradication as a cost-effective option. Even when control/containment options were very similar to eradication options in terms of cost-effectiveness, it was recognised that:

- containment/control programmes did not guarantee that *Campto* would not spread from areas of establishment to other available habitat throughout NZ over time
- the risk of Ross River Virus epidemics would increase the longer potential vectors are present in NZ
- successful eradication would ensure that Ross River Virus and other arboviral diseases transmitted by *Campto* will not become established in NZ; and,
- eradication would also obviate the need for an ongoing mosquito-control industry to manage the nuisance and health effects.

Eradication was recognised to be challenging, but still technically and scientifically feasible.

However, the initial response in Hawke's Bay generated considerable work in securing funding for the response. At that time, no specific contingency fund was available for biosecurity responses. Consequently, the funding of any sizeable and sustained response required the approval of the government to commit additional resources to the issue. This is summarised below.

Funding the Hawke's Bay Incursion Response.

When the first incursion was detected, the MoH had no funding or resources identified for such a response. On 6 January 1999, the MoH advised the Minister for Biosecurity of the establishment of *Campto* in Hawke's Bay and noted the possible financial implications. Specifically it was noted that current MoH activities were being met from within Vote Biosecurity, while the local public health unit was seeking contingency funding from the Health Funding Authority (for further information about the role of the Health Funding Authority see later in this chapter).

On 21 January, the MoH indicated it would meet the current emergency costs, but would seek reimbursement from the government's emergency response fund. By the beginning of February 1999, the MoH had identified NZ$270 000 it would transfer from its core Health funding to support the response. It also requested a further NZ$1.2 million to fund the response. However, Treasury officials recommended that this funding be found from within the Health Funding Authority's risk reserves or from predicted underspend. Discussions continued throughout February 1999 until it was agreed that the Health Funding Authority would transfer NZ$200 000 from their risk reserve, and the balance (NZ$917 000) would be new funding. This result represented a compromise between MoH and Treasury advice.

With the funding for the initial response determined, attention moved to funding the long-term response.

During March 1999, considerable activity revolved around the establishment of good cost estimates for the proposed eradication programme. Treasury's concerns mainly revolved around deviations from agreed timelines to supply them with information for analysis.

The Director-General of Health also wrote to several relevant parties, seeking to discuss possible sources of funding:

- Hastings City Council
- Napier City Council
- Hawke's Bay Regional Council
- Local Government NZ
- Employers Federation
- NZ Tourism Board (Office of Tourism and Sport)
- Ministry of Education
- Ministry of Research, Science and Technology (MoRST)
- Ministry of Fisheries
- Department of Conservation
- MAF
- Health Funding Authority

All respondents indicated that either they did not see the response to *Campto* establishment as their responsibility or they did not have available money to commit to this situation.

In early April 1999, considerable attention centred on the cost–benefit analysis formulated by John Mumford (1999) and refined by Peter Clough (NZIER) (refer Chapter 4 for further information). Particular attention centred on the projected case rates for a virgin soil epidemic, because the cost–benefit analysis was very sensitive to this figure. Treasury officials also indicated continued concerns around the costing of a local control option, the justification for minimal agency contributions, the assumption in the analysis that a further incursion was likely, discount rates and direct costs to government. In late April 1999, Treasury officials indicated that the benefit to cost ratios were 'very marginal' and that Treasury 'will be recommending against eradication'.

A paper was prepared for the government at the end of April 1999 seeking agreement to proceed with Phase Two (eradication) of the response to *Campto* with a 13-month treatment programme, based primarily on S-methoprene, a 24-month surveillance programme to confirm success and supported by national surveillance. Funding for the 1998-99 financial year had been determined but not for 1999-2000 and the following years.

The government agreed with the proposal but directed the MoH to report back on potential sources of funding for 1999–2000 and the following years. The MoH was also directed to prepare quarterly progress reports for the joint Ministers that included expenditure against budget to date, the need for contingency funding and any assistance provided by local government. The MoH was directed to negotiate with local government agencies in Hawke's Bay with a view to agreeing their level of assistance.

As a result, the Director-General of Health wrote to local government and other agencies seeking financial contributions for the eradication programme. The responses reiterated earlier reasons why funding could not be contributed to the mosquito eradication programme.

Discussions about sources of funding continued for the next several months, largely focused on whether the Health Funding Authority was able to contribute additional funding from its risk reserves. The Minister of Health felt that the programme should be funded almost entirely from new appropriations, while the Treasurer was suggesting a middle ground approach, with ~50% funding (NZ$2.5M) from Health Funding Authority risk reserves and 50% from new appropriations. In June 1999, this 'middle ground' approach was agreed by the government.

Following these initial discussions, the framework for further funding was established, using a mixture of underspend within the MoH's budget, and supported by new funding. Approval for new funding was generally able to be made through the routine government Budget processes, when funding bids are prioritised across the wider government sector.

However, it would be fair to say that the funding of the response to the mosquito incursion in Hawke's Bay was a critical and ongoing issue throughout the response period. Although all agencies with biosecurity responsibilities were funded for biosecurity, this was for routine activities and typically far too small to affect any serious response to a biosecurity emergency.

In addition, the funding chain from the MoH through the Health Funding Authority to the MRC added complexity. Funding was dependent on decisions in the government at several points through the timeframe of the response. Although the government approved each operational step in the response, the approvals were always contingent on investigation of funding mechanisms. These investigations invariably resulted in the MoH recommending that the response be funded from new appropriations and the Treasury recommending that the response be funded from within Health Funding Authority baselines. These uncertainties were translated down the contracting chain, with the MRC threatening to cease operations in the absence of financial assurances from the Health Funding Authority and the Health Funding Authority being unwilling to commit resources in advance of decisions by the government.

Support from XR-G Manufacturers

The greatest cost to the programme was the purchase of ProLink XR-G. In order for the eradication programme to be successful, the manufacturers recommended using a higher level of product per unit area (6kg/ha). The manufacturers (Zoecon and Zanus Corporation), established a special government price for this project.

David Sullivan, founder of Zanus Corporation, describes the significant price discount offered by the manufacturers to support the programme, because of its public good component:

> *'We reduced the price to the NZ MoH by 21% and by 24% for granules and pellets, respectively. The granules were priced at US$15.47/kg and the pellets were priced at US$43.99/kg. (United States prices at the time were $19.51/kg and $57.50/kg for granules and pellets, respectively). The pricing structure was established for delivery to the MoH ... Pricing was not changed during the entire [10 year] programme.'*

The role of the Health Funding Authority

At the time of the Hawke's Bay find, the complexity of the health structures created certain hurdles that had to be negotiated. Although the MoH provided the strategic direction, technical advice, statutory decision making, and liaison with government, the funding for the response and the contracts with public health units were the responsibility of the Health Funding Authority. (This reflected the wider role of the Health Funding Authority in the health sector; that is, the Health Funding Authority funded all publicly funded health services including hospitals, community health services and public health units.)

Staff in the Health Funding Authority provided active support for the response in several areas, including:

- working with public health managers to second health protection officers to Hawke's Bay to undertake delimiting surveys and other field work to support the response
- attending interagency liaison group meeting and observing TAG meetings
- contracting an audit of public health national saltmarsh surveillance
- purchasing light traps to enable and support public health national saltmarsh surveillance.

However, the split responsibilities for the response and for funding created problems with assuring the continuity of the response when funding was uncertain. On three separate occasions (in April, June and September 1999), the continued delivery of the programme was under threat because the funding had not been confirmed. On these occasions, the Health Funding Authority was unable to confirm with the Napier MRC that funding would be available to continue the programme and it may have to cease. Although funding was eventually secured, the uncertainty and risks to the programme created significant tensions, uncertainty and additional work for MoH officials and staff at the Napier MRC.

This had a direct impact on the response, by creating delays in ordering S-methoprene. On 29 July 1999, the Napier MRC asked the Health Funding Authority about the status of its contract. Without a contract, the Napier MRC had been unable to order S-methoprene on 30 June 1999. Delays in funding and contracting resulted in a belated start to the application of S-methoprene.

In addition, transactions costs were created by the public health unit contracts for services being held with the Health Funding Authority, while response activities were directed by the MoH.

A subsequent restructuring of the wider Health sector resulted in the disestablishment of the Health Funding Authority and the integration of the purchasing function into the MoH. As a result, the operational, funding, and contracting functions were aligned and such risks did not arise again throughout the course of the programme.

In addition to the MoH staff identified previously as involved in the programme, the following summarises the numbers and roles of Health Funding Authority officials involved in some way:

- governance and oversight: Chief Executive, General Manager (Public Health)
- public health unit contract management: Public Health Senior Locality Managers (4)
- mosquito response purchasing contract management: Portfolio Manager (Public Health).

Programme transfer to the MAF

When the MoH transferred the eradication programme to the MAF in 2006, it transferred all funding associated with the programme to cover the completion of the programme,

Table 3.2. *Campto* incursion sites and quantum of funding approved by government 1998–2010

	Date *Campto* detected	S-methoprene treatment started	Last adult mosquito detected	Last mosquito larvae detected	Treatment ceased	Eradication completed	Funding rounded** (NZ$M)
Napier	Dec 1998	Aug 1999	April 2000	Aug 2000	April 2001	July 2002	8.2 (incl. Haumoana)
Mahia	Nov 2000	Nov 2000	April 2001	Aug 2001	Aug 2002	Aug 2003	N/A
Tairawhiti	July 2000 (ID Oct 2000)	Nov 2000	Sept 2002	Sept 2002	June 2003	Sept 2004	3.5 (incl. Mahia and Porongahau)
Porangahau	Oct 2000	Nov 2000	June 2002	Aug 2002	April 2003	Sept 2004	N/A
Mangawhai	April 2001	Oct 2002	Dec 2002	Nil	April 2003	Dec 2004	N/A
Whitford	Mar 2002	Oct 2002	April 2002	Nov 2002	March 2004	Nov 2004	0.2
Kaipara	Feb 2001	Oct 2002	Sept 2003	Feb 2004	June 2004	June 2006	30.0 (incl. Mangawhai)
Kaipara (south)*	Nov 2005	Nov 2005	April 2006	May 2006	April 2008	Aug 2008	3.5
Whangaparaoa	Jan 2004	Jan 2004	Nil	May 2005	April 2006	May 2007	3.5
Wairau	May 2004	Feb 2005	Feb 2006	June 2008	June 2010	June 2010	16.2 (incl. Grassmere)
Grassmere	May 2004	Feb 2005	Feb 2006	Aug 2006	April 2007	Aug 2008	N/A
Coromandel	May 2006	May 2006	Nov 2006	April 2007	April 2008	May 2009	5.1M

* re-emergence of *Campto* in the southern Kaipara.
** direct costs only, not including internal government agency costs.

other than the recently detected incursion in the Coromandel Peninsula. MAF officials reviewed the operation and funding of the programme.

As a result, the NZ Government appropriated a further NZ$11.217 million between 2007 and 2010 for the eradication of *Campto*. It also approved a contingency fund of NZ$2 million per year to be used in the event of detection of additional populations or persistence of the mosquito at sites under treatment. From 1998 to 2010, the direct appropriation costs, excluding internal government agency costs, amounted to NZ$70.2 million (Table 3.2).

References

Hearnden M (1999) 'A health risk assessment relating to the establishment of the exotic mosquitoes *Aedes camptorhynchus* and *Culex australicus* in Napier, New Zealand'. Unpublished report to the Ministry of Health, Wellington, NZ.

Kay BH (1997) 'Review of New Zealand Programme for exclusion and surveillance of exotic mosquitoes of public health significance'. Unpublished report to the Ministry of Health, Wellington, NZ.

Ministry of Health (1999) 'Biosecurity section of the Environmental Health Protection Manual'. Unpublished report to the Ministry of Health, Wellington, NZ.

Mumford J (1999) 'Cost benefit analysis procedures to be applied to unwanted organisms or pest responses'. Unpublished report to the Ministry of Agriculture and Forestry, Wellington, NZ.

Thomson B, Cressey P (2000) 'Mosquito eradication process review'. Unpublished report by the Institute of Environmental Science and Research Ltd to the Ministry of Health, Wellington, NZ.

4

Listening to the experts – the advice that drove the national programme

Sally Gilbert, Andrew Forsyth, Ian Gear, David Sullivan and David Yard

The Southern Saltmarsh Mosquito Technical Advisory Group

In its *Mosquito Eradication Process Review* (September 2000), the Institute of Environmental Science and Research Ltd (ESR) found that one of the key components of the Hawke's Bay response that appeared to have been pivotal to its effectiveness was the establishment of reference/advisory groups including a Technical Advisory Group. ESR described the Southern Saltmarsh Mosquito Technical Advisory Group (TAG) as 'provid[ing] a wide range of international expertise in the absence of NZ experience'. The TAG was formed on 18 January 1999 and met for the first time on 19 January 1999.

Examination of the minutes from the TAG meetings makes it clear that all significant technical decisions associated with the Napier mosquito response were either made on the recommendation of the TAG or after peer review by the TAG. Crucially, the TAG reviewed the eradication programme against re-assessment rules at each meeting.

At each meeting, the TAG would either confirm no reassessment rule had been triggered **or** advise which reassessment rule(s) had been triggered and comment on the implications for the response to the *Campto* incursion(s).

The reassessment rules were a list of assumptions or circumstances that, if breached or incorrect, would require the NZ Government response to *Campto* to be immediately reviewed. The rules are listed below:

- *Campto* is not identified from any new sites other than the current operational zones.
- Surveillance and delimiting surveys are maintained; there is confidence that re-introductions are being detected. This refers to coverage of potential habitats, frequency of surveillance, appropriateness of method and timing, and amount of resourcing available to cover potential habitats.
- Studies of potential entry pathways demonstrate that such pathways identified can be controlled so that there is confidence that re-introductions can be minimised.
- Information on egg desiccation rates, instalment hatching and the life cycle of *Campto* is correct.

- Efficacy tests for S-methoprene demonstrate that it is effective and environmentally sustainable against *Campto*.
- The programme is not found to have an adverse environmental effect; that is, an [unacceptable] effect on non-target species.
- Information from surveillance and/or other sources suggests the eradication plans are effective; for example, zero larvae, pupae and adults are achieved after repeated applications of the control agent are completed.

In any case, eradication plans would be reviewed following any unforecast or unusual event.

The initial terms of reference for the TAG were brief , but were progressively revised, sometimes in response to concerns about the conduct of members. A key revision was to specify the expectation for members to attend meetings and accept collective responsibility for decisions.

By 2002, the terms of reference (Box 4.1) were largely complete and underwent little revision (other than updating attendance fees and expenses) while the eradication programme remained with the MoH.

The terms of reference noted that the TAG reported to the MoH's Chief Technical Officer (Health), that the MoH would provide secretarial support, and that the MoH may have observers in attendance. The terms of reference also included a media policy that prevented members making media statements unless authorised to do so.

The TAG was composed of a range of experts, identified as being needed to support the Chief Technical Officer (Health) in key decisions. A review of the mosquito eradication programme by ESR found that:

> *'The composition of TAG appeared to be appropriate for the situation being dealt with and at certain points was supplemented by the invitation of additional experts to attend specific TAG meetings (Professor Richard Russell, Dave Sullivan, Doug Van Gundy). This ability to call on a wide range of international expertise was pivotal as MoH freely admitted that they were dealing with a situation "where we do not have expertise in NZ" (Gillian Durham, letter to Treasury, 2 March 1999).'* (Thomson and Cressey 2000)

The MoH provided secretarial support to the meetings. Prior to all meetings, agenda and background papers were sent to members, and meeting notes were sent to members following the meetings. Meetings were initially held monthly, but these reduced to quarterly or half-yearly as the programme matured. The TAG was convened if it appeared any reassessment rules had been triggered, particularly the discovery of any new infested sites (including the re-emergence of the mosquito in the southern Kaipara).

The MoH's TAG ended on 30 June 2006, with the transfer of the eradication programme to the MAF. Table 4.1 shows the members and their terms of membership.

Once the eradication programme was established, a meeting agenda would typically include:

- confirmation of the agenda, notes from the previous meeting, matters arising, etc.
- progress reports from all eradication sites: treatment, surveillance, any issues
- results of national surveillance
- any scientific or research reports from overseas or NZ
- review of the programme against reassessment rules.

Box 4.1: Terms of reference for the Southern Saltmarsh Mosquito Technical Advisory Group

The Southern Saltmarsh Mosquito Technical Advisory Group will provide the Ministry of Health with high quality, independent scientific and technical advice on the response to the Southern Saltmarsh mosquito, *Aedes camptorhynchus*, including:

- the quality and completeness of information on which decisions will be made
- improvement of the quality and completeness of information
- planning of any attempted containment and/or eradication response
- assessment and review of proposals planned by the Ministry in response to the incursion
- evaluation of any attempted containment and/or eradication
- other technical, scientific and epidemiological matters in relation to the incursion as may be required.

The Southern Saltmarsh Mosquito Technical Advisory Group will report to the Ministry of Health's Chief Technical Officer (Health). It will be chaired by Dr Ruth Frampton. The Ministry of Health will service the Southern Saltmarsh Mosquito Technical Advisory Group and will have one or more observers in attendance.

You must meet the following:

- familiarise yourself with any background material sent to you before meetings and teleconferences
- participate in meetings and teleconferences as required. Where you are unable to attend a meeting, you will provide an apology and comments in advance of the meeting, and on draft notes following the meeting if there is any divergent or additional view you wish expressed. If three consecutive meetings are missed it will be taken as a resignation from the Southern Saltmarsh Mosquito Technical Advisory Group.
- undertake any additional activities agreed at the workshop (such as drafting reports, commenting on reports drafted by other members, providing material, etc.).
- assume collective responsibility for decisions.

In consideration of the provision of the services the Ministry of Health agrees to pay you a daily fee. The Ministry also agrees to reimburse you for any reasonable and actual expenses incurred in providing the services where prior approval for the expenses has been given in writing by the Chief Technical Officer (Health).

Subject to the following, the fee is GST inclusive and you have agreed to be responsible for all taxation liabilities and all other costs and expenses arising in relation to the services and the payment made to you. In relation to GST you agree and understand that if you do not provide a tax invoice acceptable to the Inland Revenue Department, you shall be paid the contract price with GST deducted.

Terms of payment are within 15 working days of receipt of a valid GST invoice following the completion of the services to the satisfaction of Sally Gilbert, Chief Technical Officer (Health).

You agree to exercise all due professional care and diligence in the performance of your obligations under this Agreement in accordance with the standards of skill, care, and diligence normally practised by suitably qualified and experienced contractors in performing services of a similar nature.

You agree that if you fail to complete the services or meet the required; or if you fail to exercise all due professional care and diligence in the performance of your obligations under the agreement, the fee payable under this Agreement may be abated or withheld by the Ministry.

You agree that you will not at any time disclose to any person otherwise than is necessary for this Agreement or as required by law, any information you acquire for the purposes of providing and completing the services. In carrying your functions as a member of the Southern Saltmarsh Mosquito Technical Advisory Group, you shall not make media statements of any kind on behalf of the Southern Saltmarsh Mosquito Technical Advisory Group unless requested to do so by the Ministry of Health. If you wish to discuss media issues, you should contact the Ministry's Corporate Communications unit in the first instance.

All physical and intellectual outputs produced for the purposes of providing and completing the services shall be the property of the Crown (for the avoidance of doubt this includes, without limitation, all reports, papers, electronic documents (including computer software), and recordings).

The Ministry may require errors, omissions, defects, or faults in the services to be corrected at any time up until 1 month after purported completion of the services.

Should, for any reason, the services no longer be required by the Ministry, you will be advised as soon as possible and will be paid for the proportion of the services provided up to when you are so advised.

You shall indemnify the Ministry in respect of costs and damages associated with any legal liability that results from your acts or omissions, where those acts or omissions were not authorised by the Ministry.

You agree not to assign, delegate, or transfer your obligations under this Agreement without specific written approval from the Ministry.

This letter sets out the entire agreement between us regarding the Southern Saltmarsh Mosquito Technical Advisory Group, and supersedes all prior oral and written representation, understandings, arrangements or agreements concerning the Southern Saltmarsh Mosquito Technical Advisory Group.

To formally record your agreement to the terms and conditions set out in this letter would you please sign and date both copies of this letter, initial all pages except this page, and return one copy to Sally Gilbert.

If a new infestation had been detected, the meeting would include reports on the response to date, delimiting survey and results, habitat assessment, results of enhanced nation-wide surveillance and options for any future response.

As early as its first meeting, on 18 January 1999, the TAG was providing crucial strategic and technical advice to the MoH to direct the appropriate response. It advised that:

- Given the assessed potential for future mosquito-borne disease, intervention is appropriate.
- Given the information and tools available, containment using *Bti* spray is the immediate/short-term response with a view to attempted eradication in the longer term.
- Containment should reduce the number of biting adult mosquitoes, help to limit the further spread of the species, and provide time to further collect and analyse

information, survey other habitat throughout NZ, assess all treatment/intervention options, assess the probability of eradication and acquire necessary resources.
- Concurrently with attempted eradication in the Hawke's Bay, national surveillance must proceed to establish whether the mosquito is established elsewhere in NZ.
- Immediate work on obtaining requisite approvals and stocks of S-methoprene is needed to provide an additional, and possibly more effective, spray agent.

At the next meeting on 29 January 1999, it:

- drafted a definition of eradication:
  - adult and larval numbers decrease to zero following treatment with control agents and/or habitat modification
  - dry period follows (at least 1 week and over 9 months depending on confirmation of the egg desiccation resistance period)
  - minimum water event to hatch eggs (habitats submerged to an identified level) and no larvae or adults found
  - water event number 2 and no larvae or adults found
  - water event number 3 and no larvae or adults found
  - one generation period (up to 1 month in winter, as short as 4 days in summer) with no evidence of adults or larvae
  - eradication successful
- identified S-methoprene as the only feasible control agent to eradicate *Campto* and noted that, while *Bti* was effective for control and containment, it was not likely to be feasible for eradication
- reported on the efficacy of aircraft disinsection at Napier airport.

Table 4.1. The Technical Advisory Group members and their terms of membership

Name	Role	Start date	End date
Ruth Frampton (Chair)	Chief Technical Officer (Forestry)	18 January 1999	30 June 2006
Brian Kay	Head, Mosquito Control Laboratory, Queensland Institute of Medical Research	18 January 1999	30 June 2006
Mark Hearnden	Department of Public Health, Wellington School of Medicine	18 January 1999	16 Nov 2001
Virginia Hope	Medical Officer of Health and Public Health Medicine Specialist (Auckland)	1 April 1999	30 June 2006
Ruud Kleinplaste	Consulting entomologist	18 January 1999	30 June 2006
Graham Mackereth	Veterinary epidemiologist (MAF)	29 January 1999	30 June 2006
Darryl McGinn	Mosquito control expert Brisbane City Council	15 November 2001	30 June 2006
Phil Weinstein	Department of Public Health, Wellington School of Medicine	29 January 1999	31 July 2002
Alistair Woodward	Head of Department of Public Health, Wellington School of Medicine	18 January 1999	1 April 1999

The approach to the eradication in Hawke's Bay was first codified into an 'eradication plan' following discussions at the third meeting on 18 February 1999 and was presented in first draft form at the fourth meeting on 11 March 1999. Subsequent meetings continued to review and revise the eradication plan as new information was received and as the eradication programme matured.

By April 1999, an operational plan had been appended to the eradication plan, defining operational procedures necessary to support the eradication plan. As the eradication programme included more sites, the eradication plan was largely unchanged as a strategic document, and each site developed its own detailed operational plan to give effect to the eradication plan.

The eradication and operational plans provided an invaluable framework for the performance of the mosquito response. They remained living documents aligned to the situation at hand and not the situation as it was a year before. It was accepted that the eradication plan would not be changed unless directed by the TAG. This ensured that any significant changes to the eradications programme, requiring an amendment to the eradication plan, were made with the full advice and endorsement of the MoH's technical experts.

In addition to advising on the eradication plan, and reviewing progress with eradication, the TAG advised on the health risk assessment and technical aspects of the cost–benefit analysis. Treasury officials were able to attend meetings, for example, to seek further advice and clarification on technical issues that underpinned assumptions in the cost–benefit analysis.

For example, at its meeting on 11 March 1999, the members discussed the estimated disease rates. A single 'virgin soil' epidemic of Ross River virus disease in NZ with a non-immune, susceptible population had been estimated to produce a case rate of 333 per 100 000 population, and there was a question as to whether this case rate in a non-immune population was realistic. The TAG cited case rates of 670 per 100 000 in a small rural population in South Australia in 1993. The most likely scenario was one initial, large virgin-soil epidemic followed by recurring, smaller epidemics. It was noted that Australia had ~7000 cases of Ross River virus disease each year, mainly in Queensland where it was endemic. During the 1983–84 epidemic in Griffith, New South Wales, the incidence rate was 127 per 100 000, though Griffith itself had a rate of 1070 per 100 000. An epidemic in south-west Australia in 1992–93 had an approximate incidence rate of 1636 per 100 000.

At the same meeting, the spread of *Campto* was discussed. The TAG recognised that *Ae. notoscriptus* had taken 50 years to colonise both islands, but it was a relatively poor flier and more limited in potential habitats. *Campto* would be expected to spread ~5 km per generation if suitable habitat is available, and so the TAG's best estimate was that, if no action was taken, the colonisation of all suitable habitats throughout NZ could take between 5 and 50 years.

Once there is vector spread, Australian evidence suggested that 7 bites/night (summer) and 14 per night (winter) would be sufficient to sustain an epidemic in an urban/suburban environment (based on computer modelling). The Australian experience suggested that there would be around three epidemics per decade plus endemic cases in intervening years. After the first virgin soil epidemic in an area, the incidence rates would drop back to base rates which could be assumed to be around the Australian endemic rates. Australian data (**not** including Queensland) would be the best indicator of likely rates in NZ because the temperate climate regions in Australia would be more comparable with NZ. This would be subject to uncertainties around environmental limiting factors (e.g. availability of habitat), meteorology data, animal reservoirs and other factors such as human population densities.

It was noted, however, that one potential vertebrate host was more prevalent in NZ than Australia (i.e. the 60 to 70 million possums in NZ located reasonably closely to population centres) so the transmission may be more rapid.

The TAG then considered the most appropriate ways to apply incidence rates for virgin soil epidemics overseas to NZ. The data from Fijian epidemics shows very high rates, but Fiji has a relatively poorly medical service and localised population, so the rates provide only upper levels for virgin-soil attack rates.

- 10 000 per 100 000 was considered feasible as an upper limit for the initial (Napier) epidemic
- 330 per 100 000 was feasible as a conservative lower estimate, assuming biting avoidance behaviour (personal protective measures) will be promoted and people will act on this advice.

Although accepted as conservative (i.e. potentially under-estimating the rates of disease), this advice was accepted by the MoH and included in subsequent health risk assessments and cost–benefit analyses. It was crucial to the decision making around the response options.

S-methoprene: where to from here?

At its meeting on 16 July 1999, the TAG was joined by David Sullivan (Zanus Corporation) and Doug Van Gundy (Zoecon) who provided expert advice on S-methoprene and its use in the United States. One of the key agenda items was to discuss the use of S-methoprene following scientific studies for the MoH that established that chemical analysis of field water samples for S-methoprene residuals was not feasible. This meant that undertaking water analysis to verify a lethal concentration of S-methoprene was being maintained in the wet habitat was not feasible. There was also little data on release curves or decay curves for the product, although Professor Kay was continuing with research on this in Queensland (see Chapter 6 for more details).

David Sullivan and Doug Van Gundy provided background to the development of S-methoprene products. They noted that there was little or no research on decay curves and release rates, and this was crucial to the application regimen against *Campto.* Apart from this, the most significant information needed was 'whether the product is killing mosquitoes when applied in the field'. The times guaranteed for the product to remain effective in the field were based on activity in the field, not on S-methoprene concentrations in water, because the American experience was that it was notoriously difficult to find it in field water samples. Chemical analyses were used only to determine the shelf life of the product.

The manufacturers guaranteed that XR-G granules will be effective for 21 days in water. If applied to dry land for 15 to 20 days before wetting, the product will still be effective for the 21 days. It was noted that, although deterioration in the efficacy of the product began after the 21 days claimed on the label, this was a gradual deterioration and would be indicated by progressive development of the mosquito before death, rather than sudden high rates of emergence of healthy adult mosquitoes. Prior to the 21 days, all pupae would die. Once the product started to deteriorate, some adults might have started to emerge from pupae before dying.

Doug Van Gundy advised that penetration of canopy was very good with the granules and his examination of the habitat in Napier suggested there would not be any problems. Clumping when damp was not a problem, and the product could be applied in the rain (if the aircraft or helicopter is able to fly). Habitat modification had already been planned, including removal of vegetation. Areas of fast flowing water, which might wash granules away, would be cleared of vegetation, have their sides straightened and the water flows maintained to eliminate them as potential habitat.

S-methoprene had not been used for eradication purposes, but had an extensive history of use as a control agent. When asked for their view of the feasibility of eradication using S-methoprene, both Doug Van Gundy and David Sullivan agreed that there was 'an excellent likelihood of achieving eradication, provided that no reinfestation occurred'.

David Sullivan noted that 'the mosquitoes are, in effect, 'eradicated' from areas treated in California, but because only small pockets of the total habitat are treated, the treated areas are recolonised from surrounding areas'. He advised that his review of the eradication plan had not detected any problem areas other than with the chemical analysis.

Doug Van Gundy advised he knew of only one instance of tolerance developing in mosquito populations. This was a genetically isolated strain of mosquitoes located on offshore islands that had been treated for 6 to 7 years with briquettes, and it was known there had been under-dosing and tidal influences.

A discussion on work to identify the release life of the granules under field conditions concluded that this was not necessary because the information it would generate would be indirect and of questionable value for the eradication programme. It would only show that S-methoprene had been released from the granules and not provide efficacy data.

The need for laboratory bioassay work to assess the product decay properties of XR-G was also discussed. It was noted that this would provide dose–response data, rather than a decay curve, and that the bioassay work already undertaken in the Hawke's Bay showed that the product was effective. The manufacturers guaranteed the product will be effective in the field for 21 days. Professor Kay was already working to provide indicative decay curve data. David Sullivan recalls the meeting:

> *'Doug Van Gundy (Zoecon) and I were invited to participate in the TAG meeting in Wellington, NZ, to review the progress of the Southern Saltmarsh mosquito eradication programme. There was a general concern that the eradication programme was in jeopardy due to a couple of concerns. First, it is difficult to detect S-methoprene in water at low levels (< 0.2 ppb), although it still controls mosquitoes at that level. Second, there was minimum scientific data available on the effects of S-methoprene on* Aedes camptorhynchus. *Doug gave a summary of the 25-year history of the use of S-methoprene, based on known effective rates on various* Aedes *spp., application rates, water conditions and temperatures. I also shared with the group my experiences with many field people who were uneasy when using S-methoprene because there is no immediate effect. I told the group that 'you will not see any immediate effect; you just have to have faith.'*

As a result of its discussions, Brian Kay's research, a review of the scientific reports commissioned by the MoH, and the information provided by David Sullivan and Doug Van Gundy, the TAG agreed that the reassessment rule requirements (i.e. that 'efficacy tests for S-methoprene will demonstrate it is effective and environmentally sustainable against the Southern Saltmarsh mosquito and provide information for appropriate dosage rates and frequency of applications') had been met.

Transferring the TAG to the MAF

The TAG continued largely unchanged for much of the eradication programme when it was managed by the MoH. However, when the eradication programme transferred to the MAF in July 2006, the MAF took the opportunity to review the terms of reference and membership of the TAG.

The existing TAG was dis-established and replaced with a TAG that focused on scientific advice and provided an overview of research to support the operational delivery of the programme. The terms of reference were for the members of TAG to provide the MAF with:

- open and frank independent scientific / technical advice on *Campto*
- advice on its known biology and ecology, and the epidemiology of disease agents vectored by *Campto*;
- advice on the potential impacts in NZ, including public health impacts from vectored diseases
- advice on possible options and their technical feasibility for responding to the incursion
- advice on the available tools and techniques for effective treatment, surveillance, eradication and/or control of *Campto*
- advice on public health risk mitigation strategies for disease agents vectored by *Campto*
- objective evaluation and comment on the likelihood of success of the management options
- advice on the courses of further research or projects to be undertaken to aid the knowledge of *Campto*, disease agents vectored by *Campto* and risk management for both
- scientific/technical recommendations on the most appropriate course of action.

Although one member of the previous TAG joined the new Group (Virginia Hope), the remainder of the membership was new. Table 4.2 shows the membership of the TAG.

The new TAG concluded that the methodology being applied in the NZ eradication programme represented best practice for mosquito control operations when compared with programmes overseas. They agreed that the application of treatment agents over two summers was sound and recognised the criteria for post-eradication surveillance was potentially critical. The TAG identified several key risks to the effectiveness of the programme:

Table 4.2. The MAF Technical Advisory Group members and their areas of expertise

Name	Organisation	Area of expertise
Matthew Stone (Chair)	MAF	Veterinary epidemiology and biosecurity responses
Virginia Hope	Institute of Environmental Science and Research Ltd	Public health medicine and epidemiology
Mary McIntyre	Department of Public Health, Wellington School of Medicine	Entomology and ecology
Scott Ritchie	James Cook University, Australia	Medical entomology
Richard Russell	University of Sydney	Medical entomology
Craig Williams	University of South Australia	Mosquito ecology and control

- the presence and impact of small pockets of missed treatment
- extended egg survival
- long distance dispersal of adults
- obscure breeding sites (e.g. freshwater pools and hoof prints)
- autogeny, either as a regular feature of the organism's biology or as a particular biological response to environmental stimuli.

The TAG also stressed the importance of the close interaction needed between surveillance in eradication zones and the national saltmarsh surveillance programme, and suggested validating techniques employed in the eradication programme, including dipping and trapping, to ensure they were optimal for the NZ environment.

Over the next 4 years, scientific research was commissioned to deal with these concerns and risks, and to ensure any assumptions made were underpinned by suitable scientific rigour.

In addition to this core work, additional research was able to demonstrate that the Wairau eradication zone, which consisted of another area raised by a historical earthquake, had a very unusual salinity gradient. Salinity mapping of the area identified hotspots that matched areas of emergence, and allowed a more targeted approach to be taken in the application of S-methoprene.

Other expert advice

In addition to the TAG, a large number of experts in a variety of fields became associated with the mosquito response at various points. The areas in which their advice or services were offered are summarised below.

Australian mosquito control and arboviral disease experts

- Bruce Haseler, Vector Officer, Australian Quarantine and Inspection Service: mosquito control and exclusion expertise
- Royle Hawkes, School of Microbiology, University of New South Wales, Sydney: arboviral disease advice
- Ian Myles and Greg Johnson, Mosquito and Pest Services, Brisbane City Council: operational advice and assistance
- Michael Lindsay, Senior Research Officer, Arbovirus Surveillance and Research Laboratory, Department of Microbiology, University of Western Australia, Perth: mosquito control and arboviral disease advice
- Peter Whelan, Director, Medical Entomology Branch, Territory Health Service, Darwin: mosquito control and arboviral disease advice

Information systems

- Peter Firns, Peter Firns and Associates Ltd, NZ: audit of information needs and analysis of spatial information
- John Kay, Team Leader (Bioinformatics and Land Systems), Landcare, NZ: mapping of potential habitat, using water salinity and vegetation that grew in saline areas (see also Chapter 9)
- Adam Lewis, Environmental GeoGraphical Services Pty Ltd, Queensland: audit of information needs and analysis of spatial information

- Peter Newsome, Programme Manager (Land Resource Information Systems), Landcare, NZ: mapping of potential habitat, using water salinity and vegetation that grew in saline areas (see also Chapter 9)
- Don Scott, Agribase Manager, Agriquality Ltd, NZ: outline of how Agriquality felt it could assist the MoH with its information management needs
- Craig Soper, Compudign Ltd, NZ: estimated the development of information management systems to several levels of specification.

Mosquitocide evaluations, supply and support

- Alan Aspell, Alan Aspell and Associates Ltd, NZ: peer reviewed ESR's report on the determination of S-methoprene in water
- Jeanette Conland, Folio Ltd, NZ: surveyed public opinion on using *Bti* in Hawke's Bay, prepared fact sheets.
- Bruce Evans, Evatech Ltd, NZ: applied to the Agricultural Chemicals and Veterinary Medicines Unit to register S-methoprene products
- Brett Freeman, Novartis Animal Health Australasia Ltd, Sydney: Australian suppliers of S-methoprene
- Tony Harris, Livestock Consultant, Agriquality Ltd, NZ: advised on S-methoprene with respect to the farm accreditation programmes
- Chris Hickey, NIWA Ltd, NZ: peer reviewed ESR's report on the determination of S-methoprene in water
- Glen Hughes, Opus Ltd, NZ: coordinated the resource consent application for applying *Bti* and S-methoprene in Hawke's Bay
- Terry Phillips, Marketing Manager, Novartis Animal Health Australasia Ltd, Sydney: Australian suppliers of S-methoprene
- Cliff Randall, ESR Ltd, NZ: validated a method for the determination of S-methoprene in water (see Chapter 6)
- Deborah Read, ERMA, NZ: peer reviewed reports on the environmental and health impacts of *Bti* and S-methoprene
- Doug Van Gundy, Zoecon, California: United States principal for S-methoprene
- Howard Wearing, Hort Research Ltd, NZ: peer reviewed reports on the environmental and health impacts of *Bti* and S-methoprene.

Community Liaison Group

In its *Mosquito Eradication Process Review* (September 2000), ESR found that one of the key components of the Hawke's Bay response that appeared to have been pivotal to its effectiveness was the establishment of reference/advisory groups including a community liaison group which provided local knowledge, support and communication. The community liaison/information groups were established for all eradication zones throughout the programme. Members typically included local Maori representatives, residents and special interest groups (Asthma, Forest and Bird).

In 1999, community liaison groups in Napier were intended to 'help the delivery of community issues, concerns, and advice about the control/eradication operations ... help the information flow from the mosquito operations centre to the community ...' . Community liaison groups also provided informal feedback on mosquito activity; for example,

Steve Garner's quarterly report (13 January 2000) comments that 'anecdotal evidence from those communities worst affected last summer is that there is no problem this year'.

Community liaison groups were also seen as a forum through which community concerns could be fed back to the Napier MRC and the MoH.

As a point of difference, a community advisory group established by the MAF during its Painted Apple Moth eradication programme, perceived itself as having an advisory and decision-making role. When it was not able to prevent the widespread aerial application of *Btk* over parts of residential Auckland, this group developed an activist role criticising the programme and mobilising protest action.

In comparison, the MoH specifically identified the role of community liaison groups as information sharing and consultation only. Although their views would be considered, members were given no expectation that these views would prevail over technical or other imperatives.

Cost–benefit analysis

The *Biosecurity Act 1993* establishes 'the need to conduct analyses of benefits and costs for national and regional pest management strategies conducted by biosecurity ministries, departments and agencies'. However, the way in which such analyses are conducted is not specified.

The various government agencies had recognised the need for a standardised decision-making methodology for analysis of biosecurity incidents. A framework was needed that included both an outline of the required ecological modelling and risk analysis and a financial model for calculation of the cost–benefit analysis. The cost–benefit analysis model would be more structured, with room to alter the variable inputs. Components would include:

- standard cost–benefit analysis model incorporating standard output formats
- suggested time horizons with their rationale and appropriateness for specific circumstances
- suggested discount rates with their rationale and appropriateness for specific circumstances
- risk analysis evaluating the impact of variances in key input variables.

Because no two infestations or outbreaks will be the same, the methodology needed to be robust, with guidelines on the suitable inputs. Outputs need to be as standardised as possible to ensure consistency of application and to prevent reinvention by different government departments and to ensure similar issues within will be treated consistently and that the government receive comparable information.

At the time of the Napier incursion, Dr John Mumford (from the Imperial College of Science, Technology and Medicine in the United Kingdom) had been contracted by the MAF to prepare a discussion paper and develop associated spreadsheets for cost–benefit analysis procedures to be applied to unwanted organisms or pest responses. This fortuitously provided the MoH with access to expert advice to prepare a cost–benefit analysis on options to respond to the *Campto* incursion.

The framework prepared by Dr Mumford (1999) provided an excellent step towards the development of a rigorous biosecurity decision-making framework. The fact that it had already been tested in 'live' situations was a further advantage.

One of the crucial areas in any cost–benefit analysis, and particularly if the costs and benefits are similar, is the treatment of intangibles. *Campto* is an aggressive day-biting mosquito that was having a significant impact on the lifestyles of those living near infested habitats. The impacts on residents unable to go outdoors in good weather, outdoor workers who suffered significant biting and New Zealanders' concerns about the invasion of exotic species and feelings of grief from the perceived 'loss of pristine environment' were not able to be costed. Dr Mumford was specifically asked by Health officials about how these costs could be included in a cost–benefit analysis. Dr Mumford advised that there were expensive and time-consuming methodologies available to develop costings, but suggested it may be more appropriate for such intangible costs and values to be identified, but not quantified. This would still allow them to be considered by the decision makers. For example, if the overall estimated costs outweighed the benefits by, say $1M, then government could decide if the impact on intangibles, such as New Zealanders' lifestyles, was worth that difference. Health officials agreed with Dr Mumford that Ministers and the government were the most appropriate decision makers to make those sorts of assessments, because they had been elected by, and represented, New Zealanders.

The Mumford spreadsheets accounted for the costs of the response, including the costs of pesticide and labour and the environmental cost associated with pesticide application (to give a cost per unit area). The costs of the mosquito included the cost of lost work and medical expenses arising from the disease and the nuisance value associated with the presence of the mosquitoes (a cost per person), based on the health risk assessment. Not all economic costs associated with an outbreak of Ross River virus disease or the presence of mosquitoes, were captured. For example, there may have been an impact on tourism because people are less inclined to visit places where they may be exposed to serious illness. In view of the NZ pure and clean image, this effect may have extended nationally.

Following on from Mumford's report, for subsequent incursions, the MoH contracted economists at the NZ Institute of Economic Research to use and develop a cost–benefit analysis framework to examine possible responses to the incursion of *Campto*. The original framework was developed to bring consistency to cost–benefit analyses of unwanted organism and pest responses. Hence, the subsequent cost–benefit analyses did not make any adjustment to that framework; rather adopted it and expanded the spreadsheets to allow analysis of a greater number of sites simultaneously (McWha 2001). In particular, the revised spreadsheets enable analysis of:

- eradication – elimination of the mosquito with S-methoprene
- containment – preventing the spread of the mosquito by spraying with *Bti* over a 2-year period. Containment is assumed to eliminate 95% of the mosquitoes.
- control – monitoring the mosquito numbers and spread and responding if it reaches set boundaries. Local control eliminates only 15% of the mosquitoes.
- secondary disease prevention – accepting the presence of the mosquito and focusing on disease control. This option includes behavioural modification programmes, disease surveillance and treatment.

The first step in a cost–benefit analysis is to establish the counter-factual, or 'do nothing', scenario. In this case, if no other action were taken, then the secondary disease response would be adopted by default. In essence, this option made no attempt to control the mosquito, rather focusing on education of the public and medical specialists, and disease surveillance and treatment. The counter-factual thus included all the medical and lost work costs of allowing the mosquito and Ross River virus disease to be distributed around NZ. It also included monitoring costs, because all options assumed some monitoring of the spread of the disease.

The cost–benefit analysis concluded that, for every 1000 ha affected, ~350 000 people needed to be exposed to the mosquito to make eradication feasible, using a 17-year time horizon. When considering the options for the Kaipara response (the largest eradication zone), the cost–benefit analysis indicated that if eradication prevented the invasion of salt-marsh mosquitoes into Auckland for at least 17 years, eradication was a feasible option. The approximation suggested that eradication would have a positive net present value over an area up to 3250 ha (McWha 2001).

Because containment and local control were relatively less effective, delaying the invasion by only 4 years and 2 years, respectively, the benefit provided by these options did not outweigh the additional cost relative to the counter-factual of a secondary disease response.

Health risk assessment

An assessment of the health risks posed by the establishment of *Campto* was essential to inform decisions about the appropriate response, and as an input to the cost–benefit analysis.

A risk assessment of the introduction of exotic mosquitoes, and particularly Dengue, Ross River and Barmah Forest virus diseases, had been undertaken by Brian Kay in 1997, before the *Campto* incursion. The likelihood of an incursion of exotic mosquito into NZ and the need to be prepared was clearly recognised: 'The general finding is that incursion of exotic mosquitoes or vector-borne diseases would seem to be only a matter of time ...' (Kay 1997).

Assessments of the environmental and health impacts of the treatment options recommended in the Kay (1997) review, namely *Bacillus thuringiensis israelensis* (*Bti*) and S-methoprene, had also been completed (Glare and O'Callaghan 1998a,b).

An initial risk assessment prepared by the Napier MRC in mid-December 1998 identified that, of the five factors needed for the outbreak of an arboviral disease, four appear to be present:

- the vector (*Campto*)
- probable animal reservoirs (domestic/agricultural and feral animals)
- weather/climate and environmental (the established nature of the infestation attests to this)
- proximate human population (the northern suburbs of Napier have reported frequent incidents of biting, particularly just before Christmas 1998, and as a generalisation the NZ population is non-immune and susceptible).

The only element missing was the arbovirus. The risk assessment found that, given the level of tourism in the Hawke's Bay and the number of local residents returning from risk areas overseas, it would only a matter of time before a viraemic individual or individuals brought in the disease.

The need for a detailed, formal health risk assessment, incorporating information on the extent of the mosquito habitat, the potential mosquito habitat and climatic outlook, was recognised.

A health risk assessment was prepared by Mark Hearnden (from the Department of Public Health, Wellington School of Medicine) and a draft of a report *A Health Risk Assessment Relating to the Establishment of the Exotic Mosquitoes* Aedes camptorhynchus *and* Culex australicus *in Napier, New Zealand* was presented to the TAG meeting on 19 January 1999 (Hearnden 1999).

In addition to review by members of the TAG, the report was peer reviewed by Virginia Hope, Richard Russell, Peter Whelan and Michael Lindsay. The health risk assessment was revised and finalised on 17 February 1999.

In its review of the mosquito eradication programme, ESR found that 'The rigour of the health impact review process with participation from NZ and Australian experts meant that the health impact assessment was a useful and robust basis for response decision-making and planning. The health impact assessment provided a significant contribution to the effectiveness of the eradication programme' (Thomson and Cressey 2000).

The health risk assessment followed the usual format for such a report and included sections on:

- hazard identification: *Campto* biology, distribution and biting behaviour, associated human arboviruses
- health effects: distribution, occurrence, transmission, outbreak ecology and diseases caused by Ross River virus
- exposure assessment: current and potential mosquito distribution and habitat
- risk characterisation: spread of mosquito, incidence rates for Ross River virus disease
- estimated costs of disease.

The health risk assessment completed for the initial incursion response was updated with each further discovery of *Campto*, but the fundamental findings did not change throughout the subsequent reviews.

The health risk assessment estimated that, if eradication was not successfully undertaken, the mosquito would spread throughout NZ within 5 to 50 years, although if the incursions were linked to the original incursion in Napier this could take a little as 5 to 7 years.

Campto has been defined as an efficient field vector of Ross River virus and implicated in the transmission of Barmah Forest virus disease (clinically similar to Ross River virus disease). It is an extremely efficient vector that doesn't need to live long or have high numbers to maintain a Ross River virus epidemic. It would become a viable vector for the pathogen in the presence of an efficient vertebrate host, resulting in disease in the animal and human populations.

The health risk assessment warned that it was also likely that *Ae. notoscriptus*, a mosquito already distributed throughout NZ, may play a role in augmenting local urban transmission of Ross River (and Barmah Forest) virus disease in susceptible areas.

Given the establishment of a competent vector for Ross River virus, the health risk assessment indicated that virgin soil epidemics were likely to occur sequentially throughout NZ over the next 0 to 50 years as the mosquito spread, with likely incidence rates of between 330 and 10 000 per 100 000. Incidence rates would be likely to drop to 10 per 100 000 in the year following an epidemic, but repeated outbreaks or epidemics with incidence rates of 100 to 330 per 100 000 were likely to occur every 3 to 5 years in endemic areas on top of this baseline rate.

The predominant clinical presentation for Ross River virus disease is epidemic polyarthritis in at least 20% of infected individuals. Symptoms include the sudden onset of an acute aching in the muscles and joints, headaches and fever, followed by a maculopapular rash (in between 40 and 78% of patients) and extensive polyarthritis. Most symptoms may settle in a few days but the effects of the polyarthritis may incapacitate an adult for 5–6 weeks. Recovery may take longer and it is estimated that 1% of cases will have joint symptoms lasting more than 5 months. Common symptoms 30 months after infection include myalgia, lymphadenopathy, headache and depression. Relapses of arthritis and fatigue may also occur after the illness has originally subsided. A chronic fatigue-type syndrome may persist in ~10% of patients.

The health risk assessment concluded that there was significant saltmarsh habitat in the vicinity of major concentrations of the NZ population, such as Auckland, Wellington, Christchurch, Napier, Tauranga and Invercargill. If *Campto* was not eradicated, its eventual spread to all available NZ habitats, and the eventual introduction of Ross River virus into the NZ environment, was inevitable.

The resident population of NZ in 2006 according to NZ statistics was estimated to be 4 138 400 people. At an estimated cost of NZ$2859 per case, this translates to an estimated direct health cost of around NZ$118.3 million for a virgin soil epidemic (using conservative virgin soil epidemic rates of 1000 per 100 000 for this example) for the treatment of disease.

Although the first reports of endemic cases of Ross River virus are likely to initiate or extend mosquito control measures, it was thought unlikely, based on Australian experiences, that these measures would significantly limit a virgin soil epidemic in an exposed population.

Following a virgin soil epidemic, the cyclical epidemics with an incidence rate of 100 to 330 per 100 000 were estimated to incur a direct health cost of NZ$11.8–39.0 million every 3 to 5 years. The intervening years would be estimated to incur direct health costs of NZ$1.2 million per annum with a background rate of 10 per 100 000. These costs did not include costs required to control mosquito populations, the economic injury to the tourism image, lost productivity and the impacts on lifestyle. The impacts on indigenous species, such as ground-nesting birds, from an aggressive, biting mosquito were not known.

References

Glare TR, O'Callaghan M (1998a) 'Environmental and health impacts of *Bacillus thuringiensis israelensis*'. Unpublished report to the Ministry of Health, Wellington, NZ.

Glare TR, O'Callaghan M (1998b) 'Environmental and health impacts of S-methoprene'. Unpublished report to the Ministry of Health, Wellington, NZ.

Hearnden M (1999) 'A health risk assessment relating to the establishment of the exotic mosquitoes Aedes camptorhynchus and Culex australicus in Napier, New Zealand'. Unpublished report to the Ministry of Health, Wellington, NZ.

Kay BH (1997) 'Review of New Zealand programme for exclusion and surveillance of exotic mosquitoes of public health significance'. Unpublished report to the Ministry of Health, Wellington, NZ.

McWha V (April 2001) 'Southern Saltmarsh Mosquitoes: Technical notes to accompany cost benefit analysis'. Unpublished report by the NZ Institute of Economic Research (Inc.) to the Ministry of Health, Wellington, NZ.

Mumford J (1999) 'Cost benefit analysis procedures to be applied to unwanted organisms or pest responses'. Unpublished report to the Ministry of Agriculture and Forestry, Wellington, NZ.

Thomson B, Cressey P (September 2000) 'Mosquito Eradication Process Review'. Unpublished report by the Institute of Environmental Science and Research Ltd to the Ministry of Health, Wellington, NZ.

5

Who knows how to do broad-scale aerial control of mosquitoes?

Darryl McGinn and David Sullivan

Australia has had a long and interesting history with control of saltmarsh mosquitoes. In most of Australia's coastal zones that support suitable habitat, the so-called 'saltmarsh mosquito', *Aedes vigilax* is the primary target species of broad-scale control operations conducted by government. Under various Australian state government systems, these control programmes are provided either by local government or at state level. Nowhere in Australia is control more concentrated than in south-east Queensland.

In 1998, there were 10 contiguous local governments all actively providing broad-scale control of *Ae. vigilax* around south-east Queensland. The largest of these was within Brisbane City Council that, at the time, maintained an operational programme for saltmarsh mosquito control using helicopters treating 22 000–25 000 ha per year, and ground-based treatment of saltmarsh and freshwater breeding mosquito species employing 37 full-time staff throughout the City. Darryl McGinn was the head of Mosquito and Pest Services and was responsible for Brisbane City Council's programme. It was by chance that in the mid-1990s, Darryl had attended a mosquito control workshop sponsored by the New South Wales Department of Health. Also attending was Henry Dowler, Deputy Chief Technical Officer (Health) to the NZ MoH. As is not unusual at such meetings, and with moderate amber stimulant, Henry and Darryl became acquainted over discussions of why was NZ so interested in mosquito control in Australia. It was an unremarkable discussion at the time but could only be described as serendipitous when recalled a very few years later.

Into the storm

On 18 January 1999, Darryl McGinn received a surprising early morning phone call from Henry Dowler. Henry first apologised for the hour, which only served to heighten Darryl's sense he was about to hear something fascinating. Had Darryl heard anything about the discovery of *Campto* in Napier? No? OK! For the next few minutes Darryl was briefed on the discovery and of the MoH's decision to start control measures while more intelligence was gathered for the NZ Government to consider an eradication response. Darryl was asked if it might be possible for him to become involved advising on control?

As an employee of Brisbane City Council, Darryl needed to seek approval from his corporate masters, but thought he could persuade Council to support MoH's request. This prediction proved correct. The next few hours went down in Brisbane City Council history as the fastest international travel approval ever to be walked through several layers of increasingly senior management and finally signed off by the Chief Executive Officer (a process required by Council's travel policy to be commenced not less than 3 weeks before the travel date). The day after Henry Dowler's call, Darryl McGinn was reclining at 35 000 feet and 460 knots heading for Auckland with domestic connections to Napier.

The Air NZ Link Saab 340 regional turboprop (an aircraft Darryl McGinn was to become very familiar with) was approaching Napier airport. Darryl was, for the first time, seeing the saltmarsh habitat of Napier and especially how much was immediately adjacent to the airport and shoreline of Hawke's Bay. Impressive.

Arrangements for being met on arrival in Napier were non-existent at departure from Brisbane, but never mind. Sure enough someone waded through the little knots of happy greetings at Napier airport to ask was this the Australian mosquito expert? Steve Garner introduced himself as a health protection officer from Hawke's Bay and said Henry Dowler had asked him to collect Darryl and give him an orientation of the area. Bags collected and stowed in the car, Darryl began a very steep journey – both the learning curve and physically getting to the Napier MRC located in the old Napier hospital perched on one of the highest and steepest hills in town.

Darryl was given a guided tour of very flat marshy terrain spreading from the wide low shingle and black sand berm of the surf beach to the foot of rolling hills, a strip ~1–3 km wide. Getting out from time to time looking more closely, Darryl noticed a wide variety of sun-bleached bivalve shells and crustacean fragments littered much of the habitat. Prompted by Darryl's puzzled expression and perhaps thinking, 'Oh God – this Aussie doesn't know where he is', Steve explained that much of the habitat was former seabed lifted by around 2 m in a dramatic 2½ min during a major earthquake that destroyed much of the town of Napier in 1931.

The town was rebuilt largely in the style of the time (mostly Art Deco – much of which remains today) and, over the next few decades, the uplifted seabed was variously ditched to allow drainage and eventually improved into pasture for agriculture. Large expanses, however, rested at just over mean high tide height and could not be effectively drained, and over time were colonised by saltmarsh vegetation including low growing *Sarcocornia* (samphire) and *Sporobolus* (salt couch grass) in the lower elevations, and taller *Juncus* (spear grass) in higher elevations. Much of the saltmarsh habitat of Hawke's Bay was reminiscent of typical saltmarsh around south-east Queensland, but without mangroves (although they are present in more northern parts of NZ).

The tour continued, and what was very obvious were the masses of saltmarsh mosquito larvae occupying almost every part of the habitat covered by water. A quick taste test to assess water salinity. 'About 25 parts per thousand' announced Darryl McGinn, 'typical for saltmarsh mosquito breeding'. A puzzled look from Steve Garner provided an opportunity to even the score – thinking 'don't you Kiwis know anything?' Darryl explained that, with experience, a taste test will give accurate salinity to within +/– 5 ppt and it was a useful indication of the likelihood of presence of saltmarsh mosquitoes.

'So – worked much with Campto*?'* asked Steve.

'Not at all' Darryl replied. *'There is relatively little known about* Campto *as it is not subject to much control in Australia. We do, however, know a lot about* Ae. vigilax *though and I think we will be starting by using it as our* Campto *model until we know otherwise'.*

Figure 5.1 Steve Garner (left) and Darryl McGinn (right) 20 January 1999 (with Steve Turpin). (Image: Noel Watson)

> *'Maybe this bloke does know what he's talking about'* was the thought bubble above Steve's head and they proceeded to the Napier MRC.

So began what was to be one of the most productive working relationships Darryl McGinn had yet experienced in his career, and typified the easy conversational style but highly motivated and rigorous flavour of how much of the operational *Campto* programme commenced (Fig. 5.1).

Napier Mosquito Response Centre (MRC)

Arriving at the Napier MRC that afternoon was an experience. About that time, various folk (health protection officers borrowed from various parts of the country) started to return from having been out in the field as part of the Napier delimiting survey. This was done partly by conducting larval sampling in habitat at increasing distances from the apparent centre of the infestation and by deployment of mosquito light traps to collect adults.

A debriefing was about to be held for gathering of information on sampling locations and what was found. Field samples were handed to Gene Browne, the MoH's contract entomologist, for identification and recording. Darryl McGinn was invited to attend the debriefing, where he was introduced by Henry Dowler as the Australian expert in saltmarsh mosquito control.

Henry explained that the MoH had approved some funding to undertake control in an attempt to contain *Campto* while the delimiting was underway and more information was gathered for urgent government consideration towards the feasibility of eradication. Henry then identified Steve Garner to Darryl as the local health protection officer he would work most closely with. Steve was later appointed the Manager of the Napier MRC and consequent surveillance and control operations in Napier. The collective question that welled up from the control room was how was control likely to succeed and how could it be done. All eyes were directed towards Darryl for his vital first words and actions. 'From what I've seen so far, there's nothing here that doesn't look straightforward to start controlling *Campto* in Napier!' 'Good start' Darryl thinks 'I seem to have their attention'. An *ad hoc* briefing on *Ae. vigilax* control in Australia follows:

- Saltmarsh mosquitoes lay desiccation-resistant eggs around the base of saltmarsh vegetation and in small cracks in soils just above the receding waterline of saltmarsh pools.
- The eggs remain viable for several months and the embryo develops relatively quickly and is ready to hatch out within minutes of the egg being flooded by rising water from tidal activity or rainfall.
- Subject to the extent of the flooding determined by water level reached, some of those eggs hatch with larvae liberated into saltmarsh pools.
- Tidal cycles produce several increasing tide heights leading up to the spring tide that progressively inundate the marsh producing several 'instalments' of hatching and larvae of varying stages of development.
- Larvae grow and moult through four larval instars, then pupate and the adult mosquito emerges from the pupa.
- For *Ae. vigilax*, in summer, from hatch to adult emergence can be as short as 5–6 days.
- For cooler conditions in Napier, for *Campto* the development time is likely to be longer (but by how much was not yet clear).
- Control is best focused at the larva stage because they are in definable habitat, relatively fixed in location, most concentrated in number and most readily treated by aerial and ground-based application of larvicides relatively specific to mosquitoes.
- The two larvicides most commonly used in Australia are *Bti* (*Bacillus thuringiensis israelensis*) and S-methoprene.
- Aerial application is generally undertaken using rotary wing (helicopter) rather than fixed wing aircraft because of a helicopter's high manoeuvrability around the irregular saltmarsh habitat, their relatively low noise and ability to operate without a runway.
- Much of what we know is currently based on *Ae. vigilax*. We will apply control agents at similar rates and observe field response of *Campto*. Specific research may follow.

Following the briefing, and further questions and answers, several matters become apparent. The most important issues were the availability of suitable control agents. *Bti* was available but in very limited supply. S-methoprene was not available at that time. There was no existing experience with aerial application of mosquito larvicides within NZ. Several agricultural aerial operators did exist and had pro-actively submitted expressions of interest towards involvement in a possible control programme.

Steve Garner provided Darryl McGinn with a file of the interested aerial applicators and requested that he contact them with a view to selecting one that could be tried for a treatment. There were several larger operators based around Wellington that were contacted by Darryl. In general, they sounded unconvincing that they could be easily torn away from flying tourists about the harbour to mount a treatment within the short response time as demanded in mosquito control.

One operator on the list was, however, Napier local Phil Deadman, owner of Helicopters Hawkes Bay. Darryl dialled and started asking questions. Phil owned a relatively small operation providing small-scale aerial application services to the fruit and vegetable industry in the Hawke's Bay area. He operated a Robinson R22 helicopter fitted with formulation tanks and spray boom. He indicated that he was very keen to be involved because he was local and did not want the Aussie mossies to spoil his area.

In talking with Phil Deadman, Darryl McGinn suggested that the programme was seeking a first treatment the following day. After the briefest of pauses and a just audible expletive, Phil said he could be at Napier airport the next morning at dawn. This was confirmed by Steve Garner after Darryl's recommendation based on Phil's proximity, availability and Darryl's sixth sense that Phil would be able to troubleshoot most issues that typically happen in aerial applications.

One negative issue was present. The R22 was a very small helicopter with limited load capacity, and was set up to treat (total volume) at 20 L/ha and likely to treat around 5–6 ha per load. Typical aerial systems then used in Australia were capable of treating 120 to 240 ha per load. Notwithstanding the limitation, availability was more important than high efficiency in the context of commencement of immediate control. Since the discovery of *Campto* in December 1998, intense media interest was following every aspect of the government's response. With high public expectation that something must be done, the media had been informed that an aerial treatment was scheduled for the following day 21 January 1999, and only 36 h after Darryl McGinn's arrival on site. Definitely, no pressure!

First strike

Preparations for the first aerial treatment were relatively simple given the total lack of 'normal' operational equipment and a limited supply of larvicide (460 L available in all of NZ at the time). 'Normal' means at least one sizeable truck capable of transporting larvicide, a mixing tank, pumps and hoses to formulate the spray mix, and fuel for the aircraft delivered to a pre-planned and demarcated landing zone. Areas to be treated are premapped for the pilots so they know where to apply larvicide. Weather information is available showing likelihood of high wind likely to interfere with treatment. The first aerial treatment for *Campto* in NZ had none of these usual niceties.

Pre-dawn on 21 January, 1999, Darryl McGinn and Rebecca Fox (one of the health protection officers on secondment to the Napier MRC) collected several boxes of Vectobac 12AS larvicide (*Bti*) and squeezed them into the boot and back seat of one of the available Toyota Corollas. The rendezvous agreed with Phil Deadman was the general aviation apron of Napier Airport.

Waiting for Phil Deadman to arrive in the R22 provided an opportunity to informally assess the adult *Campto* activity at dawn. They were highly active with Darryl McGinn estimating biting rates exceeding 30–40 bites per minute.

In the brightening dawn light, an Air NZ Link Saab 340 parked in front of the terminal building was loading passengers for the first flight out for the day. What was amazing to see, however, was that as they each crossed the tarmac, the boarding passengers were furiously waving their arms about their heads in an attempt to ward off attacking *Camptos*. This continued as they climbed the aircraft stair and ducked inside. It was all too obvious that many adult *Campto* were about to hitchhike to the next destination (see Chapter 7).

Figure 5.2 First strike 21 January 1999 – Phil Deadman and R22. (Image: D McGinn)

Darryl McGinn refocused on the task at hand but noted that transport by local aircraft was a real concern, to be raised as soon as possible with Steve Garner and Henry Dowler.

A different buzzing sound announced the approach of the R22 and it duly landed beside the stack of *Bti* boxes (Fig. 5.2). Without shutting down (which is acceptable in agricultural operations) the pilot exited and introduced himself as Phil Deadman. A quick inspection of the spraying equipment fitted to the R22 provided Darryl McGinn some relief that it would indeed be adequate for the mission planned for today. It had two ~50 L formulation tanks per side, a spray boom fitted with various hydraulic nozzles capable of being selected to adjust treatment pattern and a petrol-driven power pump to supply the boom and recirculate formulation around the tanks. What about a satellite-based GPS guidance? No? OK, a guess might be alright for dumping large loads of fertiliser on paddocks but a bit hard when the dosage rate is small as is typical for mosquito control.

Following Darryl McGinn's previous day inspection of habitat and identification of mosquito larvae around the airport itself, this seemed a reasonable place to start. Darryl advised he would go on a brief orientation flight to show Phil Deadman the areas to be treated that day. The survey flight provided additional opportunity for Darryl to scope habitat not visited on his first ground-based inspection. Aerial inspection is an immensely good use of a relatively expensive resource (if used prudently and efficiently) because it gives perspective and allows observation of areas otherwise hidden at ground level. With the proposed treatment area now agreed with Deadman, the aircraft returned for loading.

Vectobac 12AS larvicide is a liquid and is applied at rates specified for several mosquito species. *Campto* was not listed on the label. However, *Ae. vigilax* was, and it was reasonable to use it as a guide. The recommended application rate was 0.6–1.2 L/ha of the active *Bti* concentrate. Given the high density of mosquito larvae in the field, the uncertainty about exact susceptibility of *Campto* to *Bti* and lack of GPS aircraft guidance, Darryl McGinn took the decision to specify the high-end rate. Phil Deadman advised that the equipment was calibrated to deliver 20 L/ha total volume. So the calculation of how to formulate was:

- aircraft total load is 2 × 50 L tanks = 100 L
- total formulation application rate is 20 L/ha
- total area covered by 100 L at 20 L/ha = 5 ha
- active *Bti* application rate desired = 1.2 L/ha
- active *Bti* per load is therefore 1.2 L/ha × 5 ha = 6 L.

The aircraft formulation tanks were partly filled with water from a nearby standpipe and hose. Vectobac 12AS was decanted into a graduated measuring jug to a volume of 6 L and added to the aircraft tank. The spray power pump was operated in recirculation mode to effectively mix the two and further water added to top up the tanks to their 50 L graduation mark. With the application tanks filled, there was no additional payload available to carry an observer, so Phil Deadman was on his own.

Phil lifted off to commence the first treatment run. At 20 L/ha, it takes only ~10 min for the aircraft to empty its tanks and it returns to the landing zone for resupply. This cycle continued through the morning until the *Bti* allocated for this first treatment was used. In total, 120 L of *Bti* was used in the first treatment of 100 ha.

During this treatment, the media was watching, filming, photographing and interviewing. It was the first strike back against *Campto* and it was big news. Noel Watson (health protection officer) recalled the local newspaper photographer wanted a really good photograph of the first treatment occurring. The journalist advised that he wanted to stand at a particular spot on Watchman Road where the helicopter would be applying *Bti* into habitat on both sides. Thankfully, we did warn him that in respect to his expensive camera, it might not be the wisest action but he took no heed. His photograph in the paper was great, but apparently he claimed his camera did get damaged as a result and tried to seek costs from us. Due to our prior warning to him, we were not forthcoming!

This treatment was the first of three undertaken over several days and expanded to adjacent active saltmarsh habitat. It was clear that the R22 and its small payload was not going to be up to the task of providing treatment in reasonable time over this, and the apparently expanding, treatment areas. Something larger and more efficient was required.

Scaling up and overcoming habitat influences

As delimiting continued, it was becoming increasingly apparent that the *Campto* treatment area was expanding to several hundred hectares. In response to the unsuitability of the small R22 helicopter, Phil Deadman sourced a larger machine – a Hughes 500 model C. The Hughes 500 is a single engine turbine utility helicopter in common use throughout the world and they are very popular in NZ. It had a useful payload in the context of large-scale mosquito larviciding. Fitted with a 450 L formulation tank, spray boom and nozzles, the H500 could now treat more than 22 ha per load (still at the 20 L/ha total rate). Much of the identified *Campto* habitat could now be effectively treated with *Bti* from the air. Some habitat vegetation, however, provides a significant obstacle to liquid aerial treatment.

Native speargrass (*Juncus*) colonises significant portions of the *Campto* habitat. It is thick and, in some places, so thick it forms a thatched-like 'roof' above the ground pools containing *Campto* larvae. Post-treatment inspection revealed that *Campto* control in these areas was very poor. The apparent lack of larvicide penetration into these cryptic pools threatened the efficacy of the treatment. No larvae can be left alive if eradication is to be at all possible.

Figure 5.3 Dye paper in open field of *Sarcocornia* (Samphire). (Image: D McGinn)

Measuring the penetration of liquid larvicide is achieved by deploying specially manufactured water-sensitive dye paper strips that turn from yellow to blue where droplets collect (Fig. 5.3). Deploying these under the speargrass roof before the aerial treatment and collecting them after to observe droplet density and coverage (compared with those in open areas) showed a very low percentage larvicide penetration. This confirmed that the liquid larvicide was being trapped in the vegetation canopy and was not reaching the larvae below.

In order to further characterise the problem, Phil Deadman was asked to reduce airspeed and treatment altitude to around 10 knots and 2 m, respectively, while treating the speargrass in an attempt to flood the protected pools (Fig. 5.4). Dye papers deployed under the thatched roof again showed insufficient penetration and post-treatment *Campto* mortality results were very poor. *Bti* has a short activity life in the habitat and remained effective at best for a day or two. Timing of treatments was critical and must be undertaken when *Campto* larvae are most susceptible during their development from first to early fourth instar. Late fourth instar larvae become refractory to *Bti* and pupae are not susceptible at all.

Timing of *Bti* treatments is therefore strongly linked to the events that cause saltmarsh to be flooded and causing *Campto* eggs to hatch. In most cases in Australia, tidal flooding can be predicted with fair accuracy and forward planning for treatments is possible. Rainfall in the saltmarsh and its catchment zone can also flood habitat and cause eggs to hatch. For these events, a more reactive activation of treatment is normal. Flooding of saltmarsh habitat in Napier, however, was somewhat more complex due to a very interesting connectedness it appeared to have with the coastal shoreline and by controlled manipulation of water levels in the agricultural pasture.

Figure 5.4 H500 trying to treat thatched *Juncus* fields. (Image: D McGinn)

The shoreline of the coastal strip adjacent to the Napier saltmarsh consists of shingle berms and coarse black sand. It is very permeable and sub-surface water moves freely back and forth with the tides. The permeable coastal geology is highly connected with the uplifted seabed subsoils forming much of the Napier saltmarsh. This connection is observed as significant movement in surface water levels within the marsh without overland flow. The water movement is observed to be sufficient to trigger *Campto* egg hatching between spring tides, apparently in response to the always varying sea state. Rough seas with crashing surf would 'pump' up saltmarsh water levels from below.

Further complicating prediction of flooding, and so *Campto* egg hatching, across the area was a water management system of three pumps used to empty drainage channels on agricultural land. Discharge from these automated pumps influenced adjacent saltmarsh flooding.

The irregular influences on saltmarsh flooding and *Campto* egg hatching at Napier produced an untenable treatment demand on the short-acting *Bti*. The limitations of liquid treatments had been fully exposed and we all wished that we could move away from this stop-gap situation and attack with an eradication plan which we had proposed earlier (see Chapter 2). A solid granulated residual larvicide such as ProLink XR-G would probably roll off the vegetation and find hidden *Campto* breeding pools and remain active for extended periods regardless of water level changes.

Getting XR-G to NZ

Production of XR-G is based in Dallas, Texas in the United States. The active ingredient is an insect growth regulator, S-methoprene, which is sold in Australian under the name ProLink (Altosid in other parts of the world). Australian mosquito-control programmes had been using a variety of formulation types of ProLink products including locally formulated sand, ready-to-use pellets and briquettes, all providing varying degrees of residual activity. The Australian control programmes were supplied from the US via the manufacturer's local distributor's warehouse in Redcliffe, Queensland. It might have seemed like a relatively logical and easy solution to supply XR-G into NZ by sourcing Australian-based stock, but there was one major hitch: XR-G was not used in Australia, nor was it even registered there. No relatively local stockpile of XR-G existed for NZ to tap into.

XR-G granules were considered a best fit (indeed, the only viable fit) product for *Campto* eradication in NZ, but they did not fit well into existing control of saltmarsh mosquitoes in Australia. A solution must be found to make stocks of XR-G available in NZ and to have a reliable supply chain to ensure *Campto* treatments would be made in accordance with the eradication protocol. Sourcing XR-G stock from Australia, by using that country's existing ProLink supply chain, was deemed impractical.

For any control products to be brought into Australia, the Australian Quarantine and Inspection Service required that they be first registered for use there by the Australian Pesticide and Veterinary Medicine Authority. There was no such import/export approval process available that would simply allow a product to transit through Australia. Obtaining Australian registration was going to be very time consuming (more than several months minimum) and very expensive. There was not time for that and little commercial demand for XR-G in Australia. The only way XR-G was going to become available in NZ was to directly ship it in from Dallas, Texas. An urgent application for NZ registration of XR-G was made and importing processes developed.

The ordering process was mutually planned between US-based supplier Zanus Corporation, the Napier MRC (Steve Garner), NZ BioSecure (Bryn Gradwell and Steve Crarer) and MoH (Sally Gilbert). The order and supply process was agreed as follows:

- Local staff would determine sites to be treated.
- The number of treatments was established based on a 21-day cycle for granules and a 30-day cycle for pellets.
- Treatments would be based on an estimated length of the season.
- Local staff would contact Zanus/Pacific BioLogics and advise when they would require two treatments worth of product. The second treatment requirement was put in place to have product on hand for any emergency.
- Zanus/Pacific Biologics would then place a tentative order for the season and establish a preliminary shipping schedule based on a 2-month transit time from Dallas, Texas, USA to NZ.
- Zoecon/Wellmark in the US would schedule production based on the forecast.
- During the treatment season, constant communication between stakeholders kept the flow working.

John Neberz, Marketing Manager for Zoecon/Wellmark, ensured product availability for all shipments. David Sullivan and John also established a special lower government price for the products and, to their credit, this price was maintained throughout the 11 years of the eradication programme. They were aware of the importance of having product when needed and any supply problems could jeopardise the success of the project. The group worked as a team during the entire manufacturing and supply chain process.

It is one of the remarkable achievements of the *Campto* eradication programme that supply of XR-G (and ProLink pellets) was maintained uninterrupted over the course of the programme and almost exclusively by sea-based freight. On only one occasion was air cargo used to meet urgent demand as new *Campto* infestations were discovered and responded to. The level of personal 'buy-in' to the mission of *Campto* eradication in NZ by these individuals and their respective organisations is an outstanding example of how a public-health-based project can focus effective action even across half a world.

Broad-scale treatment refinement

Availability of XR-G and the determination that it would be used to attempt *Campto* eradication meant applications systems for both aerial and ground-based treatments would change from liquid to solid granulated products. Aerial application of solid products requires very different equipment to handle dry material, disperse it evenly and efficiently.

At that time in NZ, there was no equipment designed for aerial dispersal of granulated products by helicopter. Helicopter dispersal of possum-control baits were designed for large chunks of 1080-laced vegetable matter and were not considered easily adaptable for

Figure 5.5 H500 with newly constructed under-slung XR-G spreader. (Image: D McGinn)

XR-G. Helicopter systems did exist elsewhere in the world for small granule dispersal, but availability in NZ was not immediate. Systems in NZ did exist for fixed-wing aircraft to spread fertiliser to pastures and may also have spread XR-G. However, fixed-wing was not considered a suitable platform in the irregular margins of the restricted saltmarsh treatment zones.

Having already deployed the Hughes 500 for *Bti* treatments, Phil Deadman set about the task of rapidly developing application equipment that would disperse XR-G granules. One method used elsewhere is based on an agricultural spreader designed to be attached to the three-point linkage of a farm tractor. A large hopper (inverted cone) feeds solid material onto a rotating vaned disk (spinner) that spreads granules in an arc. Flow is controlled by a manual lever operated by the tractor driver. They are generally powered by the tractor's power take-off shaft. The problem for aircraft use was that the unmodified tractor unit is heavy, and it cannot be fixed to the aircraft and must be suspended underneath by a tether and be stable in flight, the spinner has to be powered somehow and the flow has to be controlled remotely from the helicopter above. Notwithstanding the limitations, Phil Deadman produced a modified spreader that was set for testing for aerial dispersal of XR-G. (Fig. 5.5)

Aerial application of solids (or any other format) depends on knowing the performance of the equipment and how much it applies. Calibration of the equipment and characterisation of the swath width and pattern of deposited material reaching the ground is essential to deciding the treatment parameters of flight lane separation, treatment altitude and ground speed.

Calibration test runs were initially conducted using a linear array of target trays coated in light grease to capture falling granules that strike the tray. Trays were 0.5 m^2 in area and deployed at 2 m intervals. The aircraft flies perpendicular to the array, releasing material. Captured granules on each tray are counted and, based on known average granule weight, the deposition pattern and application rate can be characterised. Following several such calibration runs and adjustments, the underslung spreader consistently produced near the target 6 kg/ha, and swath of 20 m at 100 feet altitude and 60 knots. The flight lane separation was set at 15 m to allow conservative overlapping of the deposit.

A critical addition to the aerial system is the specification that aircraft guidance be supported by a highly accurate aerial agriculture GPS system. This system facilitated setting of a flight lane separation to fly parallel tracks accurately by giving the pilot a virtual illuminated self-correcting centre line to follow. The GPS also records the aircraft track and logs application runs via the pilot's spray on/off control. GPS records of treatments are highly important in visualising where aerial treatments have been completed and allows ground treatment teams to treat areas missed due (mostly) to obstructions (radio masts, navigational beacons, farm stock, etc.). Application equipment can wear and drift out of calibration over time. As a standard practice, the Napier MRC conducted spot checks of XR-G deposition by field deployment of catch basins of known size. Captured granules were counted and averaged to record depositions against the treatment standard of 6 kg/ha. Any significant variation was investigated and corrected immediately.

A major problem with the initial liquid applications was the inability to effectively treat *Campto* breeding under thatched speargrass canopy. Introduction of XR-G granules should solve this, but would it? We needed to know. Measurement of granule penetration through the speargrass canopy was attempted by Gene Browne using catch trays set out in both open ground (acting as controls) and under the speargrass canopy. The aerial treatment would then follow with falling granules captured in the deployed trays. Granules were then counted and a calculation applied to convert results to an application rate based on mean granule weight. The data were then statistically analysed. In addition to assessing recovered granules, trays of *Campto* larvae were also set out to provide bioassay data as a direct measure of treatment efficacy.

These experiments were useful at demonstrating the complexity of assessing aerial treatments and producing conclusions that try to make sense of variable real-world factors such as wind, size and placement of catch trays, shedding of solid material by canopy, mechanical and aerodynamic determinants of where granules end up. It is easy to over analyse results to perhaps explain a theoretical notion and still not know the real answer. Statistical analysis (Browne 2005) concluded that only 4 kg/ha was available under the canopy even though 6 kg/ha was being applied. So where did the missing 2 kg/ha go?

Notwithstanding, the *Campto* larvae knew the XR-G granules were there and delivering lethal concentrations of S-methoprene. The bioassay results of XR-G treatments showed that, indeed, *Campto* were effectively killed by granules penetrating speargrass canopy about the time of treatment and over the 21 days required. In a relatively short time, the dense granules apparently did find their way through canopy and into the breeding sites below.

The underslung bucket spreader has several limitations, not least of which was that it could not be flown over built-up areas for risk of it detaching and causing damage or injury, it extended the risk of the aircraft striking obstructions on the ground and take-off and landing were complicated and difficult in adverse weather.

During aerial application operations for the very large treatment areas of the Kaipara Harbour eradication zone, the underslung bucket was replaced in favour of the integrated, on-board, Isolair broadcast system (Fig. 5.6). At that time also, Deadman acquired Hughes

Figure 5.6 H500D with Isolair XR-G spreaders. (Image: J Wakeford)

500 Model D aircraft to replace the H500 Cs to maintain performance of the now industrial-scale treatment capacity deployed over several *Campto* eradication zones.

By 2004–05, the use of helicopters for broadacre saltmarsh larvicide treatments in NZ was now sophisticated, using very similar equipment as used in Australia and the United States. However, the major difference in NZ remained the intensity of treatment because it was geared for eradication. This meant that all actual and potential breeding sites outside the 'broad-acre' habitat required treatment.

The wide swath of the underslung bucket and later Isolair broadcast system was not suited to aerial treatment of very narrow drains, many kilometres of which criss-crossed farm paddocks to keep them usable in the winter rainy season. Many of the drains themselves were positive habitat and therefore included in the treatment zones. Ground-based treatment of farm drains was possible, but very time consuming due to the typical network of farm fencing under which many of the drains pass. The solution was some novel and ingenious use of aerial application of S-methoprene pellets.

ProLink pellets are made from an extruded mix of carbon and the active ingredient, S-methoprene. They are relatively large compared with XR-G granules. Pellets are applied at 4kg/ha and remain active in water for 30 days. At 4 kg/ha, however, only around 4–6 pellets are present per 1 m^2. This 'point source density' is too low for the broken, cattle hoof-print pugged habitats often used by saltmarsh mosquitoes, and many individual hoof-prints would be untreated. XR-G produces a better coverage in those situations, with adequate point sources of S-methoprene so that even a single hoof-print is going to receive at least one XR-G granule, which is adequate to kill *Campto*. In open water though, where there is good connection and mixing, pellets are a useful formulation type to use. They are particularly

Figure 5.7 Pellet applicator Adrian Brocas, adapted to helicopter as a 'door gunner'. (Image: J Wakeford)

Figure 5.8 Adrian Brocas treating a farm drain with pellets. (Image: J Wakeford)

useful in farm drains that are either permanently or periodically filled with water. Their longer residual life reduced the frequency of re-application compared with XR-G.

Modified back-pack blower applicator equipment was fitted to the helicopter as a 'door gun' and operated by a 'door gunner' (Fig. 5.7) to fire a narrow stream of pellets directed into farm drains in a coordinated dance of flying and 'firing'. Linked by voice communications to the pilot and working as a unit (Fig. 5.8), the door gunner directed a stream of pellets into the drain while the pilot maintained ground speed and manoeuvred the aircraft along drains with all their irregularities and many 90 degree bends, in the most advantageous attitude for treatment. This method allowed many kilometres of farm drain to be treated in a few hours compared with several days of ground-based work. Some interesting experiences were recalled by eradication door gunners. Bruce Hammond, Technical Officer in the Blenheim eradication zone recalled:

> *'While treating at Marshlands, we used to fly low under the transmission lines and on one occasion when Wayne Spencer was the pilot and I was gunner he took the top off a cabbage tree with the rotors. No damage to the chopper but nerves and the foliage were frayed.'*

To our knowledge, this method has not been used anywhere in the world beforehand or since the NZ *Campto* eradication programme, perhaps for good reason.

The broad-scale treatment of saltmarsh in NZ over 10 individual eradication zones developed from a knowledge base of zero in late 1998 when *Campto* was discovered in Napier to a highly sophisticated, robust and fit-for-task programme a few years later.

It was initially heavily informed by existing expertise in aerial programmes for control of *Ae. vigilax* in coastal south-east Queensland, but rapidly grew a life of its own in response to the unique challenge of delivering a treatment programme designed to achieve a world first – the complete eradication of an exotic and highly invasive saltmarsh mosquito species.

From 1999 to 2010, more than 741 000 kg of S-methoprene granules and 84 000 kg of pellets were applied in NZ to eradicate *Campto*. Approximately 123 000 ha were treated with granules from the air and ~21 000 ha were treated with pellets, both aerially and from the ground.

Reference

Browne GN (2005) Biosecurity surveillance of mosquitoes in New Zealand with a case example of methods used for the eradication of the Australian southern saltmarsh mosquito *Ochlerotatus Camptorhynchus*. PhD thesis. The University of Auckland.

6

Coming to grips with ProLink XR-G at Hawke's Bay

Brian Kay, Michael Brown, Gene Browne and Barbara Thomson

Background

The persistence of different formulations of S-methoprene in aquatic environments has most often been followed through bioassay of mosquito immatures, but some practitioners have tried to measure the concentrations of methoprene in the water. The efficacy of detection methods has varied but the use of capillary gas-liquid chromatography by the US Environmental Protection Agency (Knuth 1989) and high performance liquid chromatography has ensured that recoveries around 100% occur (e.g. Ross *et al.* 1994a with 99.0+/–6.9%). Knuth (1989) emphasised the lack of reliability of single water samples from one pool and recommended that composites be taken and that quality assurance procedures were important to validate the testing system.

Persistence also varies according to the formulation used; some methoprene products are sustained release with encapsulation whereas others are not. Ross *et al.* (1994b) evaluated five sustained release and other formulations, notably Altosid liquid larvicide (5%), briquettes, XR briquettes, pellets and a 1.3% sand granule. Only the briquettes and the pellets gave similar decay curves. Because of their formulation, these products produced peak concentrations of up to 5 ppb from 2 to 10 days after treatment but the mean concentrations over 35 days were between 0.14 and 0.32 ppb. This study may have been marred by the use of poly-vinyl chloride liners in the artificial pools used, because ~60% of S-methoprene has been shown to bind to this medium, as we demonstrate later in this chapter.

Thus, there was not much useful information available that could be extrapolated to the XR-G formulation and, over the first months of 1999, we really were working on faith. The original data about the performance of XR-G in this chapter provide some surprises, and these had to be managed in the eradication program. At the onset, it needs to be said that the XR-G product was new and based to some extent off the granular formulation evaluated by Ross *et al.* (1994b). Thus, technical advice proffered by those designing the program sometimes had to be based on a reasonable expectation, in terms of total theoretically available pesticide. This is not to say that the options considered and the ultimate outcome would have been any different if those making the decisions had been aware of these deficiencies at the outset. S-methoprene still would have been recommended, because of its residual properties and track record, but in view of some of the findings listed below in relation to the XR-G formulation, it is possible that pellets (4.25% S-methoprene content) would have been used more widely to eradicate *Campto* than just in the channels and drains at Hawke's Bay.

The stability of the XR-G granule (1.5% S-methoprene) product was a key issue in the eradication plan because the strategy relied on ensuring a blanket coverage of S-methoprene at a level above that sufficient to kill all *Campto*, but we were starting with a blank page and belief in advice from suppliers! What were the possible effects of the various biotic and abiotic factors at Hawke's Bay likely to be?

Specifically, we needed to answer the following:

1. Could we measure levels of S-methoprene in the field and laboratory?
2. How much S-methoprene was available from XR-G?
3. What sort of release pattern did it have?
4. Was there enough active ingredient to kill *Campto*?
5. How did salinity, sediment and solar radiation affect its efficacy?

Degradation is influenced by ultraviolet light from sunlight (Schaefer and Wilder 1972; Mian and Mulla 1982), salinity (Pree and Stewart 1975; Floore *et al.* 1991), microorganisms and temperature (Schaefer and Dupras 1973). The last authors demonstrated that at 10, 24 and 38°C over 5 days, methoprene losses were >30, 70–80, and >95%, respectively. Of relevance to Hawke's Bay, the half-life of mixed isomers of methoprene at 4.5 and 20°C, respectively, was reported as 134 and 49 days (Pree and Stewart 1975). These authors found the residual life of methoprene to be shorter in salt (marine) compared with fresh water. However, we were comfortable with a plan which involved residual action for 21 days.

Although methoprene from encapsulated formulations has been reported to remain near the sides and bottom of ponds, charcoal-based formulations (such as XR-G and pellets) have been reported to last longer and remain near the surface (Schaefer *et al.* 1974). Thus, with the decision to base the eradication campaign on the new XR-G product, first used in mid 1998 in the USA, it was imperative that parameters were estimated in local context where possible, because a lot of money was about to be spent.

Steve Garner and Michael Brown completed an inspection of the fresh, brackish and saline mosquito-producing habitats surrounding Napier on 26 April 1999. Habitat size and salinity were found to be highly variable. Within a region, pools ranged in size from ~1 m^2 to > 100 m^2. Salinities ranged from 0 g/L to > 33 g/L. They decided that there were three major categories that constituted *Campto* habitat: (1) permanent/semi-permanent ponds; (2) temporary pools; and (3) drains. On the basis of subsequent sampling, including the measuring of abiotic parameters, these probably could be divided further. Consequently, they were unable to locate several suitable small pools containing water of the same salinity conveniently situated together in one area. Small pools of the same salinity would permit accurate hand broadcasted small-scale treatments and replication. The majority of pools was greater than 20 m^2 in area, which was too large for accurate treatment by hand. Apart from the problem of identifying discrete permanent pools/ ponds, they had little knowledge of the flooding regimen, which meant that during the course of an experiment or a monitoring program, such pools could dry or re-flood to alter the parameters.

S-methoprene trials

Accordingly, an alternative method was designed to facilitate environmental decay curve analysis of S-methoprene applications. It was proposed that four sites in each of the three habitat categories be monitored with 10 samples each, at intervals of 0, 1, 2, 5, 10, 15, 20, 25, 30 and 35 days post-treatment, if necessary. Data from these (120 samples) would form the basis for comparison with data from the proposed artificial ponds in winter (42 samples) and summer (a maximum of 180 samples). This was to ensure that the degradation patterns in the field and in the artificial ponds were similar. This gave a minimum estimate of 360 samples for decay curve testing and, on the basis of revised sample numbers suggested by Steve Garner, it allowed for ~250 samples available for random periodic S-methoprene monitoring of the aerial application program. This approach could be labour intensive and very expensive, and as realised later, was impractical!

The report, submitted to MoH on 1 June 1999, contained a chemical monitoring schedule for ensuring that S-methoprene concentrations were being maintained at levels far in excess of the dosage to cause 100% mortality of *Campto.* Although this seemed feasible on the basis of the Ross *et al.* (1994b) tank trials, in practice it was not. Although S-methoprene was detected from field treatments of Altosand at 2.7 kg/ha at Redlands, Queensland, concentrations were only of the order of 0.1–0.3 ppb for 3 days. This was to be expected because the Altosand formulation was not an extended release product.

The artificial ponds came in the form of 18 concrete 90-cm diameter stock troughs, and were placed in a secure fenced compound adjacent to the Napier airport saltmarsh at Turfery Rd during May 1999. It was recommended that the troughs be water-blasted, filled with field water and a representative substrate, and allowed to stand for ~2 weeks before the initiation of testing. After this settling period, most water from the ponds was to be drained and gently refilled with filtered habitat water to a depth of 15 cm, and a substrate provided because benthic matter can absorb S-methoprene. This was to simulate the natural habitat.

There were two concerns about using the artificial pond method. The first was that lime leeching from the cement of new troughs would increase the pH of contained water. Accordingly, the supplier was contacted and the pH of water contained in new display troughs was measured. Our readings showed that fresh water contained in new troughs had a pH of ~8. This was well within the tolerance limits of non-target fauna found in saltmarsh habitats. So far, so good. The second concern was that, after application, S-methoprene released into the troughs would bind to the cement, leaving unrealistically low levels in the contained water. Michael Brown contacted David Sullivan who said that only a small amount of S-methoprene would be absorbed by the cement, but this small fraction would be released over time.

The idea was to monitor S-methoprene levels, resulting from 4 and 8 kg/ha treatment rates first in water of known salinities ranging from fresh (< 4 g/L), to brackish (14–20 g/L), to hypersaline (> 33 g/L) in a two-way factorial design. That is, there would be treatments (4 kg/ha = 0.26 g product, 8 kg/ha = 0.52 g of product and a control of 0 kg/ha) plus salinity (as above), which could be evaluated statistically in a controlled experiment with replication. By this time, it was May and water temperatures in the troughs were 10^0 C and data from Pree and Stewart (1975) indicated that any S-methoprene would be released very slowly.

To guide water sampling, each tank was divided into four quadrants by two intersecting string lines stretched across the tops of each tank. A 250 mL sample was to be taken from the top of each quadrant to make up a filtered 1 L water sample for testing to a resolution of 0.1 ppb. From this, sample vials (43 mL) were filled completely and stored in the dark until despatch by air freight to Christchurch for analysis by the Institute of Environmental Science and Research Ltd (ESR). No special precautions were taken to prevent biological degradation in these samples. This turned out to be a problem.

The advantages of the artificial pond protocol were as follows: (1) less resource would be expended on generating data of marginal relevance to summer conditions; (2) the type of decay pattern should nevertheless be apparent (i.e. a spike release within a 5–10 day period or a constant release over 21 days or more); (3) longer residual life during winter could facilitate less frequent treatments and ensure an extra carry-over of product for use in summer (or with the 21-day regimen, facilitate a build-up of S-methoprene); (4) results could be analysed statistically by parametric methods; and (5) it provided practice in carrying out methodologies crucial to summer success.

Timing of experiments and timing of the first trial was largely determined by advice from Cabinet on whether eradication was approved. Stock troughs were treated with equivalents to 4 and 8 kg/ha on 20 May 1999. This trial had the dual functions of evaluating the concentration of S-methoprene in the water and also to evaluate any mortality of the key environmental indicator species, a mysid *Tenagomysis novae-zealandiae* identified by Assoc. Prof. Jack Greenwood of the University of Queensland as being suitable for the task.

No S-methoprene was detected and the mysids were happy!

The nightmare begins – detecting S-methoprene in water samples

There was some degree of shock when the first two sets of water samples from treated field water were without detectable S-methoprene after 24 h and 5 days, when tested by ESR during late May. It was fair to say that there would be teething problems in the early stages, possibly caused by seemingly minor changes in methodology. ESR opted to use validated in-house methodology, based on an understanding of the chemistry of S-methoprene, information from the literature, including Brown *et al.* (1985), and communication with Australian Laboratory Services (ALS) and extensive experience in trace organic analysis, rather than United States Environmental Protection Agency protocols 3510 and 8270. The latter relate to generic extraction and analytical protocols for a range of semi-volatile organic compounds, but not specifically S-methoprene. ESR was concerned that the methodology was apparently not maximised for S-methoprene extraction and that triplicate amounts of dichloromethane, a possible human carcinogen and environmentally undesirable chemical, were being used. Alternate extractants, such as ethyl acetate, sulphuric acid and mercuric chloride, gave average recoveries from 92–104.7% from spiked samples, indicating the suitability of several methodologies. Ethyl acetate was selected as the extractant of choice and, on the basis of analysis of 7 × 0.3 ppb spiked samples, the limit of detection was determined to be 0.1 ppb. On the basis of 5 × 10 ppb samples, the average return was 10.1 ppb ± 0.45% and on later comparison between ethyl acetate and methyl t-butyl ester as extractants, the former was found to have 97% efficacy. Sodium sulphate and a rotoevaporation step were added to increase the sensitivity. Percentage recoveries from samples spiked at 9.3 ppb were reported at 107 and 113.

On 19 May 1999, ESR were contracted by the MoH to provide accredited results of the concentration of S-methoprene in Hawke's Bay water samples as received in the laboratory to a sensitivity of 1 ppb (later modified to 0.1 ppb), for up to four dosage rates (3, 4, 5 and 6 kg/ha), in saltmarsh pools at Napier, for up to 35 days.

Initially no special precautions, other than minimising storage time, were taken to prevent biological degradation between the time of collection and sample analysis. However, when S-methoprene was not detected in the first field samples analysed on 24 May, biological degradation was suggested as a likely problem. Various preservatives including mercuric chloride, sulphuric acid and dichloromethane were tested. Mercuric chloride was preferred because it is non-flammable (the samples had to be airfreighted to the laboratory), non-corrosive and was compatible with the preferred ESR extraction

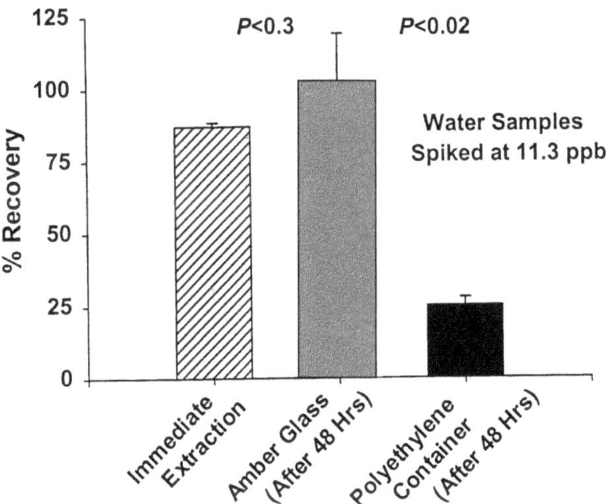

Figure 6.1 S-methoprene recovery from spiked water samples held in brown borosilicate glass and polyethylene containers ($n = 3$).

method. Polyethylene containers were considered as potentially more suitable for S-methoprene storage than glass. This idea followed from an abandonment of a sample in a clear polyethylene bottle with a 10 ppb S-methoprene water sample on a laboratory bench for 9 days. Without rinsing the container to remove any bound S-methoprene, recovery was reported at 90%. There were practical advantages for this. The containers were lighter and less fragile than glass. Polyethylene containers were subsequently examined analytically at ALS/QIMR by comparing water samples spiked with S-methoprene at 11.3 ppb (Fig. 6.1), which clearly showed that, even after 48 h, amber borosilicate retained a significantly greater ($P < 0.02$) proportion of the original spike than did opaque polyethylene.

ESR noted the ability of S-methoprene to adhere to glass, but this method of storage was thought to be adequate if an extractant was also included or the glass vessel was subsequently rinsed with solvent before analysis. This had implications for the testing of the active life of the product. At QIMR, 20 L glass aquaria were treated with XR-G at 4 kg/ha rate to monitor both S-methoprene concentrations and their efficacy against the salt-marsh mosquito, *Ae. vigilax*. Remember that this product was said to have a release life of 21 days or a bit more. At 62 days after treatment, the trials were terminated, the aquaria washed, wiped with paper towels, rinsed with deionised water and refilled. The mortality of third instars introduced into the new water averaged 90 ± 7% (Fig. 6.2). Thus laboratory trials gave us a totally different picture of biologically effective S-methoprene because glass was involved, but this activity would not be so prolonged in the field because substrates were different. This made it difficult to model field performance from laboratory studies.

Subsequent to the Hawke's Bay field trial, glass aquaria containing deionised water were spiked with S-methoprene at a rates of 4 or 8 kg/ha on 8 June 1999. Separate aquaria were located outside (exposed to the UV) and in the laboratory at ESR. Samples were analysed immediately at 24 and 48 h, to avoid any potential biodegradation issues. Maxima of 0.1 ppb and 0.2 ppb S-methoprene were detected in one rooftop aquarium spiked at 4kg/ha. At the higher application rate of 8 kg/ha, no methoprene was detected in any of the aquaria.

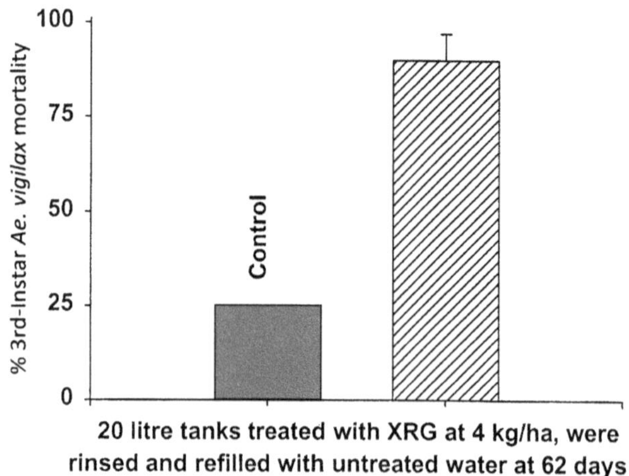

Figure 6.2 Bioassay of S-methoprene activity against third instar *Aedes vigilax* in glass aquaria, washed, wiped and rinsed with deionised water, 62 days after treatment with XR-G at a 4 kg/ha rate.

Subsequent to the Hawke's Bay field trial, on 16 June 1999 at the Gold Coast, Michael Brown treated 10 discrete pools by hand with XR-G at rates of 4 and 8 kg/ha and 9 × 100 mL composite water samples were put into brown borosilicate bottles containing 100 mL dichloromethane. Field water was monitored for 10 days. At both dosages, no S-methoprene was detected by ALS.

Needless to say, certain team members were getting rather frustrated at not seeing S-methoprene in either deionised or field water!

At the TAG meeting of 16 July 1999, the decision was taken to abandon chemical analysis of field water samples in favour of surrogate bioassay. Ministry of Health officials wore furrowed brows as they lined up on one side of the conference table. Their boss, Gillian Durham, believed that serious reassessment was needed as attempts to detect S-methoprene in the Hawke's Bay saltmarshes were far from satisfactory and could not inform dosage and frequency of application specifications locally. Both industry representatives of Wellmark International (Doug Van Gundy) and Zanus Corporation (David Sullivan) reassured the Ministry personnel and its technical advisors that sufficient S-methoprene would be in the field water at Hawke's Bay to have reasonable probability that eradication of *Campto* would occur after persistent applications at 6 kg/ha, despite not detecting S-methoprene in the water trials.

Do aerial applications of granules reach the ground?

If we could not detect the product in the water, then could we see XR-G granules on the saltmarsh? This is akin to counting soot specks on a dark night! In theory, satisfactory applications should be reasonably even but the variation in the number of granules across what is known as the swath width will have a coefficient of variation in the order of 25%. We placed 0.5 m^2 greased boards (or, alternately, catch buckets) out around selected pools for evaluation. The actual dosage was estimated by averaging the number of granules stuck to 10 replicates of 3–4 greased boards. Previously in the laboratory, we had weighed and counted 10 replicates each of 0.3 g (3 kg/ha) to get 60.6 granules +/− 9.22 standard deviation (range 48–75) granules per m^2. Thus at the designated 6 kg/ha treatment rate with batches supplied for New Zealand, we should expect to apply 120 granules/m^2 or 20 granules/kg/ha. At least we knew the XR-G granules were having an impact on saltmarsh habitat but did we know if the granules were releasing their lethal payload?

Characteristics of the XR-G product

The specimen label (Wellmark International, January 1999) indicated 1.5% S-methoprene content. The XR-G product is not a solid granule but S-methoprene and unspecified protein coated onto sand, which is then coated with carbon. The carbon is used to protect S-methoprene from ultraviolet radiation as the conjugated 1, 3-diene moiety in S-methoprene can undergo cycloaddition reactions catalysed by UV light.

Because both ESR and ALS did not detect S-methoprene in the field and in some aquarium samples, one possibility was that that the batch as supplied for trials was defective. Alternatively, advice from a chemist at QIMR indicated that: (1) because carbon is used as a common filter, excessive binding may occur; and (2) dichloromethane extractant might actually bind free S-methoprene to the protein rather than release it. Accordingly, 2 g replicates of the New Zealand test granules were analysed as four batches: two replicates of crushed granules and two replicates of intact granules by putting them directly into dichloromethane extractant (Table 6.1). The table shows that the manufacturers were generous because the specified 1.5% of active ingredient was actually 1.83–2.19%. So far, so good.

At the 6 kg/ha application rate adopted by Technical Advisory Group 6 (1999) in pools averaging 15 cm depth, the average daily release rate, assuming 100% availability of active ingredient, is 2.9 ppb over the suggested 21-day period of activity. Previously at the 4 and 8 kg/ha application rate, this theoretical figure was calculated at 1.9 and 3.8 ppb, respectively.

In order to obtain an estimate of the degree of binding and actual daily release rates of S-methoprene, the performance of crushed and uncrushed XR-G was compared in 20 L aquaria with salt water at 20 g/L at 25 ± 1°C in the QIMR insectary in Brisbane. Two different batches of product were obtained (NZ test, batch # 980317949, manufactured in March 1999: #98AMCA, possibly 1997) and each treatment at a 4 kg/ha rate was replicated twice. One-litre composite samples, comprising 4 × 225 mL water samples plus 100 mL dichloromethane, were collected at 24 h post-treatment and every 5 days for 40 days and analysed by ALS.

Two assumptions were made for this experiment: (1) that crushing of XR-G would free up the S-methoprene for release into water for mosquito control; and (2) that the half-life of the free technical grade component is short enough to not significantly bias the estimates, via a build-up of residues. The total S-methoprene (ppb) was summed for all positive sampling days; in the case of the NZ batch, two crushed replicates for 10 samples over 45 days (58.7 and 83.7 g) and for the uncrushed XR-G (3.16 and 38.6 g). Note that with an

Table 6.1. Percentage S-methoprene content of Altosid (ProLink) XR-G

Treatment	Replicate	Percentage S-methoprene
Crushed	1	1.83
	2	2.19
Uncrushed	1	2.05
	2	2.00

assayed mean 2.0% content, instead of the specified 1.5%, this means that at a 4 kg/ha rate, the total S-methoprene that theoretically should be available is 80 g/ha, which, in relation to 15 cm depth (1 500 000 L), was 53.3 ppb total. The explanation for the apparent over-abundance of total S-methoprene in the second crushed replicate could be due to the water non-replacement sampling policy for this experiment. Nevertheless, the estimates from this batch suggested S-methoprene availability of perhaps as little as 5.4 to 46% for this formulation. Estimates for the #98AMCA batch were 24 and 10.2% (Table 6.2). This was a trap for young players!

We had assumed that if the label specified an active content of 1.5%, then that is what we should have. Wrong! So, now we knew how much S-methoprene we had to play with, but Brian Kay at least felt that we had been short changed. However, he did know that the active ingredient had to be protected from the sun.

The second consideration related to its release pattern (Table 6.2). For three of four uncrushed batches, heaviest release occurred within the first 24 h, with activity being recorded up to day 45 with the NZ test batch, but only on day 1 for the 98AMCA batch. There may be true differences in activity between these two batches because the latter was at least 2 years old. The variation between replicates illustrates possible inter-batch variation in S-methoprene content or, more likely, the difficulties of accurately sampling small batches in units of ppb, or part thereof, even in aquaria. Nevertheless, these assays by ALS (Brown *et al.* 1985) indicated that positive assays from 4 kg/ha rates were possible and that, for the NZ batch at least, the active component was being released up to 45 days. Hallelujah!

To examine the release pattern further, Brian Kay and Michael Brown designed an approach to measure what was left in the XR-G granule over time, rather than measuring amounts released into the water. Batches each with 2 g of XR-G were placed into 200 mL perforated glass beakers that sat in 2 L of reverse osmosis water in glass beakers. The residual S-methoprene content was analysed at ALS on days 1, 5, 10, 15, 20, 30 and 40 by gently lifting the beaker out and putting the remains of the granules directly into dichloromethane. Two replicates were done for each sample. In contrast to the previous experiment, this was done at ambient conditions (29–31°C) with UV exposure, and the fresh water was replaced after each sampling period to maintain the volume at 2 L.

Table 6.2. Total availability of S-methoprene (in ppb/day) with crushed and intact XR-G granules, in 20 L aquaria of salt water

			Time (days)									
Batch	**Treatment**	**Rep.**	**1**	**5**	**10**	**15**	**20**	**25**	**30**	**35**	**40**	**45**
NZ	Crushed	1	8.4	6.5	3.8	3.7	5.6	1.0	1.9	2.3	1.9	23.6
		2	7.5	10.2	6.9	7.5	4.0	2.6	1.2	8.1	2.6	33.1
	Uncrushed	1	0.4	0.4	0.4	0.8	0.8	< 0.2	< 0.2	< 0.2	0.3	< 0.2
		2	< 0.2	0.2	< 0.2	< 0.2	< 0.2	< 0.2	< 0.2	< 0.2	< 0.2	0.4
98A	Crushed	1	2.3	4.5	2.0	1.1	< 0.2					
		2	1.5	1.0	< 0.2	1.3	< 0.2					
	Uncrushed	1	0.4	< 0.2	< 0.2	< 0.2	< 0.2					
		2	0.4	< 0.2	< 0.2	< 0.2	< 0.2					

Table 6.3. Residual S-methoprene content of ProLink XR-G granules for up to 40 days of being held outdoors in freshwater microcosms

Time (days)	0	1	5	10	15	20	30	40
Residual S-methoprene (%)	2.03±0.04	1.2±0.08	1.49±0.41	1.16±0	1.36±0.01	1.18±0.03	1.21 ±0.01	1.24±0.1

As shown (Table 6.3), the XR-G formulation released 40% (21.3 ppb) of the S-methoprene within the first day and from day 5 to 40, minimal release occurred. The carbon coating was extremely efficient in the presence of UV, because 61% of the S-methoprene in the granule was still bound after 40 days of exposure. This substantiated the earlier trial about the availability of active ingredient.

To summarise the situation graphically (Fig. 6.3), we could not detect S-methoprene in field water after 4 and 8 kg/ ha doses, but could do so in the laboratory. It was also apparent that the XR-G product did not fully release the lethal ingredient because it needed to be protected from adverse environmental factors.

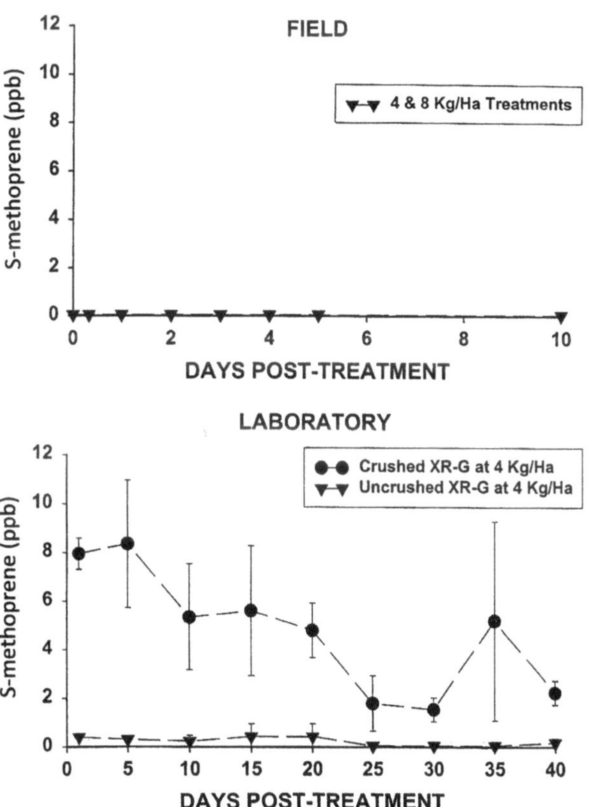

Figure 6.3 Comparison of concentrations of S-methoprene, following XR-G treatments in the field and in the laboratory ($n = 2$).

Was there enough XR-G released to kill mosquitoes?

We then decided to conduct bioassays in two 20 L salt water aquaria at 20 g/L treated with 4 kg/ha XR-G. Fifty third-instar *Ae. vigilax* were introduced into the tanks on days 35, 50 and 62. The S-methoprene content from water samples analysed at ALS had fallen below the detectable limit of 0.1 ppb from days 25, 30 and 35. Despite this, 100% mortality occurred (Fig. 6.4) on all test days whereas mortality in two untreated tanks was acceptable and significantly different ($P < 0.01$) from those treated.

Given the apparently poor release of S-methoprene from the XR-G formulation, and noting that the LC_{95} for both *Campto* and *Ae. vigilax* has been estimated between 0.1 and 0.5 ppb (Kay 1999), 100% control still occurred at undetectable concentrations. Larval susceptibility tests have traditionally been done by diluting Altosid liquid larvicide because of the huge volumes necessary to accurately dilute expensive technical grade material. Traditionally, LC data are not corrected for the fact that only 20% S-methoprene is said to be released on day 1, rather than to the total amount released over a further 4 days. Given its short half-life and non-accumulation (see later), these susceptibility data must represent maximum concentrations. Furthermore, the data are not adjusted for the insolubility of S-methoprene in water, although it is conceivable that some larvae could ingest undissolved pesticide. Thus, we now believe that the LC_{100} value for *Campto* is in the order of at least five times lower (i.e. 0.02–0.1 ppb). This should offer extra security that concentra-

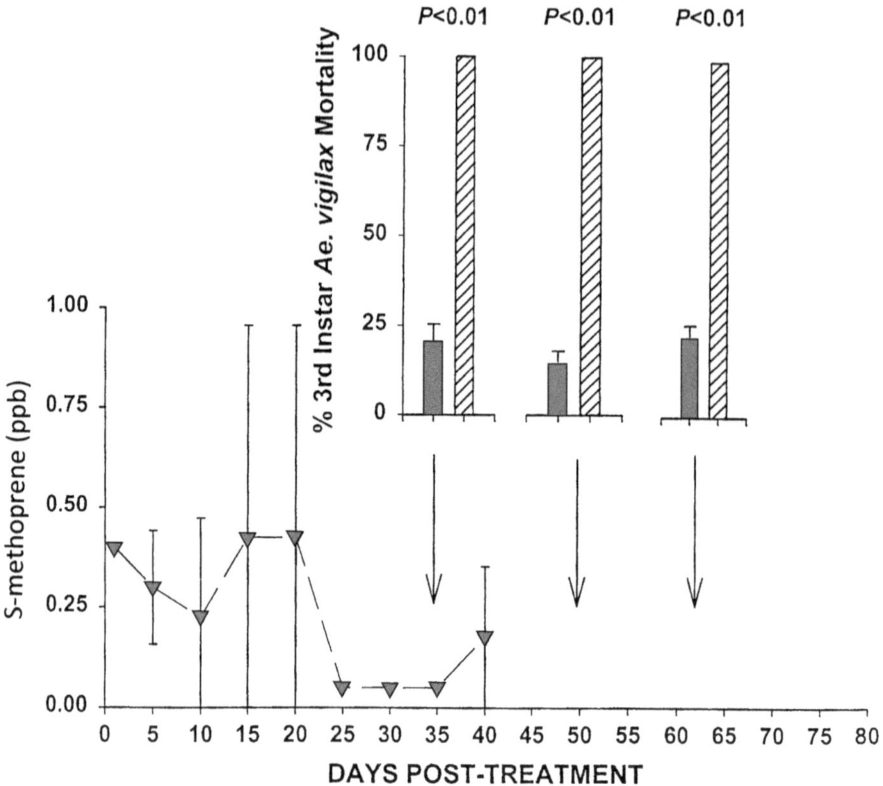

Figure 6.4 Laboratory decay curve of XR-G at 4 kg/ha and subsequent bioassays at days 35, 50 and 62 with *Aedes vigilax*.

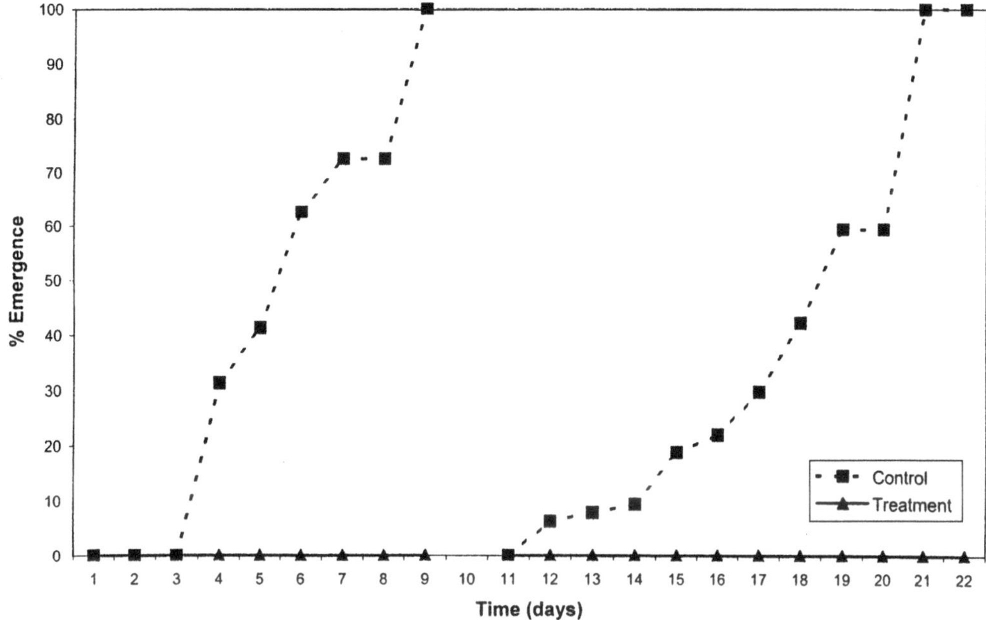

Figure 6.5 Emergence of *Aedes camptorhynchus* in laboratory aquaria at 25°C, Napier Emergency Response Centre, following treatment with XR-G at 4 kg/ha.

tions in the outbreak area after treatment are in excess of requirements, although it is disconcerting that such low concentrations cannot be measured.

During April 1999, laboratory and field bioassays (Figs 6.5 and 6.6) were done with Napier water and *Campto* using 20 L aquaria and 4 kg/ha treatment rates (Browne 2005). Three aquaria were held within the Exotic Mosquito Response Centre whereas three were kept on a balcony on Napier hill, exposed to sunlight. Tanks were filled with field water at a salinity of 14 ppt. Two of each set were treated and one left untreated as a control. In the laboratory tanks at 25°C, two batches of third instars were added at 0 and 10 days and emergence and physical and behavioural characteristics noted for 22 days. Larvae became pale and were noticed to swim abnormally, whereas pupae often were not fully developed with some larval features persisting or were pale with unusual appendages. Only one deformed male from 375 test subjects emerged but, in reality, mortality can be counted as 100%.

Of those untreated, 80% emerged successfully 3–10 days later at 25°C. Under the oscillating ambient conditions, development was slower from 11 to 21 days in the untreated tank but 100% control was recorded in the treated tanks, with one deformed male emerging out of 200 larvae tested. These data indicated that, at a 4 kg/ha rate, XR-G gave 100% control and that the product was effective enough to kill any larvae which may emerge several days after treatment.

Environmental variables

This section examines the effect of salinity, sediment and solar radiation (UV) and how it might affect detectable S-methoprene concentrations, based on the literature (Cooney 1995). To examine the effect of salinity and sediment, four aquaria were filled with 20 L of

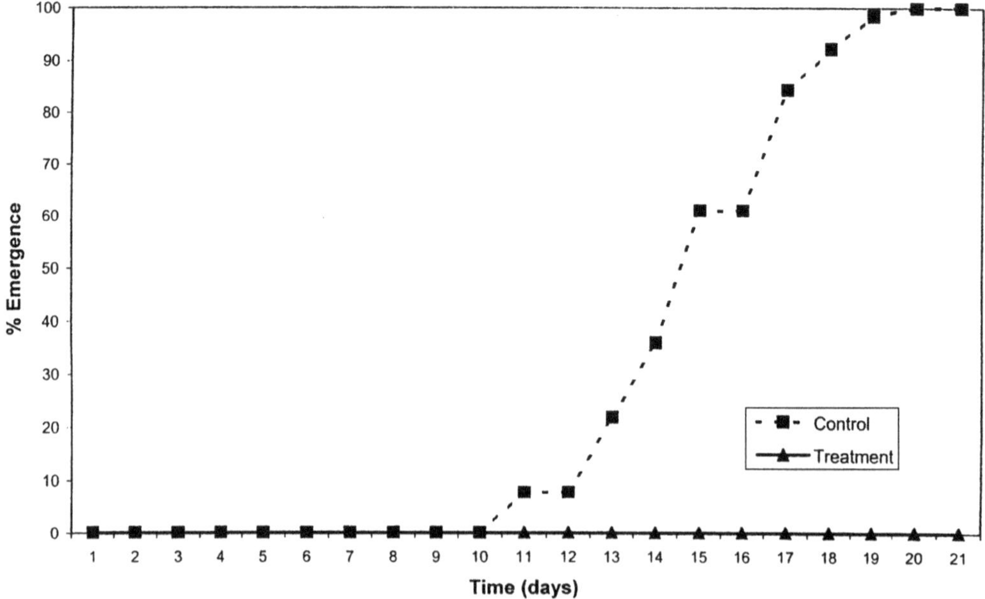

Figure 6.6 Emergence of *Aedes camptorhynchus* in aquaria at ambient conditions on Napier hill, following treatment with XR-G at a 4 kg/ha rate.

salt water at 36.5 ppt and a 5 cm deep organic-sand substrate was added to two. Four aquaria were filled with fresh water and the same substrate (which had been washed to remove salt) was added to give a final salinity of 0.3–0.5 ppt. These aquaria were placed in direct sunlight with a photo-period of 14 h light: 10 h dark. Analysis of the substrate indicated it was 90% silica, with 64% of the particle size being in the range of 212–600 µm. Each tank was spiked with 22.6 ppb of technical grade S-methoprene, the ingredient in XR-G. Daily samples were taken for 5 days with the water volume being replaced.

The results (Fig. 6.7) indicated that, at the 22.6 ppb treatment rate, ~90% of S-methoprene degraded within 24 h, regardless of the conditions. After 2 days, the concentrations were significantly less ($P < 0.01$) than the spiked level in tanks with or without sediment, regardless of salinity.

So what about the effect of solar radiation? Because 90% of the decay of S-methoprene occurred within the first 24 h, aquaria were spiked with 22.6 ppb of technical material at 0800 h and water samples taken every 3 h for 24 h. Because the type of water and presence of sediment had relatively little impact on the stability of S-methoprene during this time, the trial was done with fresh water and no sediment. To eliminate solar radiation, cardboard boxes were placed over two aquaria whereas the other two were left exposed.

These data (Fig. 6.8) indicate that solar radiation during daylight hours had a major effect on the stability of S-methoprene and degradation slowed down considerably during the second 12 h period (after nightfall). For technical grade S-methoprene, which is released from XR-G, the latter two studies demonstrate that S-methoprene has a half-life in the order of 3–8 h and solar radiation seemed to exert the most influence on decay, compared with the other environmental factors. To increase the efficacy of XR-G in Hawke's Bay, the MoH could consider the practicalities of aerial applications in the late afternoon/early evening to maximise the available concentrations in the water. This may not be practical,

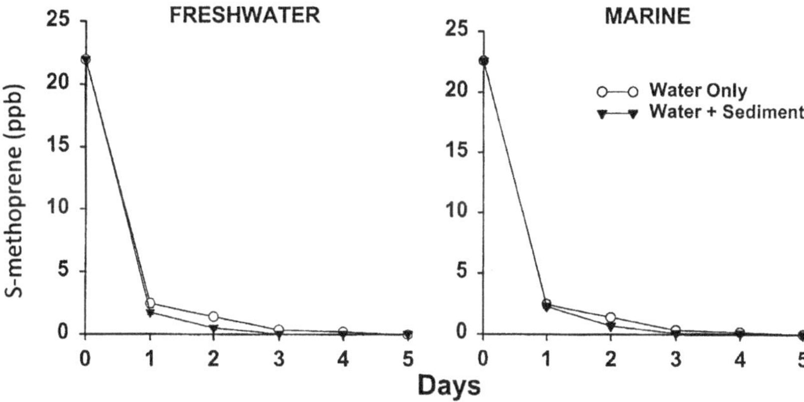

Figure 6.7 Decay curves for technical grade S-methoprene in fresh and marine water, with and without sediment.

however, because increased wind velocities in the afternoon can influence the evenness of the granule distribution pattern.

So, after all this, we were ready to go and had evidence that the ProLink XR-G product would kill *Campto* even at 4 kg/ha. To allow for environmental imponderables, we elected to apply the product at 6 kg/ha to install some degree of security. The distribution pattern on the greased boards indicated that we could expect around 120 XR-G granules for every m^2 of breeding habitat, but some of these would get caught up in heavy undergrowth (Browne 2005). We could not detect S-methoprene in the field situation, but, as, David Sullivan said, 'The mosquitoes know it's there'. The limit of detection of S-methoprene was 0.1 ppb, but we think that environmental factors were limiting detection in field sites. This supports our belief that the amount of S-methoprene required to kill off 100% of *Campto* was certainly less than the estimated 0.1–0.2 ppb and possibly 5-fold less. So, with respect to the best time for treatment in order to avoid solar degradation, the answer probably is that it does not matter because *Campto* are extremely sensitive to S-methoprene.

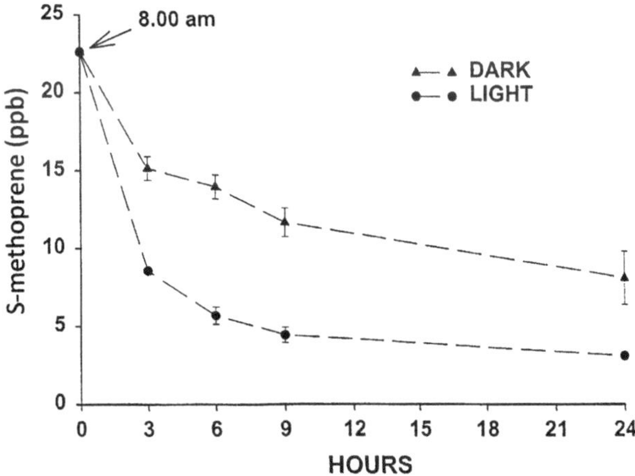

Figure 6.8 Decay curves for technical grade S-methoprene in fresh water, in dark and light conditions ($n = 2$).

Acknowledgements

We thank David Sullivan from Zanus Corporation for product and constructive advice, and also David Walker of Pacific BioLogics Pty Ltd for supplying product for testing.

References

Brown MD, Kay BH, Blow P (1995) Development of a mass spectrum gas chromatography method for determination of methoprene concentrations in natural waters. *Local Authorities Research Committee Annual Report* **1994–1995**, 125–128. [QIMR, Brisbane.]

Browne GN (2005) Biosecurity surveillance of mosquitoes in New Zealand with a case example of methods used for the eradication of the Australian southern saltmarsh mosquito *Ochlerotatus Camptorhynchus*. PhD thesis. The University of Auckland.

Cooney JD (1995) Factors that modify the toxicity. In: *Fundamentals of Aquatic Toxicology* (Ed. GM Rand) pp. 94–97. Taylor and Francis, Washington DC.

Floore TG, Rathburn CB Jr, Dukes JC, Clements BW Jr, Boike AH Jr (1991) Control of *Aedes taeniorhynchus* and *Culex quinquefasdatus* emergence with sustained release Altosid sand granules and pellets in saltwater and freshwater test plots. *Journal of the American Mosquito Control Association* 7, 405–408.

Kay BH (1999) 'Laboratory evaluation of the efficacy of S-methoprene and three 1200 ITU/mg *BTI* products for control of *Aedes camptorhynchus,* compared to other Australian vectors'. Report to the Ministry of Health, Wellington, NZ.

Knuth ML (1989) Determination of the insect growth regulator methoprene in natural waters by capillary gas -liquid chromatography. *Chemosphere* **18**, 2275–2281.

Mian LS, Mulla MS (1982) Biological and environmental dynamics of insect growth regulators (IGRs) as used against Diptera of public health importance. *Residue Reviews* **74**, 27–112. doi:10.1007/978-1-4612-5756-1_2

Pree DJ, Stewart DKR (1975) Persistence in water of formulations of the insect developmental inhibitor ZR 515. *Bulletin of Environmental Contamination and Toxicology* **14**, 117–121. doi:10.1007/BF01685609

Ross DR, Cohle P, Ritchie BP, Bussard JB, Neufeld K (1994a) Effects of the insect growth regulator (S)-methoprene on the early life stages of the Fathead minnow *Pimephalas promelas* in a flow through laboratory system. *Journal of the American Mosquito Control Association* **10**, 211–221.

Ross DH, Judy D, Jacobson B, Howell JR (1994b) Methoprene concentrations in freshwater microcosms treated with sustained-release Altosid® formulations. *Journal of the American Mosquito Control Association* **10**, 202–210.

Schaefer CH, Dupras EF Jr (1973) Insect development inhibitors. 4. Persistence of ZR515 in water. *Journal of Economic Entomology* **66**, 923–925.

Schaefer CH, Wilder WH (1972) Insect development inhibitors: A practical evaluation as mosquito control agents. 1. *Journal of Economic Entomology* **65**, 1066–1071.

Schaefer CH, Wilder WH, Mulligan FS, IIIDupras EF (1974) Insect development inhibitors: Effects of Altosid, TH6040 and H24108 against mosquitoes (Diptera: Culicidae). *Proceedings and Papers of the Californian Mosquito. Control Assoiation* **42**, 140–145.

Technical Advisory Group 6 (1999) *Aedes camptorhynchus* infestation in the Hawke's Bay. Agenda for the Sixth Meeting, 16 July 1999. Ministry of Health, Wellington, NZ.

7

Pathways of entry and mosquito dispersal

Andrew McFadden, Graham Mackereth, Ruud Kleinpaste, Robert Sanson, Nigel Beebe, Bryn Gradwell, Mark Bullians and Ruth Frampton

Introduction

In order to carry out effective biosecurity, it is important to understand and mitigate the pathways for entry and spread of exotic species that have the potential to establish. If pathways into a country are identified, steps can be taken to reduce the likelihood of the species entering the country.

The pathways can be countered at source, with pre-embarkation or pre-export procedures, or by interception at the border on arrival (interception is the process of detecting a species at the border). An incursion is considered to have occurred where exotic species are detected post-border in circumstances where they are found to be breeding.

In NZ, an exotic mosquito detected in a surveillance trap at a port of entry is most likely to be considered an interception rather than an incursion. Surveillance at international ports and airports is undertaken routinely and mosquitoes are intercepted every year. It is possible for mosquitoes to get sequestered with personal effects or sealed into containers and bypass residual insecticides applied to the cabins in planes or disinsection processes on ships. Despite measures taken, pathways still exist for the entry of mosquitoes, particularly species that lay desiccation-resistant eggs.

Entry of mosquitoes does not necessarily mean that a population will establish. Whether it does depends on several biological factors, such as the type, number, viability and sex. However, a single gravid female will pose an establishment risk. In some situations it can be difficult to determine whether the detection of a mosquito in a surveillance trap is the result of an interception or an indication that a small population has established in the immediate vicinity. This was the case in a biosecurity response initiated for Asian tiger mosquito, *Ae. albopictus*. One adult male was collected from a CO_2-baited sentinel surveillance light trap at the Port of Auckland on 2 March 2007 (Holder *et al.* 2010). In this particular instance, a biosecurity response was initiated due to uncertainty as to whether it was an interception from a recently arrived ship or an indication of a recently established local population.

On finding a mosquito at a port, it is necessary to determine the significance of the find by considering the manner in which it was found, the likely means of entry, association with specific goods or vessels, the previous ports those vessels visited, the life stage of the mosquito and the nature of any regular surveillance in place. The find may indicate that mosquitoes have gained entry and been conveyed to other sites within the country by various means. Detection indicates that border measures are operating but does not necessarily reduce uncertainty about the effectiveness of measures for mitigating mosquito incursions. Often we do not know for certain how often entry has occurred without establishment. However, one way of assessing the biosecurity system is by looking at how many interceptions are made or how often incursions have occurred. Thus, it can be difficult to estimate the true number of incursions that would have occurred without any measures applied at the border.

Those mosquitoes that get through the border must find suitable habitat for breeding. This is not difficult for container-breeding mosquitoes, but is less likely for saltmarsh mosquitoes. Surveillance activities in NZ are tailored to detect both container- and saltmarsh-breeding species near ports and airports. In addition, resources allocated for surveillance should be applied with an understanding of the risk of entry and likely breeding sites of the various exotic species.

Once a population is established, it has many and varied potential pathways of spread. All stages of a mosquito's life cycle can be inadvertently spread through human activities, but adult mosquitoes can also move themselves. Local spread will occur and is characterised in terms of the utilisation and invasion of nearby habitat. Mechanical spread to distant sites can occur by a variety of means and poses the greatest challenge for eradication. The most significant long distance spread is associated with movements in cars, planes, boats and other vehicles. Just as with entry at the port, there may be many translocations of mosquitoes that do not result in new breeding populations. In assessing the risk of new sites being invaded, one must consider the abundance of the source population, their interaction with human activities that may result in movement, the time taken to translocate and the conditions at the new location.

In this chapter, we tell the story of the introduction and spread of *Campto* in NZ, pieced together from anecdotal and scientific evidence.

Pathways for entry

Aedes camptorhynchus (Southern saltmarsh mosquito), when introduced, was the most recent addition to NZ's mosquito fauna, comprising just 12 native species (Laird 1995; Macfarlane *et al.* 2000; Weinstein *et al.* 1997) and three introduced species. As the common name suggests, it is a saltmarsh species (Macfarlane *et al.* 2000) and not, as might be expected from global trends of mosquito spread, a container-breeding species. Prior to the discovery of *Campto* in NZ, there had not been an exotic mosquito establishment since the early 1960s. The earliest establishment of an exotic culicid involved *Cx. quinquefasciatus*. It arrived in NZ (Weinstein *et al.* 1997) and Australia (Lounibos 2002) during the early years of European settlement. Notably, three of the four species of exotic mosquitoes that have established in NZ are native to Australia, reflecting its close proximity and associated pathways for entry.

Campto does not have a history of global spread. It established in NZ despite the probability of establishment being very low compared with container-breeding mosquitoes. Unlike the container-breeding mosquitoes, the more probable (albeit unlikely) pathways of entry involve adults (i.e. in cabins or in the holds of ships, internal contamination of shipping containers or on aircraft arriving from other countries). In the main, such pathways provide low propagule pressure, with the recent notable exception of more than 12 individuals of a species of *Culex* (Ministry of Agriculture and Forestry 2003).

Campto has rarely been intercepted at the NZ border. MoH records show four detections of dead adults: one from a container in 2004; two from decks of a used vehicle-carrying vessel travelling via Australia on separate occasions in 2009; and another somewhat exceptional interception from inadequately cleaned goods (mosquito traps from Australia) also in 2009. Lounibos (2002) suggests that propagule pressure and past success are the best predictors of the invasiveness of a mosquito invader. Based on these predictors, *Campto* would not be expected to be invasive; exerting very low propagule pressure through the more probable pathways of entry and, before this present introduction to NZ, having no past success. It's good, however, that at least some thought eradication was possible (see Chapter 1)!

Recent successful mosquito invasions internationally almost exclusively involve container-breeding species possessing a desiccation-resistant egg stage (e.g. *Ae. albopictus*, *Ae. atropalpus* and *Ae. japonicus*). Although there are documented records of long-distance dispersal of mosquito adults unassisted by humans (i.e. natural dispersal), far from their larval habitats, resulting in short-term colonisations that temporarily extend the range of a species, most recent successful invasions of mosquitoes have resulted from human transport of immature stages. The duration of survival (hatching viability) of such desiccation-resistant eggs can be in the order of some years (up to 4 years recorded). Furthermore, the spread of container-breeding species is easily effected through the transport of immature stages (desiccation-resistant eggs and/or larvae) in artificial containers, their natural habitat. Pathways of entry for container-breeding species (e.g. used tyre imports, used vehicle and machinery imports) are well known and, consequently, measures to prevent the entry of mosquitoes via these pathways have been identified. The importance of various pathways of entry varies with different species. The possible pathways of entry of *Campto* into NZ are presented, and qualitatively ranked, in Table 7.1.

The global spread of the more cold-tolerant, container-breeding mosquito species, *Ae. albopictus* and *Ae. japonicus*, during the last two decades, as well as the ongoing threat of *Ae. aegypti* in tropical areas, suggests that few countries will ultimately be immune to the invasion of one or more of these species. Even those countries, such as Australia, France and NZ, with rigorous biosecurity systems in place targeting mosquito species are frequently challenged. As noted by Lounibos (2002) regarding propagule pressure, it is noteworthy that most successful mosquito invaders have arrived by ship. Mosquito arrivals on aircraft are typically adults consisting of only a few individuals of any given species. In contrast, ships, especially modern container vessels, can themselves harbour, as well as transport cargo, which carries a large number of propagules, especially of the immature stages of mosquitoes. For example, the transport of desiccation-resistant *Aedes* eggs in tyres appears to account for the establishment of container-breeding species such as *Ae. atropalpus* in France (Schaffner *et al.* 2013) and Italy (Lounibos 2002; Snow and Ramsdale 2002), *Ae. japonicus* in France (Schaffner *et al.* 2013) and the United States (Fonseca *et al.* 2001; Lounibos 2002; Snow and Ramsdale 2002) and *Ae. albopictus* almost worldwide (Ayres *et al.* 2002; Lounibos 2002; Schaffner *et al.* 2013; Snow and Ramsdale 2002).

Lounibos (2002) further states that 'the dominance of a few species among successful mosquito invaders suggests that previous success may be a potentially good predictor of vector invasiveness'. Although one cannot fail to agree that such a statement applies to the aforementioned container-inhabiting species, the most recent and only new mosquito invader to NZ for over four decades has had no previous success. In addition, a substantial

Table 7.1. Possible pathways of entry of *Campto* into NZ

Pathway of entry	Demonstrated *Campto* pathway of entry (Y/N)	Likelihood of *Campto* entering NZ by this pathway (probable, possible, low or negligible)	Likelihood of *Campto* establishing in NZ as a result of entering by this pathway (probable, possible, low or negligible)
Used tyre imports (containerised and non-containerised)	N Has not been recorded to breed in containers such as tyres	Negligible	Possible
Used vehicle and machinery imports (including any accompanying accessories)	N Has not been recorded to breed in containers	Negligible	Possible
Water pooled on the deck, items on deck or deck cargo on ships, fishing boats and yachts	N Has not been observed to breed in containers such as items on deck	Low	Possible
Water-storage containers or bilges of ships, fishing boats and yachts	N	Negligible	Low
In cabins or in the holds of ships	Y (Two **dead** adult females: MoH, 6 May 2009 and 3 June 2009)	Possible	Negligible
Internal contamination of containers (including empty containers)	Y (**Dead** adult male: MoH, 30 September 2004)	Possible	Negligible
External contamination (including water collected in sagging canvas 'soft tops') of loaded and empty containers	N	Low	Possible
Imports of plants or plant products	N	Negligible	Negligible
On aircraft arriving from other countries	N	Low	Negligible
With passengers' baggage (e.g. within a rolled up tent)	N	Negligible	Negligible
Deliberate illegal (human-instigated) introduction	N	Negligible	Probable
Wind dispersal	N	Negligible	Negligible
Migratory birds	N	Negligible	Negligible

proportion of recent establishments seem to originate in countries that are near neighbours. Although the 'journey' may be a rough one, the close proximity of Australia nevertheless enhances the probability of survival during the short trip of ~2000 km, whatever the mode and propagule pressure.

Whether or not one takes account of the measures associated with the more probable (albeit unlikely) pathways of entry (i.e. in cabins or in the holds of ships, internal contamination of shipping containers, external contamination of open shipping containers and on aircraft arriving from other countries), because of the low propagule pressure, it is not clear how *Campto* got to NZ. Moreover, it is difficult to envisage such a rare event occurring more than once.

Nevertheless, a couple of the pathways warrant further discussion, if for no other reason than it has been speculated that they may have provided the immigration route for *Campto*. The first of these pathways is trans-Tasman wind dispersal, particularly to the Kaipara Harbour, situated on the west coast of the North Island. There is good evidence, and some hard data, indicating that several species of moths (both macro- and micro-Lepidoptera) and aphids have been carried across the Tasman Sea from Australia to (colonise) NZ (Graham Walker *pers. comm.*). For example, several entomologists were involved with running a large light trap at Pukekohe (near Auckland) over a 10-year period (1981–1991). All the catches from this trap were identified and the very large dataset is held by Plant and Food Research awaiting analysis. While acknowledging that this trap was set up primarily for monitoring moth and butterfly populations, it is interesting to note that Graham Walker (*pers. comm.*), of Plant and Food Research, confirmed that there were no mosquitoes among the insects caught. Moreover, several factors weigh against wind-borne transmission as the means of entry. The limited NZ mosquito fauna is evidence of isolation from the diverse mosquito fauna of Australia, suggesting wind has not previously been a pathway of entry. The index site in NZ was likely to have been Napier on the North Island's east coast. The infested habitat was close to Napier's sea- and airports.

Approximately 600 vessels from overseas and other NZ ports enter the port of Napier each year. It was noted that 22 vessels from Australia had first-ported in Napier over the 3 years before 1999, and six of these were livestock vessels from Freemantle/Adelaide.

The introduction of *Campto* in association with imported sea containers has also been mooted as a possibility. There are thousands of container devanning sites throughout NZ (Ministry of Agriculture and Forestry 2004). The MAF maintains a publicly accessible register of MAF-approved transitional facilities for sea containers (Ministry of Agriculture and Forestry 2004). In examining the *Campto* incursion at Kaipara, it was thus determined that four transitional facilities were registered in the Helensville area. Based on the number of container devanning sites in the Helensville area, and the information on imported containers obtained from the importers, it was determined that container traffic into Helensville (the urban centre closest to the *Campto* infested area at the southern part of the Kaipara Harbour) was minimal and an unlikely pathway of entry.

Pathways for spread within NZ

Several possible pathways were identified as being associated with spread subsequent to entry and establishment of *Campto* in NZ (Table 7.2). Nine possible means of spread were identified, some of which will be further discussed below. In decreasing order of probability, the most probable means of spread involves adult flight from an infested area, (in combination with) wind dispersal of adults, as adults inside vehicles or caravans with the road transport of people or livestock, deliberate illegal (human-instigated) spread and the carriage of immature stages in water receptacles. Although there is no evidence supporting the deliberate illegal (human-instigated) spread of *Campto*, this means of spread cannot be ruled out.

Adult flight and wind-assisted dispersal

In reality, it may be impossible to separate adult flight from wind-assisted dispersal. To date, *Campto* adults have been shown in mark–recapture studies to disperse distances of up to 6 km (Russell 1988; Mike Lindsay (with reference to Cameron Gordon's PhD studies) *pers. comm.*). For wind-assisted dispersal, however, both *Culex* and *Aedes* mosquitoes have been recorded travelling in the faster geotropic flows for maximum displacements of over 600 km (Kay and Farrow 2000) and *Campto* has established (perhaps step-wise) in inland saline areas such as Mildura (and further east at Lake Tyrrell near Swan Hill) in the Murray Valley of Victoria, more than 400 km away from coastal saltmarsh habitat (Lee *et al.* 1984; Russell *pers. comm.*). Given frequent dispersal of *Campto* up to 48 km through the woodland coastal belt of Victoria, distances between some of the infested sites in NZ can easily be explained by wind-assisted dispersal.

With reference to wind direction and speed data (presented as wind roses) provided by the National Institute of Water and Atmospheric Research (Tony Bromley *pers. comm.*), wind-assisted dispersal may well have contributed to the spread of *Campto* from Kaipara to Mangawhai, Kaipara to Whangaparaoa Peninsula, and Wairau estuarine area to Lake Grassmere. For instance, readings taken from 1976–1981 at Oyster Point, at the southern end of Kaipara Harbour, indicate that easterly, south-westerly and westerly winds were experienced most frequently. Although wind readings were not available for a relevant site at or near Mangawhai, it is noteworthy that Mangawhai lies ~30 km north-east of South Head at the southern end of Kaipara Harbour. Furthermore, for the period 1994–2004, the prevailing winds recorded were westerly at Whangaparaoa, which lies almost due east of the southern end of Kaipara Harbour.

Similarly, Lake Grassmere (the smaller of the two South Island areas of infestation) is located some 30 km south-east of the Wairau estuarine area (~10 km east of Blenheim). Westerly and north-westerly winds prevailed at Blenheim from 1996–2004, while at Cape Campbell (the closest but more exposed weather station near to Lake Grassmere), northerly, north-westerly and southerly winds were recorded most frequently.

Although less likely because of the distance involved, such wind readings may also be seen as supporting the possibility of wind dispersal of *Campto* from Napier to Mahia. From 1994–2004, south-westerly and westerly winds were those most frequently recorded at Napier. During the same period, south-westerly, westerly and northerly winds prevailed at Mahia, which lies ~95 km (across Hawke's Bay) to the north-east of Napier. However, adult flight and wind-assisted spread is not likely to have been involved in the spread of mosquitoes to several the infested areas, including Kaipara, Blenheim and Coromandel (Table 7.2).

Adults in aircraft

Another possible means of spread involves the transport of adult mosquitoes in aircraft. Reports of mosquitoes in aircraft are numerous (Gratz *et al.* 2000). Obviously, given the presence of airports (cf. air fields) near Blenheim (Wairau estuarine area), Gisborne and Napier, there is the possibility that adult *Campto* may have arrived directly from Australia and established in these areas (where infestations were subsequently discovered). However, none of these airports are approved places of first-arrival in accordance with the

Biosecurity Act 1993 (but obviously mosquitoes don't know this). Furthermore, it was determined that seven to 10 corporate jets arrive at Napier each year, and mainly from southern Australia where *Campto* occurs, associated with the vegetable or wine industries and a golf course development venture (Ian Jarvis *pers. comm.*), and commuter jets fly from Napier to Gisborne. Many, including Darryl McGinn, commented on the antics of passengers embarking on aeroplanes in the early days of the Napier infestation (see Chapter 5).

Consequently, based on the small numbers of international aircraft arrivals at Napier airport and the comparatively poor invasion success of mosquitoes arriving on aircraft, one might be tempted to discount the possibility of aircraft being the means of spread of *Campto*. However, it has been reported that aircraft disinsection procedures were less than optimal in 1998 because Napier was not recognised as an international port of entry. On arrival, of course, a female *Campto* satiated with passenger blood, or not, could simply fly or be blown a few metres into suitable habitat to begin a new lineage of NZ *Camptos*.

The spread of *Campto* via domestic aircraft travelling from Napier Airport to Kaipara Harbour (e.g. the airfield at Parakai), and Kaipara to Blenheim (Blenheim Airport and Omaka Airfield), however, requires further consideration. All commercial flights departing from Napier from January 1999 to December 2000 were disinsected. Furthermore, aircraft disinsection was instigated for flights departing from Gisborne Airport in October 2000. Also, arrivals of private aircraft at Omaka Airfield are few and far between. Most private flights (where flight plans are not required and therefore there is no formal record) from Northland to the South Island involve a refuelling stop at Paraparaumu. At most, one or two aircraft a month arrive from Northland at Omaka Airfield (Kevin Wilkey *pers. comm.*). In addition to the comparatively poor invasion success of mosquitoes arriving on aircraft, measures such as aircraft disinsection taken during the relevant time periods may have further reduced the possibility of *Campto* spreading via domestic aircraft.

Unintentional spread by birdwatchers or duck shooters

The unintentional carriage of *Campto* eggs from site (e.g. Kaipara Harbour) to site (e.g. Wairau Lagoon) by birdwatchers and/or duck shooters was suggested as a possible means of spread. However, there was no concrete evidence supporting this as a means of spread, even though duck shooters were ultimately responsible for bringing the presence of *Campto* at Wairau Lagoons to the attention of the MoH. The duck shooting season is relatively short (May to July) and more often than not, opening day is the highlight of the season for duck shooters who have a favourite site from which to shoot. As a result, duck shooters are unlikely to be going from site to site. It is not unreasonable to surmise that birdwatchers are similarly inclined. At most, a particular birdwatching expedition may involve time at different sites in relatively close proximity to one another but is unlikely to involve, within a short period, visits to sites located as far apart as Kaipara Harbour in the north and Wairau Lagoons at the top of the South Island.

Added to this is the fact that egg hatch of floodwater mosquitoes, such as *Ae. vigilax* and *Campto*, typically occur by instalments and is associated with reduction in oxygen concentrations in the water following immersion. The eggs of floodwater species can survive at least 4 months of dry conditions and, therefore, might be spread through inadvertent carriage in mud on the boots of bird watchers or duck shooters.

Migratory birds

As with unintentional spread by birdwatchers and/or duck shooters, to date there was no evidence supporting the idea that migratory birds may spread mosquitoes. The relevant

Table 7.2. Possible means of spread of *Campto* within NZ.

Means of spread	Demonstrated *Campto* means of spread (Y/N)	Likelihood of *Campto* spread occurring by this means
Adult flight from an infested area	**Y** *Campto* adults may disperse distances of up to 6km as shown in mark–recapture studies, and possibly up to 48 km into Victorian coastal woodlands, but probably aided by winds.	**Possible** Distance too great to explain spread by active flight of *Campto* from Napier to Porangahau, Napier to Gisborne or Mahia, Napier to Kaipara, Gisborne to Kaipara, Kaipara to Mangawhai, Kaipara to Whitford, Kaipara to Whangaparaoa, and any of the North Island areas of infestation to the Wairau estuarine area in the north of the South Island.
Wind dispersal of adults	**Y** Wind dispersal up to 600 km recorded for *Culex* and *Aedes* in upper air flows, and *Campto* found 400 km from coastal breeding habitat in Victoria (also long distance dispersal of *Ae. vigilax* to Toowoomba, Mount Isa and Barrier Reef islands).	**Probable** Analyses of prevailing winds (as in wind roses) provided by NIWA and the distances between infested sites suggest it is possible that wind dispersal led to the spread of *Campto* from Napier to Gisborne, Napier to Mahia, Kaipara to Mangawhai, Kaipara to Whangaparaoa Peninsula and Wairau estuarine area to Lake Grassmere. The spread from Napier to Porangahau, Napier to Kaipara, Kaipara to Whitford or any North Island area of infestation to the Wairau estuarine area/Lake Grassmere in the South Island is unlikely to be due to prevailing winds.
Deliberate illegal (human-instigated) spread	**N**	**Negligible**
Immature stages in water receptacles (e.g. used and/or spare tyres) transported between an infested area and an uninfested area	**N**	**Negligible** *Campto* has not been recorded to breed in small readily transported containers such as tyres
On light aircraft flown from an infested area to an uninfested area	**Y** *Campto* reported on commuter aircraft at Napier	**Possible**
As adults inside vehicles (cars, trucks) or caravans with the road transport of people or livestock	**Y** *Campto* reported in vehicles and caravans from Napier and Mahia	**Possible**
As adults inside the cabins of boats moved from an infested area to an uninfested area	**N**	**Low** Reports of local fishing boats going from Hawke's Bay to Kaipara Harbour to visit families
Unintentional spread by birdwatchers or duck shooters	**N**	**Negligible**
Migratory birds	**N**	**Negligible**

category of birds to consider is referred to as 'migrant' (i.e. those that move annually and seasonally between breeding and non-breeding areas, either within NZ or between NZ and other countries). Spurr and Sandlant (2004) list several species that fall into this category, including the little egret (*Egretta garzetta*), turnstone (*Arenaria interpres*), three species of tern (*Sterna* spp.), three species of dotterel (*Charadrius* spp.), cattle egret (*Bubulcus ibis*) and two species of plover (*Pluvialis* spp.). There may well be other birds in the 'migrant' category and whether any of these migrants move between the known areas of *Campto* infestation was not examined further.

Understanding human-mediated methods of spread

The Coromandel Peninsula was the last major area identified with *Campto*. The finding of *Campto* in Coromandel was of concern not only because it raised issues around the effectiveness of the eradication programme, but because of the distribution of affected areas within the wider Thames-Coromandel district. There were five locations where *Campto* were found (Waiaro, Kennedy Bay, Waikawau, Colville and Whangapoua). Waiaro was the first location where *Campto* had been identified and was likely to have been the index site for introduction to the district. This supposition was based on the large larval biomass present compared with Colville, Waikawau and Kennedy Bay. At Waikawau and Colville, no further larvae were identified subsequent to initial detection and treatment of susceptible habitat. At Whangapoua, only four larvae (on just one occasion) were ever detected.

The steep topography separating the five infested sites was likely to have prevented natural spread through wind dispersal. Some of the infested sites were also very remote and so it was not clear how spread had actually occurred. To understand how spread was occurring in the area, a postal survey of Thames-Coromandel properties located within 5 km of *Campto*-infested sites was carried out (Fig. 7.1). Households were asked to describe their movements by cars, movements of livestock and movements of plant material and soil. These conveyers were selected based on their likelihood to be associated with mosquito spread. Spread was hypothesised to occur through *Campto* eggs being transported among plant material and soil and adults attracted to human hosts in cars and livestock hosts in animal transport vehicles. For this survey, selected households were asked to provide information on the movement frequency and destination of these conveyers for the previous 12 months.

Designing a survey to get maximum response, and thereby eliminate bias in the results, can be an art form, particularly with a survey of this complexity that required considerable effort by the respondent to complete. A careful strategy was developed to maximise exposure and interest from the public. The strategy involved communicating the survey through the local media, letters to residents and creation of visually pleasing design to the survey form itself. Over 1000 chocolate bars were sent out with the survey forms as a further inducement. In addition, half a dozen university graduates were employed to contact non-responders to encourage them to complete their forms.

After this considerable effort, 68% of households responded with completed questionnaires. This high rate of response reflected not only the survey design and procedures adopted, but highlighted the high level of community concern regarding *Campto* and its potential role in vectoring disease, as well as its considerable nuisance value.

This study provided some interesting insight into the movements of people, and conveyers of *Campto*. The output of a frequency distribution of distance travelled presented a

THE SOUTHERN SALT MARSH MOSQUITO HAS INFESTED SITES IN COROMANDEL

Has it spread to any other places?

Please fill in the questionnaire and help us find them

This mosquito can play a part in the spread of several human and animal diseases!

- The New Zealand government aims to eradicate the Southern Salt Marsh mosquito because of its potentially harmful effect on public and animal health.
- There is evidence that adult mosquitoes may become trapped in vehicles and spread to non-infested places.
- Biosecurity New Zealand will use the travel data that you provide to identify other places infested with the Southern Salt Marsh mosquito.

All information gathered is confidential

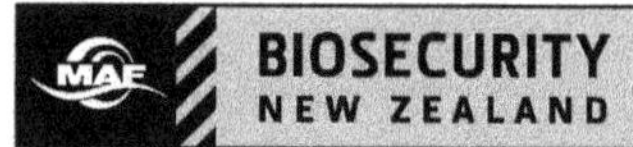

October 2006

Figure 7.1 Survey cover page carried out to understand movements of *Campto* conveyers in Coromandel.

means of guiding the amount of surveillance effort given to susceptible habitat for any future incursions of *Campto*. Potentially, an estimate of local travel to surrounding towns could be predicted from population density statistics as the frequency of travel to towns within 10 km of infested sites was correlated with the population density of the town. Inter-

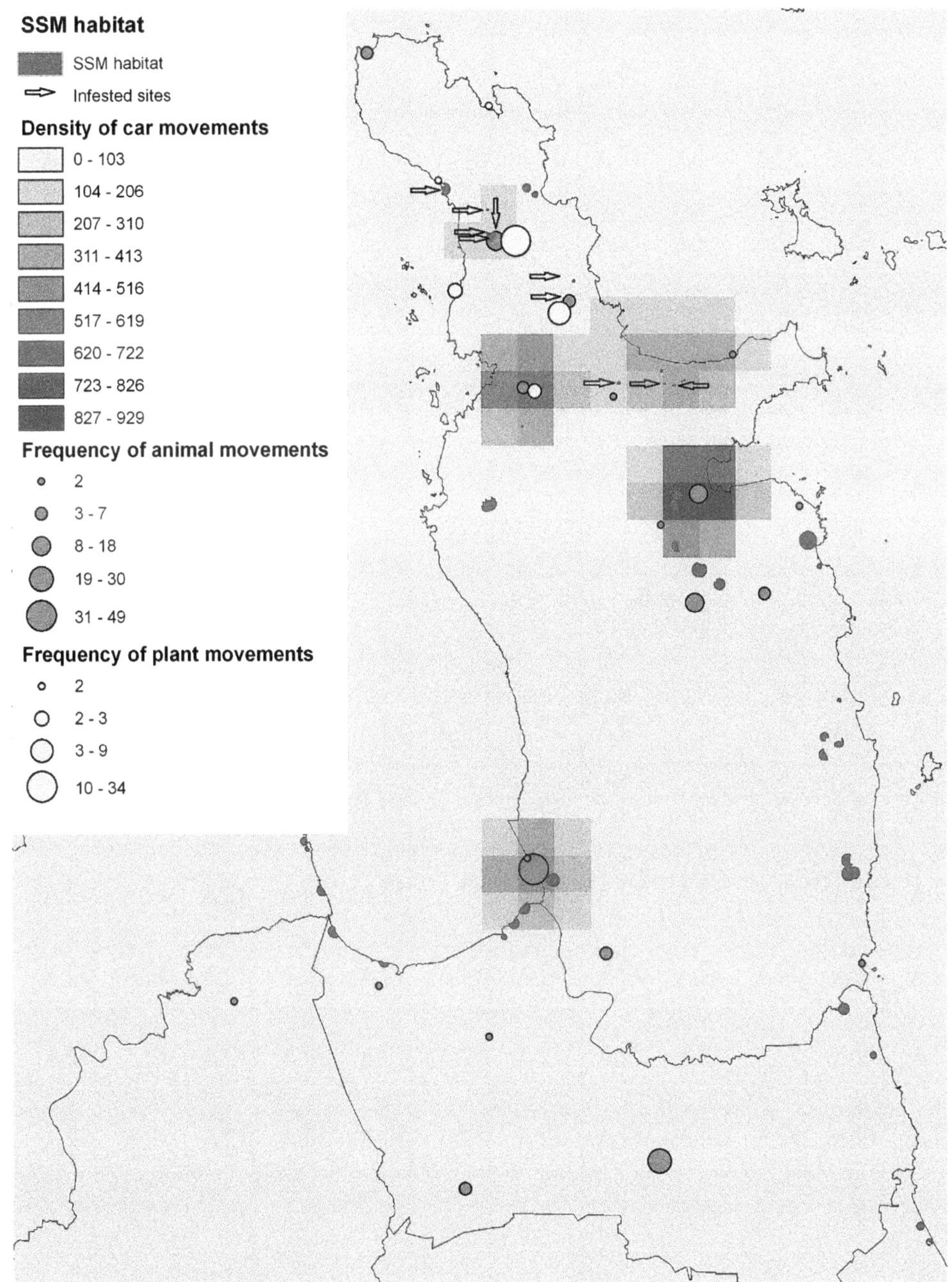

Figure 7.2 A kernel smoothed density plot of vehicle movements from surveyed locations to other places in the Thames-Coromandel district with overlying symbols representing the frequency of movements of plant/soil material and animals.

estingly 90% of travel carried out by those surveyed was less than 58 km from the respondent's property location (McFadden and Bullians 2008).

Within the Thames-Coromandel district, there was a high frequency of car travel between *Campto*-infested sites, particularly between Kennedy Bay and Waikawau/Whangapoua/Colville and between Colville and Waikawau (Fig. 7.2). Although this did not necessarily mean that cars were responsible for transfer, it did show that a suitable network existed between *Campto*-infested sites within the Thames-Coromandel district. It also highlighted that there was more travel between some reasonably remote areas than was originally appreciated.

Other factors were identified from the survey that showed how important it is to have good local knowledge and undertake careful investigation when working through potential risk pathways. The managed Waikawau camping ground was in close proximity to infested habitat. The risk of adult *Campto* transfer from this site was likely to have been high, given that adult mosquitoes may have become trapped in tents and camping equipment or cars and caravans during the process of packing to leave a campsite.

A high proportion of respondents within a few hundred metres of *Campto*-infested sites carried out activities resulting in plants and soil being moved. The majority of these movements were over short distances. Nevertheless, long distance movements were identified. One property that was fortunately not in the immediate vicinity of susceptible habitat had received a substantial number of plants from a variety of locations both within and outside the district. The purpose of these movements was to create new habitat for NZ wildlife and recreate salt water marshlands, an ideal habitat for *Campto*. Although this property did not represent an immediate risk, it does show how important it is to identify these sorts of activities during an eradication response to an organism like *Campto*.

Although the study did show that in some cases there was a relationship between the frequency of car travel and infestation status, it did not explain why habitat from some locations was not populated despite being the destination for a high frequency of travel from properties in close proximity to *Campto* populated habitat. It is therefore clear that the occurrence of a pathway is not the only factor governing establishment. It also suggested that establishment may be a low probability event.

Mitigating risk of pathway spread

There were no controls imposed on the movement of personnel or vehicles off a known infested habitat, or restrictions imposed on people entering infested land. Advice on limiting spread was provided through media communications and publicity, as well as through personal communications by operational personnel when interacting with affected people. As publicity got out that there was a vicious daytime biting mosquito at the Wairau infested sites in Blenheim, there was almost a self-imposed restriction of people entering the Department of Conservation land habitat, which is popular for duck shooting and fishing. Possibly this may have reduced spread in the area.

The movement of cattle trucks was a suspected pathway. Farmers with livestock located in infested areas were asked to provide details of when livestock were being moved. Later, information was provided that dairy stock had been transported from the Kaipara to Mangawhai farms, providing a clear pathway for spread. Unfortunately, there was no control over the contractors who used earthmoving machinery to clear drains and fill flooded low lying land in *Campto*-infested locations. These pieces of excavation equipment were known to have moved freely around the countryside and were recognised as a risk for spread.

There was well-organised residual insecticide spraying of the small aircraft using the Napier and Gisborne airfields located adjacent to infested habitat (Box 7.1). The MoH provided the insecticide and the airport authorities ensured that treatment was carried out.

Box 7.1: Aeroplanes as conveyers

When an exotic mosquito turns up in the wetlands surrounding a busy regional airport, the threat of it 'hitching a ride' in one of the departing planes is very real indeed. Aeroplanes viewed from the perspective of a mosquito are nothing more than aluminium barrels, filled with sweaty and breathing humans, creating carbon dioxide and all sorts of attractive odours. Indeed, aircraft can be the perfect mode of transport for insects in general, and mosquitoes in particular, especially over long distances.

Over the past several decades or so, researchers have focussed on technologies to prevent the world-wide spread of unwanted vectors of human and animal diseases through aircraft traffic; the term 'aircraft disinsection' covers the whole range of activities inside (and outside) aircraft to stop unwanted or exotic invertebrates using airliners as their private jets.

At the time of the discovery at Napier airport, aircraft disinsection was a familiar term around the biosecurity systems in both NZ and Australia. Indeed, the most recent techniques were developed 'down under' in collaborative programmes with MAF, Australian Quarantine Inspection Service, Air NZ and Qantas under the guidelines and auspices of the World Health Organization, Geneva.

Originally, disinsection was a simple affair: as soon as an international aircraft had landed and was 'on-blocks', a quarantine officer or two would enter the front door, armed with several aerosol cans, filled with insecticides. Yes, DDT was the earlier model of active ingredient (they called it the 'DDT aerosol bomb'), followed by the more benign pyrethrins (in deodorised kerosene!) and synthetic pyrethroids, of which d-phenothrin was the star performer, recommended by the World Health Organization for use in cabins and cargo holds.

The on-arrival disinsection was always terribly unpopular with tourists and New Zealanders alike, but especially with visitors from the USA. Passengers had to sit in the insecticidal fog for 10 min before the quarantine officers gave the all clear for disembarkation. One major improvement was the 'blocks-away' disinsection procedure, carried out as the aircraft was taxiing from the gate to the end of the runway before take-off. The idea was to kill the interlopers at the port of departure, giving the invertebrates usually plenty of time to die, before arrival in NZ. The great advantage was that passengers were strapped into their seatbelts, leaving the aisles clear for the cabin crew to do their insecticidal run through the aircraft, while the captain switched off the air conditioning systems, so as to not lose any insecticide from the cabin prematurely. Passengers also had their minds on the impending take-off and journey, so the psychological impact of the disinsection was not as great as when it was carried out on arrival. The disadvantage was that cabin crew had to be trained and monitored to do the disinsection job adequately and according to the quarantine manual. A further improvement saw the cabin disinsected by crew at 'top-of-descent', some 20 min before landing. At that point in the flight, crew were able to find some time in their busy schedule to spray the cabin, while passengers were distracted by the imminent arrival in NZ.

Box 7.1: (Continued)

In the early 1980s, Pat Dale commenced preliminary research into what is now known as 'residual disinsection' (Dale 1980). This involved the aircraft being sprayed with permethrin every 8 weeks in the absence of passengers, usually after a regular maintenance check or engineering audit. The residual deposit dried after an hour or so and caused no harm to mammalian passengers, according to the recommendations of the World Health Organization experts in Geneva. Flies, mosquitoes and other invertebrates, however, were quickly knocked down, once they came in contact with the permethrin deposit (Dale and Kleinpaste 1986). Residual disinsection becomes economically viable for airlines when the treated aircraft frequently visit airports in countries that require disinsection under public health or quarantine laws. Aircraft that do so haphazardly are more suited to one-off treatments as required.

As the 'on-arrival', 'blocks-away', and 'top-of-descent' treatments became to be considered to be out of date, and causing increasing amounts of passenger discomfort and complaints, the quarantine authorities of NZ and Australia decided to initiate research into alternative ways of cabin disinsection in 1995, aiming to develop a technique by which aircraft are disinsected by aerosols in the absence of passengers. This technique was named 'Pre-embarkation' disinsection (Kleinpaste and Walker 1996). The rationale of this technique was to treat the cabin after all catering was finished, just before passengers boarded the plane, and within an hour from departure or closure of the main entrance door. The active ingredient in the aerosol cans was 2% permethrin, propelled by non-CFC propellants. The spray was carried out according to regular World Health Organization recommendations, aiming to disperse 10 g of the aerosol product per 1000 ft^3 (28.3 m^3). The toilets, cupboards, flight deck, lockers, and so on were sprayed as well. In general, the procedures and pace of 'pre-embarkation' disinsection are similar to the 'top-of-descent' or 'blocks-away' techniques. The permethrin aerosol will kill insects that may be present inside the cabin at the time of disinsection. It will also give a light and patchy coating of insecticide residue designed to knock down and kill any vectors that may enter the cabin in the time span between disinsection and departure. Moreover, the permethrin residues will continue being lethal to stow-away insects (including hard-bodied specimens) during the flight. With air-conditioning packs switched off, the 10 g permethrin per 1000 ft^3 has a proven efficacy.

When *Campto* had descended on the brackish wetlands around Napier airport in the late 1990s, it became quite clear that 'pre-embarkation' disinsection was the preferred method to treat the small aircraft that were using Napier airport and, since the method had been officially adopted in the menu of disinsection options in NZ and Australia from the 1 January 1999, the timing was perfect too. To ensure that no mosquitoes were surviving a trip in the cargo holds of the aircraft, these small luggage spaces were treated with the same aerosol pre-departure. Seeing 'pre-embarkation' disinsection procedures were tailored for wide-bodied, international aircraft, the adaptation of the technique for small, regional planes took a few weeks of fine tuning by measuring internal cubic footage and auditing ground crew performance.

Communication with Air NZ and Napier aero clubs resulted in most – if not all – aircraft being treated before departure from the infested airfield and audits continued on a regular basis. As *Campto* dwindled over the subsequent months, the TAG considered the possibility of suspending aircraft disinsection from Napier under specific conditions. A formal proposal to do so was adopted at the meeting of 13 October 2000, almost 2 years after the discovery of the mosquito.

It was very important that operational activities did not result in further spread of *Campto*. NZ BioSecure had several standard operating procedures in place to ensure that staff and vehicle movements working in infested habitat did not inadvertently cause spread. Several actions took place subsequent to visiting infested habitat:

- thorough cleaning of all vehicles using high-pressure hoses, water-blasters and, in the case of the Kaipara incursion, a specifically designed wash-point
- all field footwear (gumboots, boots and safety footwear) was required to be cleaned upon return to base
- no vehicle was to move from within the eradication zone without residual insecticide treatment of the inside of the vehicle. This same precaution was taken when the operational helicopter landed with doors opened in known infested areas.

Another important operational procedure was ensuring 'safe' packaging and transport of *Campto* larvae from the field to the laboratory for confirmed identification. For the technicians sampling in the field, special dispensation and authority was sought and approved under the *Biosecurity Act 1993* to collect and handle an unwanted species and taking them alive to an area outside the eradication zone. The accepted protocol for preparation of the specimens for post or courier transportation of the adults and larvae was established by NZ BioSecure Entomology Laboratory. Adults were frozen overnight before being placed into plastic test tubes and *Campto* larvae first had boiling water added before being put into a test tube of ethanol.

Understanding spread through molecular analysis

The discovery of *Campto* in NZ in 1998, and its subsequent spread to numerous localities in both the North and South Islands, raised the question of whether multiple incursions from Australia were responsible for these dramatic expansions. A genetic investigation was initiated to attempt to identify the dynamics of the incursion and whether the working hypothesis that the *Campto* incursion into NZ and subsequent spread stemmed from a single introduction to Napier (mosquito or egg batch) was correct. Evidence of multiple introductions into NZ would have had a negative effect on any eradication strategy and may have compromised attempts to keep the mosquito out of NZ.

Genetic investigations to attempt to determine the dynamics of this incursion began in 2006. To this end, mainland Australian *Campto* material from its home range needed to be compared with the NZ material through DNA-based genetic analyses. However, and unfortunately – thanks to the early successes of the eradication campaign – the extent of field material available from NZ was limited. This meant that instead of undertaking a potential microsatellite-based approach that would permit the use of a vast array of nuclear DNA molecular markers, an alternate approach – using the maternally inherited mitochondrial DNA (mtDNA) cytochrome oxidase I (COI) marker – was used instead.

However, one of the advantages of working with the maternally inherited mtDNA marker is that it can be used as both a population genetics marker and as a proxy for the physical movement of females. In terms of the latter, the rationale is that the mtDNA sequence of each female will be identical to that of her offspring, allowing insights into the diversity and movement of females contributing to each population. For example, if the initial NZ population of *Campto* had been established by a single female or egg batch, all future offspring would carry her haplotype sequence. Alternatively, if several *Campto*

Table 7.3. Summary of collection sites for the southern saltmarsh mosquito in NZ and in Australia

Country	Group	Town	Location	Sample code	Haplotype
NZ	Group A	Napier	Napier	Nap-2	H1
			Hector	P9	H1
			Indigo North	P06	H1
			Hector	P29–1	H1
			Hector	P29–2	H1
			Mahia	M03-a	H1
			Mahia	M03-b	H1
			Mahai	M02-a	H1
	Group B	Blenheim	Wairau	B14_1	H1
			Wairau	B07–2	H1
	Group C	Kaipara	South Head	K16	H1
			South Head, Carter's Bach	K08–1	H1
			South Head, Carter's Bach	K08b	H1
	Group D	Coromandel	Kennedy Bay Dent01	C07	H1
				C11	H1
	Group E	Kaipara recent	Thorpes	K21	H1
Australia	South Australia	Wellington	Coastal	W1	H6
				W2	H1
				W3	H2
				W4	H3
				W5	H3
		Poltalloch	Coastal	P1	H4
				P2	H4
				P3	H1
				P4	H3
				P5	H1
		Berri	Inland	B1	H1
				B2	H1
				B3	H1
				B4	H1
				B5	H1
		Jervis	Coastal	J1	H1
				J2	H1
				J3	H5
				J4	H4
				J5	H1

(Continued)

Table 7.3. (Continued)

Country	Group	Town	Location	Sample code	Haplotype
	New South Wales	Sydney	Olympic Park	SPO1	H1
				SPO2	H1
				SPO3	H1
				SPO4	H1
			Shellharbour	Sh1	H1
				Sh2	H1
	Western Australia	Perth	Woods Road	WA1	H4
				WA2	H4
				WA3	H4

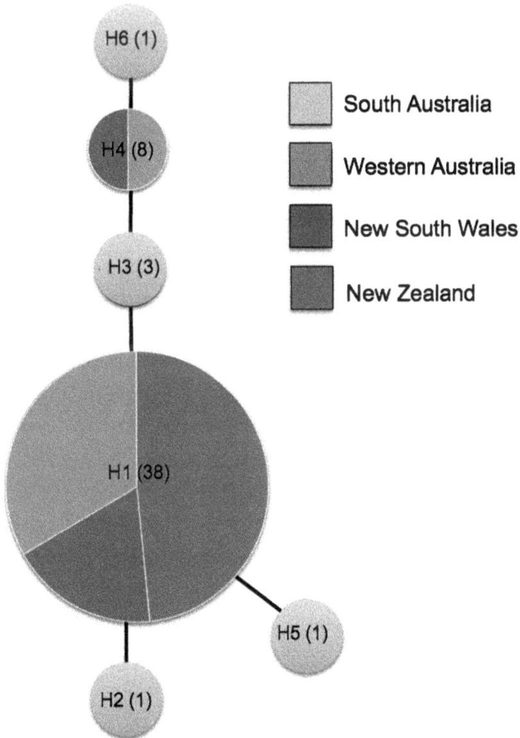

Figure 7.3 Mitochondrial DNA COI haplotype network of *Campto* material from mainland Australia and NZ. Each circle represents a different mtDNA haplotype sequence with connections representing single mutational steps between sequences. Of the six haplotypes identified, only the most common (H1) exists in NZ, and although this is consistent with the hypothesis of a single incursion event from Australia to NZ, it is not conclusive evidence of such activity.

haplotypes were discovered in NZ, then we can assume that multiple females must have contributed to these populations, indicating either multiple incursions or one large initial incursion of multiple females.

Although using this marker as a proxy for female movement would ultimately provide an underestimate of the number of females contributing to the population, given that different females of the same sequence cannot be distinguished, an incursion that originated from a single event should reveal only one mtDNA haplotype in the NZ material. As stated above, this information on the occurrence and movement of one or multiple females would have serious impacts on the design and feasibility of any control strategy, because evidence for multiple females may suggest that any eradication may be compromised by more additional future incursions unless the mechanism of the incursions could be halted.

Mosquito larvae were collected from 10 sites in NZ and seven sites in southern Australia, spanning New South Wales, South Australia and Western Australia and representing the wide distribution of *Campto* (Table 7.3). Larvae were collected from oviposition sites and adult specimens were morphologically identified using the keys of Lee *et al.* (1984). Following identification, mosquitoes were stored dried or in 70% ethanol. The predicted presence of siblings at each larval site – which would contain the same mtDNA haplotype – meant that minimal samples from each site needed to be used.

Mosquitoes (either partial or whole) were thoroughly ground and processed to produce a 580-base-pair fragment of the mtDNA COI gene. This was then amplified using a polymerase chain reaction (PCR), and the amplified products were purified and sequenced at the Australian Genome Research Facility, based at the University of Queensland in Brisbane, Australia. Chromatograms for all sequences were analysed and edited, and alignments were then generated to identify the diversity of DNA sequences in NZ and Australia. A total of 59 mtDNA COI samples was sequenced (16 individuals from 10 sites in NZ and 35 individuals from seven sites in southern Australia: see Table 7.3).

In this study, aligned COI sequences revealed six distinct haplotypes (different sequences) among all *Campto* samples. Only one haplotype (H1 – the most common haplotype) was found in the NZ samples, which was consistent with the incursion being small and a single event (or multiple events from the same Australian site or other sites where that haplotype existed) (Fig. 7.3). We also found six different haplotypes in the Australian material, with samples from South Australia being the most diverse – this increased diversity may be the effect of the limited sampling that took place in Australia. Had multiple incursions taken place, genetic analyses were more likely to show multiple haplotypes (representing different mtDNA sequences) suggesting numerous females were contributing to the population, unless the multiple incursions were coming from the same site or multiple sites in eastern Australia where the same haplotype existed.

Despite the relatively small sample size used in this study, several key lessons were learned. First, an mtDNA-based approach, building on maternal inheritance, can provide valuable insights into biosecurity breaches differentiating a small incursion in terms of a single female or egg batch from a large incursion/range expansion. Second, all material collected during the course of an incursion should be kept and stored for later molecular studies. If *Campto* reappears in NZ, the material and data already gathered from this study can be used to determine if a new incursion has manifested or that the original eradication campaign was not 100% successful (but only if it is a different haplotype).

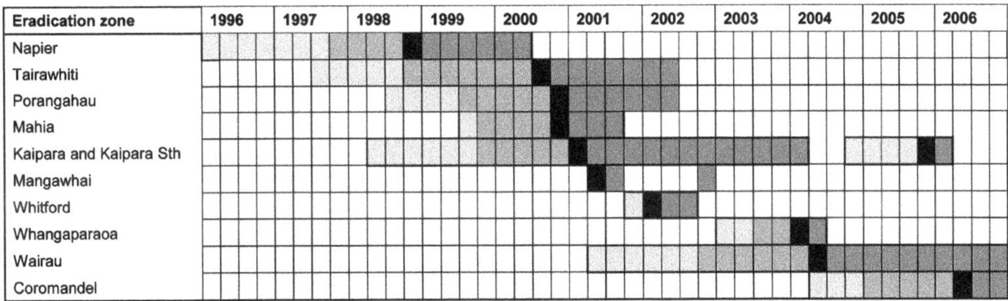

Key to Figure 7.4	
Estimated backward risk period	
Detection	
Forward risk period pre-detection	
Forward risk period post-detection	

Figure 7.4 Composite timeline showing estimated backward and forward risk periods.

The sequence of translocations

If we make the assumption that the distribution of *Campto* resulted from a single incursion and subsequent translocation from infested sites to distant sites, then the sequence of events can be pieced together by considering the likely time each eradication zone became infested and the likely time mosquitoes spread from these zones.

The length of time each eradication zone had been infected before its discovery was estimated by considering:

- the abundance of adults and larvae present
- the degree of habitat utilisation as assessed by ground staff during delimiting surveillance
- the amount and type of affected habitat
- published and unpublished data on the *Campto* life cycle
- the extent of local spread within the eradication zone (i.e. greatest distance between affected habitats, assuming 3 km expansion per mosquito generation with an estimated 10 generations per year)
- importantly, the opinion of those working in the programme at each site.

These considerations were also used to estimate how long the site may have been 'infectious' for other sites.

Figure 7.4 shows the time of detection for each eradication zone and the estimated backward and forward risk periods, where backward risk period refers to the time during which a zone may have become infested and forward risk period refers to the time when it may have become a source for other zones.

Forward risk appears to reduce significantly at 3–4 months after eradication starts and was likely to have been minimal within 6–8 months after initiation of treatments. S-methoprene is generally not effective on late-fourth instar larvae, pupae and adult mosquitoes. As an insect growth regulator, it acts to disrupt the normal progression of immature development and prevents the adult mosquito emerging from the pupa, which is non-feeding and eventually dies of starvation. Once treatment has begun, the product will not affect those late fourth instars, pupae and adults already developed. The adults will continue to contribute to the egg mass and where the larvae and pupae successfully emerge as adults they too will potentially contribute to the egg bank.

For this reason, we see in Figure 7.4 a noticeable decline in adult numbers within the first 1–3 months of treatments (no further adults emerging), and very low numbers for the following 3 months. The end of forward risk could be considered to be 6 months after treatment or when the last adult was detected, whichever came later.

Figure 7.4 is useful in considering which zones were sources for other zones. The risk of long distance translocation from Napier probably started in the summer of 1997–98, although possibly the year before. This coincides with the likely time that Tairawhiti, Porangahau and Kaipara were first infested. Mahia is more likely to have been infested by translocation from Tairawhiti in the summer of 1999–2000.

Mangawhai and Whitford could have been infested by translocation from Kaipara. More mysterious is the origin of infestations for Whangaparoa, Wairau and Coromandel. Unless Wairau had low levels of infestations for a long time, it was unlikely to have been infested by translocations from the geographically closer East Coast sites Napier, Tairawhiti, Porangahau, and Mahia. Possibly, it was infested from Kaipara, although by what mechanism we do not know.

The means of translocation

Spread to distant sites requires the presence of mechanisms that pick-up and transport viable adults or eggs (or larvae) to new locations. The locations must have suitable habitat for the survival and establishment of a population. Factors considered were the presence or absence of:

- geographic barriers
- public useage
- airports
- residents (rural and urban)
- livestock
- land use (agriculture and forestry)
- biosecurity measures used by those undertaking surveillance or control
- specific information concerning risky events or movements.

The magnitude of spread risk from each zone

In addition to the known means of translocation, the sites with the most potential to have been the source of *Campto* spread to other sites were those with large numbers of adults or large amounts of infested habitat. Semi-quantitative methods and expert elicitation placed Napier, Tairawhiti, Kaipara and Wairau as the four main likely sources of mosquitoes for other sites. Napier and Tairawhiti had the most mosquitoes, and Kaipara and Wairau had the most infested habitat. The lower biomass and habitat at Mahia, Porangahau and Coromandel meant they were much less of a spread risk than these four sites, and the other eradication zones were so small as to be of near negligible forward risk.

Habitat at risk of infestation

Table 7.4 shows the amount of suitable habitat at different distances from Napier, Kaipara and Wairau. The tables reveal thousands of hectares and hundreds of sites that did not become infested. The estimated number of successful translocations is shown. Most of the translocations were within 100 km of the source; however, some must have skipped over large amounts of suitable habitat to establish 300 or 500 km from source. Combining the

Table 7.4. Amount and number of habitats at given distance bands from Napier, Kaipara and Wairau eradication zones and the estimated number of successful long-distance translocations

		Distance from eradication zone (km)							
Eradication zone	Habitat at risk of infestation	0–100	100–200	200–300	300–400	400–500	500–600	600–700	700–800
Napier	Number of suitable sites	58	58	110	256	100	148	22	13
	Hectares of suitable habitat	1035	724	3406	5993	1059	3222	405	56
	Successful translocations	1 or 2?	1	0	1	0	0	0	0
Kaipara	Number of suitable sites	193	200	63	72	63	100	12	22
	Hectares of suitable habitat	3594	5378	1538	1342	1037	2574	42	102
	Successful translocations	3	0	0	0	0	1?	0	0
Wairau	Number of suitable sites	95	69	43	67	149	191	127	86
	Hectares of suitable habitat	2543	915	668	1147	3212	4052	2148	2162
	Successful translocations	1	0	0	0	0	2?	0	0

data from these three zones, the risk of successful translocations from high-risk sites within 100 km was in the order of 1 per 70 sites or 1600 ha of suitable habitat. One to 200 km from the zones, the risk dropped to 1 transfer per 327 suitable sites or 7000 ha of suitable habitat. The data suggest that translocation is a rare cryptic, albeit important, event.

The means and timing of the spread of *Campto* into and within NZ remains a mystery. There were sites that one would expect to have become infested, and others that were a complete surprise. Other mysteries remain: the means of spread to Kaipara and Wairau; the lack of spread from Wairau; the means of spread to Whangaparoa and Coromandel (but see page 188). These mysteries gave rise to speculations such as the presence of undetected sites or freshwater breeding, or to doubts about the effectiveness of surveillance or biology of the mosquito. Time has shown that incursion and translocation are simply rare and unpredictable events. Although there remains uncertainty in regard to the infestation network, it is plausible that the distribution of *Campto* in NZ arose from a single incursion to Napier in 1995–97 and this remains our best explanation.

Modelling spread

Entomologists are used to working with mosquito biology, life cycles and population growth models dynamics. However, the transfer of the mosquito to distant sites around NZ was more analogous to the problems faced by epidemiologists in the control of animal diseases. In studies of infectious diseases, epidemiologists understand the need to detect an infected individual or herd and contain spread early enough such that other individuals or herds do not become infected.

Very early after the handover of the eradication programme from the MoH to the MAF, operational staff from the MAF were keen to develop a spatial simulation model to further explore the feasibility of eradication, and to evaluate the impacts of varying the surveillance and control programmes. InterSpread Plus (ISP), a stochastic simulation modelling system developed at Massey University (Sanson 1993), was used.

There were several stages involved in the development of the model. First, data on the location of suitable *Campto* habitat in NZ, compiled by NZ BioSecure, was converted into a format required by ISP. The data available included the centroid of each habitat area and an estimate of the size in hectares. Circular polygons were constructed of the appropriate habitat size centred on each centroid. Overlapping polygons were merged and then intersected by the coastline of NZ, to exclude open sea. These polygons were then tessellated into 50 × 50 m cells, to represent individual *Campto* habitat sites within the model. Cells within 500 m of an airport were classified as 'airport senders'. These were areas assumed to be capable of transporting mosquitoes to other locations via long-distance (aircraft-assisted) movements. Cells greater than 500 m but less than 3000 m from an airport were classified as 'airport receivers'. Remaining cells were classified into high- or low-density vehicular areas, which could act as both senders and receivers of 'hitchhiker' mosquitoes. To do this, human population data from the 2001 census was combined with data on total vehicle licences as at June 2003 from Land Transport NZ to estimate the likely number of vehicles per km^2 in each cell, using 1 vehicle per km^2 as the threshold.

Second, the life cycle of *Campto* was represented and the temporal increase of mosquito numbers at each newly infested site. Using the advice of entomologist colleagues and information from the literature (Barton and Aberton 2005; Disbury 2006), an @RISK spreadsheet model was built to represent four distinct stages in the *Campto* life cycle, including the time from the emergence of an adult until it could start laying eggs, how long it might take for a suitable inundation event to occur allowing an egg to hatch, the larval stage and the pupal stage. The output of this model was a distribution of time in months from initial infestation of a new site to when it could function as a new source for further spread – a kind of site-level latent period. To represent the subsequent spread potential of a site over time, the build-up in mosquitoes was modelled in each site using a Verhulst logistic growth curve, as follows:.

$$\frac{\Delta P}{\Delta t} = rP\left(1 - \frac{P}{K}\right)$$

Where:

P = the population at a given time step
r = rate of growth
K = maximum carrying capacity
t = the time step

From the experience already gained in NZ across the known affected sites, it was believed that mosquito numbers reached peak carrying capacity after ~18 months. A set of values for r and K were determined such that the maximum growth rate in the population occurred ~18 months after the emergence of the first crop of adults, and peak densities reached in just over 24 months. This growth curve, converted to a proportion of peak population was used as a global infectivity parameter within the model, which influenced the probabilities of transmission associated with each spread mechanism.

Third, the main spread mechanisms and their parameters needed to be both biologically plausible and consistent with the literature and NZ data. These included:

- local diffusion based on the behaviour of an adult female searching for a blood meal before laying eggs
- longer distance wind-assisted spread
- vehicular transport
- aircraft-assisted transport.

The approaches taken and assumptions made to parameterise each of these mechanisms are summarised below.

Females were known to fly up to 5 km from a hatch site while searching for a blood meal (Dobrotworsky 1960; Howard 1973; Lee *et al.* 1984; Noiton *et al.* 2000). This behaviour was simulated using the local spread mechanism in ISP, which acts as a spatial kernel, using an exponential decay curve with probability of spread values from 0.5 at 1 km to 0 at 5 km.

For wind-assisted spread, it was assumed mosquitoes could be blown up to 120 km down-wind. ISP was parameterised to represent the prevailing wind conditions, with a likelihood that wind conditions would be suitable during any time unit.

The Coromandel survey provided valuable information on the frequency of trips by New Zealanders. Most of the Coromandel habitat areas were classed as high density (≥ 1 vehicle per km^2). The geometric mean number of vehicular journeys of greater than 5 km in length per month originating from localities within 5 km of high-density areas in the Coromandel was 528. It was assumed that one in 10 of these might carry a mosquito (once *Campto* numbers within a habitat area had reached a peak) which meant that ~53 mosquitoes per month could be transported to new locations via vehicles. If a journey end was adjacent to suitable habitat, it was assumed that the likelihood of a transported mosquito finding that habitat and laying eggs was between 0.2 and 0.4. In the absence of precise vehicular movements for other areas of the country, it was assumed that similar movement numbers would originate from other high-density habitat areas throughout NZ.

To estimate the corresponding movement frequencies for low-density habitat areas, data on visitor numbers to different regions around NZ (sourced from Tourism NZ) were analysed. There was a three-fold magnitude of difference between the median of the top half most visited regions versus the median visitor numbers of the less popular regions. It was therefore assumed that the number of transported mosquitoes out of low density habitat and environs was one third of the comparative figure for high-density sites (that is, 18 movement events per month).

Fourthly, the types of surveillance programmes conducted within NZ to detect *Campto*, and their relative sensitivities of detection, were defined. Three surveillance mechanisms were defined, including: the likelihood of the public reporting in habitat areas that were previously free of *Campto* ('public' surveillance); national water (larval) sampling, with a baseline level of 20% of habitat areas surveyed each month; and an intensive monthly sampling of all habitat areas within a 130 km radius of each newly detected area for a period of 24 months.

Fifthly, ISP was parameterised to be able to represent the control strategies employed to try and eliminate infestation in each newly discovered area. A large amount of field data was available to show larval counts as a function of the number of months since the start of S-methoprene or *Bti* treatments. This data was transformed into a function that progressively reduced the ability of an infested area to spread *Campto* to new areas once treatments were begun.

The final step before the model could be used was to validate the model against the NZ experience. *Campto* was seeded into one of the areas known to have been infested near Napier airport at Timestep 1 (equivalent to January 1998), with adult emergence set to occur in Timestep 6, and detection in Timestep 12 (December 1998). Controls were set to start 9 months after detection and the model run for a total of 108 Timesteps (9 years), for 40 iterations. The patterns of infested habitat areas predicted by the model were then compared with the known areas of infestation as at December 2006.

To our amazement, the model was remarkably predictive in terms of numbers of infestations simulated, and there was one particular model iteration that produced a spatial distribution almost identical to the actual NZ situation (see Fig. 7.5). This gave us a high degree of confidence in the base model. Crucially, the model showed that eradication was feasible, using a combination of existing surveillance and control programmes.

The model was then used to explore the effects of varying the probability of spread associated with vehicular and aircraft movements, the proportions of habitat areas

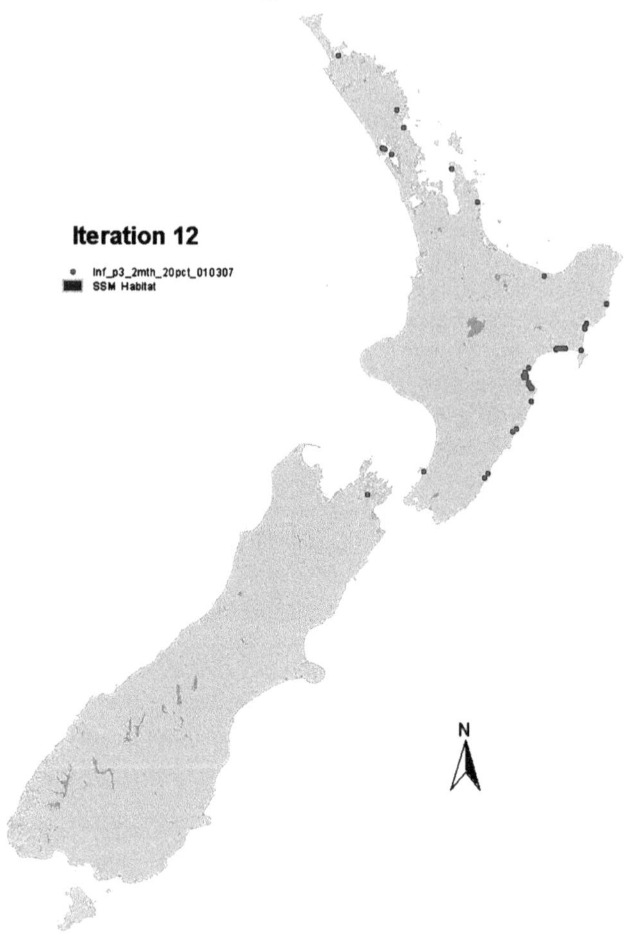

Figure 7.5 Map of NZ showing locations of sites predicted to be infested with *Campto* at the end of one of the simulated 9-year periods by the base model.

participating in surveillance programmes, and delaying the introduction of control measures following each new detection. Analyses explored the differences in the predicted numbers of infested sites and times to eradication.

There were several conclusions reached from the modelling work. First, although the overall model was highly complex, each component was parameterised using the best information or data available. Second, increases in the probability of spread via vehicular or aircraft movements increased the predicted number of infested habitat areas over the 108-month simulation period, but the relationship of this parameter with predicted epidemic duration was not as well defined. Third, progressively increasing the proportion of habitat areas participating in national surveillance from 20 to 50% each month reduced the predicted number of infested sites over the 108-month simulation period. Establishing a 130 km zone around identified infested habitat areas and conducting intensive surveillance in those areas (while continuing with the national sampling programme of 20%) was just as effective as increasing national surveillance participation to 40%. Finally, undetected or uncontrolled but known infested habitat areas acted as potent sources of infestation for other non-infested areas. These findings supported the premise that infested habitat areas should be detected and controlled as quickly as possible.

Acknowledgements

The authors of this chapter are indebted to several key individuals and organisations who participated either directly or indirectly to the work documented in this chapter. Individuals who had a major contribution to the work include: Matthew Stone, Cameron Webb, Craig Williams and Cheryl Johansen. The organisations who contributed included: the Ministry for Primary Industries (MPI) and the other agencies contracted by MPI, AsureQuality and Massey University.

References

Ayres CFJ, Romao TPA, Melo-Santos MAV, Furtado AF (2002) Genetic diversity in Brazilian populations of *Aedes albopictus*. *Memorias do Instituto Oswaldo Cruz* **97**, 871–875. doi:10.1590/S0074-02762002000600022

Barton PS, Aberton JG (2005) Larval development and autogeny in *Ochlerotatus camptorhynchus* (Thomson) (Diptera: Culicidae) from southern Victoria. *Proceedings of the Linnean Society of New South Wales* **126**, 261–267.

Dale PS (1980) Use of residual insecticidal coatings for killing insects in aircraft. *New Zealand Entomologist* **7**, 116–119. doi:10.1080/00779962.1980.9722347

Dale PS, Kleinpaste RH (1986) Residual disinsection: A simpler technique for killing insects in aircraft. *WHO Chronicle* **40**, 160–162.

Disbury M (2006) *Aedes (Ochlerotatus) camptorhynchus* (Thomson), the Southern Saltmarsh Mosquito. New Zealand BioSecure Entomology Laboratory, Wellington, NZ.

Dobrotworsky NV (1960) The subgenus *Ochlerotatus* in the Australian Region (Diptera: Culicidae). III. Review of the Victoria species of *Perkinsi* and *Cunabulanus* sections with descriptions of two new species. *Proceedings of the Linnean Society of New South Wales* **85**, 53–74.

Fonseca DM, Campbell S, Crans WJ, Mogi M, Miyagi I, Toma T, *et al.* (2001) *Aedes (Finlaya) japonicus* (Diptera: Culicidae), a newly recognized mosquito in the United States: Analyses of genetic variation in the United States and putative source populations. *Journal of Medical Entomology* **38**, 135–146. doi:10.1603/0022-2585-38.2.135

Gratz NG, Steffen R, Cocksedge W (2000) Why aircraft disinsection? *Bulletin of the World Health Organization* **78**, 995–1004.

Holder P, George S, Disbury M, Singe M, Kean J, McFadden A (2010) Biosecurity response to *Ae. albopictus* in Auckland. *Journal of Medical Entomology* **47**, 600–609. doi:10.1603/ME09111

Howard GW (1973) *Aspects of the Epidemiology of* Eperythrozoon ovis *in South Australia. Agricultural Sciences.* University of Adelaide, Adelaide.

Kay BH, Farrow RA (2000) Mosquito (Diptera: Culicidae) dispersal: implications for the epidemiology of Japanese and Murray Valley encephalitis viruses in Australia. *Journal of Medical Entomology* **37**, 797–801. doi:10.1603/0022-2585-37.6.797

Kleinpaste RH, Walker J (1996) Pre-embarkation disinsection: efficacy trials on a new disinsection technique. In *Report of the Informal Consultation on Aircraft Disinsection.* Pp. 48–54. World Health Organisation (IPCS) Geneva.

Laird M (1995) Background and findings of the 1993–94 New Zealand mosquito survey. *New Zealand Entomologist* **18**, 77–90. doi:10.1080/00779962.1995.9722010

Lee DJ, Hicks MM, Griffiths M, Russell RC, Marks EN (1984) *The Culicidae of the Australasian Region.* School of Public Health and Tropical Medicine Monograph Series, Entomology Monograph No. 2, Vol. 3 Australian Government Publishing Service, Canberra.

Lounibos LP (2002) Invasions by insect vectors of human disease. *Annual Review of Entomology* **47**, 233–266. doi:10.1146/annurev.ento.47.091201.145206

Macfarlane RP, Andrew IG, Sinclair BJ, Harrison RA, Dugdale JS, Boothroyd IKG *et al.* (2000) Checklist of New Zealand Diptera. http://www.ento.org.nz/Diptera.htm

McFadden AMJ, Bullians M (2008) Movement patterns of potential conveyers of *Aedes camptorhynchus* in the Thames-Coromandel district. *Surveillance* **35**, 4–10.

Ministry of Agriculture and Forestry (2003) 'Sea container review'. Discussion paper no. 35. MAF, Wellington, NZ.

Ministry of Agriculture and Forestry (2004) 'MAF-approved transitional facilities for sea container devanning (current as of 26 October 2004)'. MAF, Wellington, NZ, <http://www.maf.govt.nz/biosecurity/border/transitional-facilities/sea-containers/facility-list.pdf>.

Noiton D, Brady H, Shulmeister J (2000) *Southern Saltmarsh Mosquito – Literature Review of the Mosquito Ecology and the Transmission of Ross River Virus.* Ministry of Health, Wellington, NZ.

Russell RC (1988) The mosquito fauna of Conjola State Forest on the south coast of New South Wales. Part 4. The epidemiological implications for arbovirus transmission. *General and Applied Entomology* **20**, 63–68.

Sanson RL (1993) The development of a decision support system for an animal disease emergency. PhD thesis. Massey University, Palmerston North, NZ.

Schaffner F, Medlock JM, van Bortel W (2013) Public health significance of invasive mosquitoes in Europe. *Clinical Microbiology and Infection.* Online Early. doi:10.1111/1469-0691.12189.

Snow K, Ramsdale C (2002) Mosquitoes and tyres. *Biologist (Columbus, Ohio)* **49**, 49–52.

Spurr EB, Sandlant GR (2004) *Risk Assessment for the Establishment of West Nile Virus in New Zealand.* Landcare Research Science Series No. 25. Manaaki Whenua Press, Lincoln, NZ.

Weinstein P, Laird M, Browne G (1997). 'Exotic and endemic mosquitoes in New Zealand as potential arbovirus vectors'. Occasional paper. Ministry of Health, Wellington, NZ.

8

Camp Kaipara – a mosquito or programme death camp?

Bryn Gradwell, Monica Singe, Shaun Maclaren, Steve Crarer, Jessica Taylor and Mark Disbury

Kaipara – the initial find

According to the Kaipara-Kumea Catchment Management Plan, the Kaipara Harbour is a remote windswept waterway with some 800 km of coastline. Identified habitat was first assessed at a heart stopping 22 000 ha, soon to be reassessed to around 2710 ha of positive habitat. This was a potential 'deal-breaker': the successes of Napier, Porangahau, Mahia and Gisborne were gained over smaller and more-accessible sites. Was it the death knell of the eradication programme? No one had eradicated a mosquito from 2700 ha of God-forsaken saltmarsh, let alone 22 000 ha! The TAG determined that it was feasible to embark on an eradication programme based on the Napier model. The political will was found and the Kaipara eradication programme implemented with a budget of over NZ$30 million.

The TAG convened in Wellington on 26 February 2001, following the find during routine public health staff surveillance of *Ae. camptorhynchus* (*Campto*) on 18 February 2001 and the delimiting survey undertaken by the Auckland public health unit. In their tabled report to TAG, they stated that there were still locations around the Kaipara that needed to be checked and that the affected area may well increase beyond the 22 000 ha already identified as requiring treatment in any eradication programme.

We gasped for air! The TAG agreed that it did not appear feasible to eradicate *Campto* from the Kaipara because of the area of possible infestation. They reiterated that they still considered eradication from Hawke's Bay and Tairawhiti to be feasible, but that the Kaipara infestation was on a scale of greater magnitude. Treating 22 000 ha with S-methoprene was technically feasible but not economically realistic. They also agreed that containment/control was not an option, given the close proximity of suitable habitat in Northland and Auckland, because surviving mosquitoes could easily be blown into these areas. Reluctantly, TAG recommended a change in focus from primary disease prevention (i.e. by excluding the mosquito) to secondary disease prevention (i.e. prevent cases through early detection of disease and control measures to reduce epidemic rates). Were Kiwis really going to face a future of mosquito-borne disease?

Was preparation for inevitable Ross River virus epidemics all that Kaipara residents could hope for? The secondary disease prevention response could be achieved through improved mosquito surveillance around populated areas and by defining the spread of any virus activity. Early notification of any cases was important, but it was also fair to say that NZ did not have experienced arbovirus diagnosticians. Silent infections could be monitored in animal hosts. Given an alert, mosquito control would be instituted and residents told to use repellents and take evasive action.

Habitat reassessment

On 27 March 2001, the Auckland Public Health Unit engaged the services of Darryl McGinn, now of Mosquito Consulting Services Pty Ltd, to provide a report to review the adequacy of the information so far obtained and intended for use in decision making on a response to the *Campto* incursion in Kaipara Harbour. He examined the potential habitat for *Campto* in Kaipara Harbour and the delimiting survey of distribution of adults and larvae.

At the same time Steve Garner, Manager, Napier MRC and his technical field team were asked by the MoH to assist in conducting an aerial survey of potential Kaipara habitat in order to produce a more robust estimate of *Campto* habitat in the harbour. The aerial survey provided a map of habitat at a macro level, while the assessment of the delimiting data would provide a picture of distribution of the insect within the mapped habitat. The MoH would consider this information and develop options for government decision making.

The habitat size was estimated using a combination of data recorded from GPS, video, photography and notes taken during survey flights. This information was collated and assessed using GIS software. Mark Disbury and Jessica Taylor, two MRC technicians with considerable field experience and mapping knowledge, were selected to undertake this work.

Turning the mountain into a molehill

It was ironic that just as the eradication work in Napier was winding down, notification came that the largest natural harbour in NZ was found positive for *Campto*… and it potentially contained 22 000 ha of habitat, in the southern half alone.

The logistical implications were huge. Transport around the harbour for surveillance … how many trucks, quad bikes and helicopters would be needed? Would the programme need to use airboats? All-terrain vehicles? The tidal range was 5 m and this amplitude would inundate a lot of saltmarsh. Cost of treatment over an area that size would be phenomenal. Was the programme doomed? Was all the hard work to be undone by this monster of a harbour? The actual size of the habitat needed to be known. And fast!

The Auckland Public Health Unit estimated 22 000 ha based on Landcare map layers, applying a 1- km buffer zone to potential saltmarsh areas. Clear aerial photos were needed. These were the days before Google Earth was readily available, but the Napier MRC had been allowed access to some great aerial digital photography held by district councils in the past. These clearly showed vegetation, land use, and so on, and gave a clearer idea than the standard topographical map as to what habitat potential each area contained. Public health unit staff established that there were no recent photos available at the time.

If the potential habitat couldn't be accurately determined, the eradication might be written off due to the astounding cost of each treatment, based on the initial habitat estimate. The Napier MRC had a helicopter contractor who was a very experienced eradication programme pilot. The helicopter had built-in differential GPS, used to accurately map the eradication treatments.

Napier MRC had staff who were used to hunting *Campto* down to the last mosquito. They knew the best spots to find them and where they were likely to spread out to if the numbers were high. They also knew, more importantly, where they were unlikely to breed. Armed with a video camera as well as digital cameras, pencils and paper, Mark and Jess felt they could do a pretty good job of estimating habitat potential from the air.

Accurate estimations of habitat would lead to more accurate costings for surveillance and treatment. It would also give a good idea of where to start the delimiting survey. From this, staff could ground-truth each area and delete unsuitable areas off this list as they went along. However, the MoH would need to know the maximum potential area, in order to have a worst-case scenario for budgeting purposes, and the all-important cost–benefit analysis would have to be done.

Painting by helicopter

Mark and Jess got their basic mapping kit together and set out with Phil Deadman, the pilot. Excited about the prospect of getting under way to map this monster, they went out to Parakai airfield. They had watched the rippling biceps of the Auckland Public Health Unit flex since the initial *Campto* find. Did everyone need this bureaucratic arm wrestle? Auckland wondered why some gum-booted hillbillies from Napier were muscling in on their patch. The Napier MRC team wondered why they couldn't get stuck in and do the job they knew best – not wanting all their eradication sweat to be for naught.

One short flight (with the topographical maps) for the Auckland Public Health Unit in the Hughes 500 cleared the air, better than any desktop discussion could. Suddenly the scale of the work and its difficulties became clear. Moreover, the specialist helicopter and pilot Phil could not only do the job less expensively and quicker, but they could map the harbour out with a far greater sense of accuracy.

Having on board a couple of people who had worked on the ground for a few years helped too. Vegetation was a great indicator for this species. The pointless arm wrestle eased into a handshake; teams could now work together and combine their strengths. 'Operation Molehill' was under way.

A rough estimate, taking into consideration mangroves, saltmarsh and drains as layers from Landcare maps available at the time, suggested *Campto* habitat could be far less than 22 000 ha. After flying for 6 days round many sharp, nose-diving corners to get accurate boundaries mapped with the DGPS, the likely maximum habitat available was reduced to 2728.8 ha.

Eradication was now potentially feasible. A glimmer of light was beginning to line that portentous Kaipara dark cloud. Perhaps NZ could still get rid of this uninvited, irksome Australian. Later, after extensive surveillance during the eradication, it turned out the largest treatment ever was 1800 ha and that followed a significant wet weather event.

After 2 weeks of assessment and evaluation, Darryl McGinn and Steve Garner produced a briefing paper for the MoH (McGinn and Garner 2001):

> *'The extent of potential habitat for* Campto *has been largely identified through GPS assisted aerial survey and ground-truthing. Notwithstanding some refinement of data, the assessed habitat within the harbour is around 2,700ha (subject to the caveats in the Napier MRC Kaipara report). Given the technology that exists to control*

saltmarsh mosquitoes elsewhere, development of a control/containment programme is considered a practical option. The data relating to the delimiting of the current distribution of the insect within the available habitat has yet to be adequately completed.'

MoH response

The MoH had been directed to report back to government no later than 31 May 2001 on possible options for the future of the response to the *Campto* incursions in Napier (650 ha), Gisborne, Mahia and Porangahau (190 ha), and Kaipara and Mangawhai (potential habitat of 2710 ha).

The MoH's preferred approach was a combination of options that included the eradication activities in Napier, Gisborne, Mahia and Porangahau and to contain/control in Kaipara and Mangawhai and undertake secondary disease response in Auckland and Northland, and enhance the MoH's capacity for biosecurity work. The MoH was of the view that this would ensure gains on the East Coast were not lost in the short term, help to prevent the spread of the mosquito from Kaipara and Mangawhai, and reduce the risk of Ross River virus disease for the people of Auckland, Northland and the eastern North Island. On 16 August, Sally Gilbert (2001) said:

'The government agreed to contain/control the mosquito in Kaipara and Mangawhai and to strengthen the quality and reliability of surveillance and to seek information on entry pathways before making a decision to eradicate or to involve local government in any changing response in the Kaipara and Mangawhai.'

The TAG members were provided with the MoH's background paper and a meeting was convened in Wellington on 27 August 2001. The TAG were informed that the potential habitat for Kaipara and Mangawhai was now estimated to be 2710 ha and that the MoH's preferred option was identified on the basis of the revised information and cost–benefit analysis, but had to be without direct advice from TAG.

The TAG advised that the review of surveillance needed to include a retrospective review of all positive sites as well as surveillance of potential habitat.

In response to the draft plan for responding to the *Campto* incursions at sites in NZ, which had be circulated before the meeting, the TAG agreed that, although the eradication programme on the east coast could not be deferred, it would leave limited resources to respond to the Kaipara and Mangawhai incursions. They pointed out that the Kaipara habitat needed to be delimited properly, when conditions were appropriate, before containment/control measures could be appropriately implemented.

NZ BioSecure was contracted by the MoH to provide a national response to *Campto* incursions and were working diligently on scoping the extent of the infestation in the Kaipara, including the determining of larval densities. The TAG anticipated that there would be local involvement including ongoing surveillance of sentinel sites, stakeholder consultation and involvement and resource consent issues (i.e. getting permission to treat). They recommended that the priority was to delimit and characterise the distribution of the mosquito, using hydrological data and to concurrently apply for permission to apply control chemicals to the environment.

At this meeting, it was decided that NZ BioSecure should have a draft plan to present to the TAG in early December for approval. There was some concern that this could be too early, with insufficient tidal flooding events to hatch eggs, but the draft could be refined as more information became available. The final plan was to be ready for implementation on 1 January 2002.

Delimiting the Kaipara

On 24 September 2001, the first of two comprehensive delimiting operations was conducted in Kaipara Harbour when optimum conditions prevailed, using the field and management expertise from the Napier and the Tairawhiti personnel, who by this time were all employed by NZ BioSecure. Prior to mounting this operation, management had liaised with the Auckland Public Health Unit, Rodney District Council and the Biosecurity Officers of the Auckland Regional Council to identify all stakeholders and landowners who were likely to be affected by, or interested in, such an activity.

The database for landowner's property boundaries and contact details held by the Auckland Regional Council provided a valuable tool in identifying who needed to be approached.

Delimiting was undertaken by five teams of two, thoroughly examining the identified habitat for *Campto* larvae, starting from the southern end of the harbour, moving up the South Head on the west coast, and then concentrating on the eastern side where there was far more potential habitat. This effort was completed in five long days, after the northern reaches of the harbour were examined and potential habitat characterised.

The terrain was difficult to access and, where four-wheel-drive utility vehicles and all-terrain vehicles were unable to gain access, swamp, stream and thick undergrowth was traversed by foot. Radio transmission communications equipment used on the east coast operations failed in the Kaipara and cell phone coverage was intermittent.

The team was based and accommodated at 'Black Pete's' a small camping ground with the reputation of having the finest steaks in Parakai – a settlement outside Helensville the largest of the community settlements in Kaipara Harbour. At the end of each day in the field, debriefings were held to assess the information gathered, *Campto* larvae were identified along with their stage of development. Tomorrow's plans were decided.

The second full-scale delimit commenced on 18 November when the weather was very warm, the days were humid and sunny. *Campto* larval densities were very high in some places. When debriefing occurred, a mud splattered Mark Disbury said 'The water was black with them …'.

NZ BioSecure Report to the MoH

While consideration about eradication was ongoing, the Kaipara Report and Control Strategy was presented to the MoH in December 2001, proposing progressive review of existing field data and assess them for completeness and accuracy. This included on-site field assessments to determine larval densities, to establish larval surveillance sites and larval indexes that would trigger treatments.

NZ BioSecure had developed an electronic sampling database and mapping system to be used alongside the aerial photographs to direct field surveys. These were based on colour aerial digital photography and additional surveillance data reported to MoH in September and November from the delimiting surveys. From the field surveys that followed, 36 positive zones were identified. Some zones were quite large and contained extensive networks of drains and runnels. It was estimated that 200 ha of drains alone may require treatment. Larval densities varied from common (5–20 per dip) to abundant (>20 per dip) in some sites.

Further surveillance was undertaken in the southern part of the harbour during the week of 17 December 2001. This survey followed several weeks of significant rainfall in the area. Advanced larvae and pupae were recovered from most sites and sentinel sites were established around the harbour at South Head, Parakai, Hoteo, Tapora, Tinopai, Taingahe and Mangawhai. Sites were surveyed after water events. In addition to these areas, NZ BioSecure also developed a network of property owners who offered to advise them when areas become inundated.

At the 16 November 2001 TAG meeting, treatment of all sites when any *Campto* larvae were found was recommended. This was achievable, but not within the current resource levels. The MoH was advised that a treatment strategy to achieve containment and control depended on availability of larvicide and finance. Comparisons were made with the control by Brisbane City Council, not dissimilar in size or habitat type, but budget constraints currently ruled out such an option.

Bti was the larvicide available for use. S-methoprene pellets and granules would require ministerial approval, resource consent and additional funding. However, control and containment using *Bti* would require regular applications because of rapid larval development, reducing efficacy on advanced larval instars and lack of residual action.

Treatment trials on the extensive network of drains and runnels demonstrated that ground treatment was not feasible without a massive increase in manpower and equipment. Penetrating the density of the vegetation with *Bti* was a complicating factor, but granular or pelletised S-methoprene would overcome this problem. Although using a helicopter to fly the drains would save manpower, treatment with *Bti* was unlikely to be successful in vegetated hotspots without changing to a granulated formulation.

An indicative treatment budget for the Kaipara was attached as part of the NZ BioSecure tender proposal, with a note that, until all actual on-site assessments were made, it was not possible to forecast these costs accurately. This budget was based on treating around 300 ha with *Bti* over 10 tidal inundations, but this was never going to achieve eradication. Furthermore, the MoH was advised to review the actual costs of applying product within this area by helicopter and ground equipment delivery platforms because they would be higher than at Hawke's Bay due to the extreme transit distances.

At their November meeting, the TAG recommended treating all sites whenever larvae were present. This was likely to exceed the 300 ha estimate and may require weekly treatments in some areas. NZ BioSecure faced a worse case scenario that the Kaipara treatment budget could be expended in less than 10 weeks! This could occur by 30 June even using a rationalised scenario based on larval averages of 1 or greater over 10 dips. These were desperate times!

Resource Management Act 1991

Before a control and containment programme could commence, a resource consent or permission to apply control products to land and water was required. Close examination of the District and Regional Plans and early liaison with the consenting authorities planning staff was necessary to prepare for an application. The exhaustive consultative process involved is outlined and discussed in Chapter 12.

The outcome sought had been achieved: consent to allow the control and containment programme to proceed. The consents were issued with manageable conditions. Along the way, management had gained valuable knowledge about the people, communities and the environment of the Kaipara, which proved very beneficial as the programme later moved towards eradication. Huge emphasis had been placed on completing a thorough consultative process and to avoid creating an adverse public reaction, as had been experienced in West Auckland with the painted apple moth eradication programme. From the early planning sessions, an extra effort was made to ensure anyone considered affected was identified and kept up to date with progress.

NZ BioSecure created a regulatory stakeholders database, a landowner/occupier database, a media contact database, and contact was made regularly via email, newsletter or media release. Agreement was reached with Council Environmental Health Officers to direct mosquito complaints from around the Kaipara to NZ BioSecure and, in return, advice was provided on mosquito nuisance complaints and equipment was loaned out when requested.

Trust from the community seemed to have been gained and when the control phase of the operation changed to eradication, with a new larvicide product and far more intense presence of staff and activity, it was a smooth transition and, without exception, there was full community support. However, on 17 December 2001, before the resource consents commenced, the Minister of Biosecurity used the powers of Section 7A of the *Biosecurity Act 1993* to provide an exemption from the *Resource Management Act 1991* so that aerial applications of *Bti* could commence, in an attempt to eradicate the *Campto* infestation at the onset of the summer breeding season.

Establishing the team

During the control and containment phase of the Kaipara operation, only two NZ BioSecure full-time staff undertook surveillance (adult trapping and dipping for larvae) and ground-treatment activities (spraying applications of *Bti*). However, they were supported by staff from the east coast eradication programmes and Helicopters Hawkes Bay (the subcontractors providing aerial transport for the field teams to access difficult terrain and to apply S-methoprene treatment product) when required, usually following king tide events or when treating inaccessible locations. Bryn Gradwell (manager of the Tairawhiti eradication programme) and Regan Courtney (field technician from the Napier MRC) provided the full-time presence.

When the NZ Government approved eradication, NZ BioSecure management planned for the recruitment of staff. There wasn't a pool of trained field technicians out there waiting to be hired, but experience gained from the recruits employed and trained with the east coast operations provided the basis for what types of people were ideal for this work.

To assist with 'buy in' from the affected communities, NZ BioSecure targeted local newspapers when advertising for staff. The advertisements asked for fit, energetic people, who preferred to work outdoors, interested in the care of the natural environment, comfortable working in a team or unsupervised and with attention to detail. A plan was devised to establish a workforce headed by an Operational Manager (Bryn Gradwell) and Technical Manager (Monica Singe), assisted by a Logistics Coordinator and field teams consisting of Technical Officers (TO) as team leaders and Technical Assistants (TA).

The Operations and Technical Managers, one TO and two TAs already existed, originating from the Napier and Gisborne MRCs, but all other appointments were to come from the recruitment strategy. The advertisement seeking potential staff gave details of the work associated with the eradication programme for the Kaipara and the proposed locality of the operations base. A dozen staff appointments were on offer, 50 applications were received and 20 applicants were selected for interviews. They came from a variety of backgrounds, male and female, different work experiences, some with extensive tertiary qualifications and others who were early school leavers. The Logistics Coordinator was selected from some well-qualified applicants: Shaun Maclaren was an ex British Army instructor with vast field experience and business management background. Two TOs and seven TAs were initially appointed – selection was about the skill sets they brought with them and team dynamics of the group selected was a consideration.

To fit the criteria the interview panel adopted, potential candidates had to:

- be reasonably fit
- have a current clean drivers licence
- be able to read a topographical map
- be happy to work outdoors in all weather conditions throughout the full year
- be able to travel to other locations as required
- have legible handwriting and could work as part of a team.

As the programme progressed, many people's natural aptitudes surfaced and were used where possible. In the mix selected, we had skill sets and additional knowledge in computer programming, welding, building construction, engineering, safety management and field craft/orientation. Steve Crarer, our East Coast Operations Manager from Napier said:

> *'Some interviews were a lot like an encounter group, as it stirred some people up emotionally. I remember Adam Mason's interview where subsequently I told his Mum he was successful but suggested that he tie his hair back and remove his facial jewellery while on the job, but she told him to cut his hair off. I think he was a bit grumpy with me for a while about that.'*

When the Kaipara eradication programme commenced, the team comprised of 14 personnel, several with past military experience from within NZ and overseas. Maybe we remembered that the greatest mosquito campaigns ever were led with military precision by the likes of General Gorgas and Dr Fred Soper (Duffy 1977). Others had interests in environmental and conservation issues, most important to our objectives and government policy of 'first do no harm'.

Kaipara eradication programme operational planning

Because of the vast amount of territory and varying terrain under surveillance within the total eradication zone, supervision of personnel, transport movement, communications, flight liaison and the passage of information and instructions took on a military-style control. Management staff planned with a practical concept of operations, which was passed on to the team leaders, who in turn delegated tasks and responsibilities among their team members. A hierarchy was established within the workforce. Leadership qualities were identified and development encouraged, with the 'carrot' of promotion upon a successful personal development review.

Management promulgated routine orders on a weekly basis following the last debriefing of the week. This information was displayed on notice boards, with any daily changes written onto the whiteboard located in the central meeting room location. There would be regular weather forecasts, planned treatments, scheduled liaison meetings, planned training forums, HR updates, staff reviews and reports received from the parent company, the MoH and stakeholders.

Within the main office area, each field team had a designated breakout room where they had their own lockers, personal protection equipment, desks and computers. Upon returning from the field, team members would undertake their allocated responsibilities: cleaning vehicles, specimen preparation in the laboratory and preparing equipment for the next day's tasks. They would then gather in their designated rooms to enter surveillance track data and prepare reports for debriefing. Following the debriefing of the full team, team leaders would meet with the Logistics Coordinator for their daily planning session using the knowledge gained from the debriefing. Team members would use this time on personal administration or maintenance/cleaning of equipment and the workplace. Teams would then return to their rooms for instructions from the team leaders.

The Logistics Coordinator met regularly with the Operations and Technical Managers for input into the planned surveillance and treatment events. Strategic plans were developed and passed on to the technicians for action. These plans were put together following immediate observations and weather/climate conditions, but did not deviate from the overriding MoH eradication plan sanctioned by the TAG.

Training requirements

Immediately following staff selection, all new appointees signed their Individual Employment Agreements. Schedule A outlined the key tasks for their positions of TO or TA.

TOs and TAs were staff who undertook larval monitoring activities and provided input into planning mosquito surveillance activities. They recorded relevant environmental data in the field and in a computer programme. They were required to liaise with relevant landowners as to field activities and were to provide input and advice at daily debriefings in order to maintain accurate records describing the status of each treatment area.

These technicians would visit light traps at their field locations with the documented schedule to remove trapped mosquitoes. They were to ensure accurate identification for later despatch to the laboratory. Traps were maintained with an adequate supply of CO_2 and octenol bait. Staff would replace batteries, check for malfunctions, carry out repairs where necessary and prepare and maintain the light-trap programme in conjunction with management. Importance was placed on the maintenance of accurate records containing data relevant to mosquito populations and hatching levels.

Technicians were to ensure adequate and accurate data were recorded as they related to: 'wet events'; data capture including recording tide heights; rainfall measured at trapping locations; and weather forecasts with wind speed, direction and frequency. This environmental surveillance was to enable anticipated hatches and required treatments. They were to be proficient at control activities, including hand/backpack application of treatment product, ground treatments from mobile platforms and assistance with the aircraft control activities.

Routine tasks also included ensuring the workplace was safe, compliant with health and safety procedures, and cleaning and organisation of the workplace, including plant, laboratory equipment, offices, stores and workshops. They all had to undertake the administrative duties according to the standard operating procedures, such as activity reports, timesheets and employment-related documentation.

All staff members were issued with a comprehensive Induction Pack. The induction process required every new staff member to become familiar with procedure (the Company Quality Manual), identify the competencies required for their job and meet key staff appointments throughout the organisation. As a group, they were deployed to Napier for a full week's intensive training by experienced senior NZ BioSecure staff. During this time, they were exposed to on-the-job training in the field and laboratory, and classroom instruction on the technical matters concerning treatment products, *Campto* and the eradication programmes.

NZ Biosecure had set standards for core competencies – those skill sets that all field operatives were required to achieve. The measure of success for attaining the standards was subject of professional development reviews. The criteria and assessment methodology was given to staff through the induction process, standard operating procedures and their individual employment agreements.

The core competencies included: planning and organisation; effectively scheduling time, tasks and activities; and establishing a course of action to accomplish specific goals. They were to have a thorough understanding of theory and practice in the field and have appropriate job knowledge and skills to perform in their position. Staff were trained to communicate effectively and convincingly to present sound arguments to influence others and negotiate mutually acceptable solutions. They should be able to work constructively by developing effective relationships and networking with people within the company and external clients. Staff needed to show an awareness of, and act to meet, client needs and strive to continually improve service quality.

They were to identify the causes of problems, and determine and organise facts related to a given problem, and to make consistent, accurate and timely decisions based on thorough analysis of the situations concerned. We were looking for leaders: those who could articulate their commitment to the goals and direction of management through their actions and communication. We needed staff who were team players and who could inspire and energise others to pursue the company goals. We also wanted those who were able to contribute and make recommendations to improve records, policy and standard operating procedures. Upon accepting new responsibilities and duties, those selected as leaders contributed to developing new services and roles, and adopted and developed the necessary knowledge and skills to perform new tasks. To ensure that the new staff members were able to perform their respective key tasks to the expected standard, intensive in-house training was carried out using established NZ BioSecure expertise and external colleagues. Instruction on mosquito identification and screening techniques, specimen preparation and packaging for transport was provided by the NZ BioSecure Entomology Laboratory by Mark Disbury and Rachel Cane. There was also intensive training in the field – identifying most suitable habitat and demonstrating techniques for sampling. Initially, the raw staff spent time 'in camp' at Napier working alongside experienced staff from the Napier and Gisborne MRCs. This is where they learnt the three main components of field work: sampling, trapping and treatment. The calibration of treatment equipment was an essential skill and instruction was undertaken by experienced staff under the supervision of Steve Crarer.

Computer skills for the data entry were quickly learnt because the majority of the new workforce had previously used computers extensively at school. However, senior staff provided instruction on the use of GPS units, radio-transmitting devices, map reading, photography and weather forecasting.

Health and safety concerns required a programme of ongoing safety instruction. Fire evacuation plans were drawn up, instruction on extinguisher use and fire fighting was provided by the local fire service authority and fire drills exercised. All staff attended four-wheel-drive and all-terrain vehicle driving courses conducted by a credited commercial organisation. Helicopters Hawkes Bay pilots produced a Safety Manual for travelling, alighting and loading the helicopters. Instruction was also provided by senior staff on: the safe carriage, handling and charging the 12 V light trap batteries; the handling of all chemicals and fuel; the wearing of correct footwear, safety glasses, sunscreen and life-jackets when required.

Quality management

The company already had a quality management system based on ISO9002, which included a Quality Assurance (QA) Manual, Human Resources (HR) Manual and Health and Safety Manual, but, with the advent of the Kaipara programme, there was now three eradication programmes several hundred kilometres apart and three distinct teams of field personnel. The need to formalise some of the surveillance and operational methodologies became apparent to ensure a consistent approach to all aspects of the techniques and activities being undertaken. Joanna Christie, located in Napier, was crucial in the documenting of standard operating procedures for the NZ BioSecure Manual. The Manual was developed in sections, including Laboratory, Workshop, Administration, Field and Forms, and later a National Surveillance Programme (NSP) section, ensuring conformity and best practice.

In addition to operational policies and procedures, the HR manual had in place performance review appraisal systems for all employees. To ensure the large number of personnel now on the ground among the various eradication programme localities were capable of carrying out surveillance with an attention to detail, a series of core competencies was developed and attached to job descriptions and a regular performance review system. Staff reviews were carried out on an annual basis. Core competencies had to be signed off by experienced staff and management, and were documented in individual personnel files commonly referred to as their 'Induction Book'. Although most staff detested the documentation and the sometimes onerous system of performance review, goal setting and competency assessment, none would deny that it was necessary.

Of course, the HR Manual and the Individual Employment Agreements also included procedures in place for issuing warnings regarding poor performance, but we had a team of keen eager mossie hunters who took their jobs seriously and enjoyed the thrill of the catch (albeit a couple of millimetres of invertebrate). Management worked hard at keeping staff turnover to a minimum to ensure expertise was retained. All staff undertook a defensive driving course, thereby minimising vehicle accidents and incidents. Relationships with contractors, especially the pilots of Helicopters Hawkes Bay, including Phil, Peter, Tim, Frank and Wayne, were great. Some of the team members enjoyed the occasional pig hunt and fishing trip and, of course, an exciting form of transport for treatments, surveillance and light-trap maintenance.

With performance review comes professional development, and NZ BioSecure ensured all staff were given the opportunity for such development. Communication and liaison was an important part of the programme's success and linkages were built among council staff, community groups, NZ Biosecurity Institute and Auckland Pest Liaison Team via meetings. This interaction with the Auckland Regional Council biosecurity personnel at the Institute also resulted in our mossie hunters reporting Bathurst Burr and other plant pests found throughout the Kaipara to their colleagues.

Additional training/education undertaken

As the programmed progressed, some skills and knowledge gaps appeared and, to complement the advances made by members of the workforce, additional training and education courses were organised. A retired NZ Army Colonel, David Roseveare, was contracted to conduct a leadership course for the TOs, and media liaison and communication training for all field staff. St John's Ambulance was hired to conduct CPR and first aid training for everyone, while a selected number were taught about the radio transmission and communications network.

Health and safety legislation required trained staff to manage the company plan and selected personnel attended courses to meet this requirement. TO Gareth Southcombe, who had an RAF firefighter background, excelled in this capacity. Health and safety training for all staff was given top priority because of the nature of the field work and the potential dangers staff faced every day. Behaviour around helicopter operations was under constant scrutiny; unlike most vehicle safety, the danger areas of a helicopter are at the rear. However, two very close calls eventuated when Adam Kellian jumped up onto the vehicle deck while refuelling the helicopter and came close to losing his head when he straightened up right under the revolving rotor. He realised his predicament at ~100 mm from contact. On another occasion, Eliane Lagnaz, who didn't tie back her long curly mop of hair, was inside the helicopter when her hair was sucked into a vent and tore at her scalp before a team member alongside fought her free ... scary stuff!

Some staff who had left school early required assistance with writing reports and in comprehending technical publications, and attended adult education classes to develop such skills.

Because our public image was so very important, management sometimes encouraged some staff to improve their life skills, by attending time and behaviour management training, while others took opportunities outside the workplace to improve public speaking and *marae* (Maori meeting house) protocol. Many others competed in active sports, which assisted in the maintenance of physical fitness – a necessity for the field operative.

Training outcomes

With all this training, management were trying to achieve a competent, and eventually expert, workforce who had an understanding of environmental issues relating to mosquito surveillance and control activities, with a methodological approach to sampling, data recording and entry. There was a focus on accuracy and quality, attention to detail and knowledge of the local environment as it related to topography, effects of rainfall and tides, and the people who lived in the operational areas.

The team was striving to attain enthusiastic, articulate staff who were able to meet and confidently discuss issues with all stakeholders. It was soon realised this team was very capable, and proud to wear the company shirt and jacket and be identified as the Kaipara MRC field staff (Fig. 8.1).

Team organisation

The team was initially established with a management group of two, a team coordinator and three operational teams that specialised in surveillance sampling, light trapping and ground treatments. All members were cross-trained to be competent in times of absenteeism.

Figure 8.1 From left, Shaun Maclaren, Mark Dellow, Adam Kellian and Dean Singe. (Image: courtesy of *Rodney Times*)

Field teams were normally four strong, led by a technical officer and supported with a second in charge, who was the most experienced remaining member of the team. This format allowed teams to work in pairs and gave the programme increased flexibility without jeopardising operational standards and safety requirements.

Box 8.1: NOR-WEST NEWSBRIEF, Thursday January 23, 2003 (Page 5)

Mossie team sets up in Parakai. By Michelle Hyland

'Parakai will host the control centre for a $13 million, four year programme to eradicate the southern saltmarsh mosquito from the Kaipara Harbour. NZ BioSecure has set up camp on Greens Road, opposite Parakai airfield. The command centre has 17 staff, a helicopter, computers, testing devices, chemicals and a fleet of four-wheel drive vehicles. Most of the staff are working as technical officers and assistants.' When we were looking for staff for the operation we did all our advertising locally. We got some really talented people and they were predominantly from around the Kaipara,' says Bryn Gradwell, Regional Manager of NZ BioSecure. Employees were mainly in their 20s and from Muriwai, Parakai, Warkworth and other parts of Rodney. Jobs include collecting mosquito larvae and setting traps, which are carried out on foot, in helicopters and in areas accessible only to four-wheel drives. Testing is at the Parakai base and specimens are then sent to Napier for positive identification by four NZ BioSecure scientists.'

Figure 8.2 The Kaipara MRC *Campto* Eradication Team, June 2003. (Image: NZ BioSecure)

Within 3 months of starting the programme, our workforce was expanded (Box 8.1) – a fourth field team was established, an administrative clerk was appointed, three TAs were promoted to TOs and a helicopter pilot permanently joined the team.

By June 2003, just 8 months into the Kaipara eradication, the team had expanded to 22 permanent staff plus two pilots living on site (Fig. 8.2). There were four field teams of four technicians, the logistics coordinator, operations manager, technical manager, information systems coordinator, two administrations clerks and the two pilots.

Establishing the base

The property at 73 Greens Rd, Parakai was on the outskirts of Helensville, the town servicing the southern half of the Kaipara. It had recently been used as a stud farm: a facility for the training of 'trotting' horses. With the space, house and outbuildings, it was an ideal site from which to operate the programme. The only drawback was that the main house needed 'sprucing up' before it could be occupied and it required a fire rating upgrade to commercial use standard. So as not to delay or affect the programme, a sizeable 12 m long portable accommodation unit was installed while renovation/repair of main building was being carried out. The 'portacom' became the hub of the programme for a management and staff of 16 persons. It was the office, briefing and debriefing area, the cafeteria and the place where the processing of samples and initial identification of adult mosquitoes was carried out.

The 'sprucing up' involved the clearing and cleaning of the farm house and nearby garage – a task that one would have thought should not be too taxing. However, after the removal of 9 m^3 of rubbish, renovations were able to commence. The plan was to use the house as office accommodation, the portacom as the screening laboratory, the barn as a

Figure 8.3 The Parakai Operations Centre looking like it should! (Image: Helicopter Hawkes Bay)

workshop and battery shed, and the stables for storing product, vehicles and spray equipment.

Helicopters Hawkes Bay established a portable hanger able to house two Hughes 500 helicopters, and with shipping containers providing service equipment storage.

As soon as the rubbish was cleared, 75% of field staff resumed mosquito surveillance operations, in particular familiarisation of land and farm properties around the Kaipara, while the remaining 25% began preparing team break-out rooms, offices and the debriefing area. Local electricians were used to upgrade the electricity supply and services to the main building, portacom, battery shed, stables and nearby helicopter hangar, while IT technicians wired the main building to establish a computer network and telephone communications.

The small barn close to the house was converted into the battery shed and used as a workshop for vehicles/quad bikes and light-trap maintenance. The battery shed was equipped to store a minimum of 80 batteries and charge 16 batteries at any one time. A separate area within the battery shed was used to store up to 60 CO_2 cylinders for use with the adult light traps. The workshop was equipped with a full range of maintenance tools, a compressor and welding equipment (Fig. 8.3).

Being the largest building on the property, it was planned that the stables be used to store the tens of thousands of bags and containers of treatment product required for the programme. We were to discover that that these stables bore abundant evidence of their former use: it seemed the previous owners had not bothered with cleaning for several years!

Field supervision

It was important that supervisory systems be adopted from the outset to ensure that quality was maintained at an operational level and that health and safety of staff remained paramount.

This was achieved not only by the careful selection and make up of team skills, but through the adoption of standard operating procedures, practical training on the use of such tools as the hand-held GPS, a monitored communications network and an inclusive debriefing session at the end of each day's operations.

Although the logistics coordinator had the responsibility for the quality maintenance in the field, management would take every opportunity to visit the teams in the field and examine progress. Throughout the duration of the Kaipara programme, various external and NZ BioSecure audits were carried out.

Radio communications/repeater station

Effective communications while operating around the vast area of the Kaipara were problematic and a concern for management from the outset. None of the major cell-phone networks was able to provide total coverage of the eradication zone. Management believed that the health and safety of staff when working in remote locations could not be jeopardised by poor communication and, as a result, set about establishing a practical solution. It became apparent that the only suitable type of radio over such distances was VHF. Icom radios were fitted to all vehicles and this method was compatible with the helicopter communications. Every team member was trained in the use of the equipment and radio procedure. The administration desk back at base acted as the communications nerve centre and there was constant radio traffic as information was relayed back and forth.

VHF radios operate by line of sight, and unfortunately the terrain in places was severe and communications suffered. To remedy this problem, the installation of a 'repeater station' in Woodhill forest certainly helped and, in cases where the signal was weak, one of the helicopters would be tasked to re broadcast any messages back to base. The emergency services for the Kaipara region were aware of this communications capability and close liaisons were formed with yet another organised group within the community.

Daily briefings and debriefings

Before each field team deployed each morning soon after daybreak, they would be briefed by their team leader as to the location of their activities that day. The team leader would have received his instructions from the team coordinator the previous evening. These briefings would specify the activities to be undertaken, mode of transport, any particular equipment requirements and timings for return to base. Briefings were conducted while staff sipped hot coffee and devoured slices of hot toast.

Operational debriefings involved all staff and took place at the end of each working day as part of the daily routine. This was held in the main meeting room equipped with overhead projector and class seating arrangement. With teams already having downloaded their GPS tracks, everyone was able to follow exactly where ground treatment or sampling surveillance had been undertaken while being able to listen to a commentary from those involved. Each team nominated a representative to deliver their report. Everybody was rostered to take their turn and, over time, they all grew in confidence and became proficient at public speaking. This holistic approach meant that everyone in the programme, including management, was up to date with what had occurred but also allowed input from those being briefed. Tracked flight paths of the aerial treatments were able to be viewed to see if any areas had been missed that would result in further treatments being carried out aerially or by a ground crew.

These daily debriefings also provided an opportunity for management and administrative staff to pass on HR information and feedback from field staff. When the debriefing was completed, team leaders would meet with the coordinator to prepare for team and individual tasks and discuss the weather, tides and any administration for the next day.

On occasions, an urgent situation arose and all staff would be forewarned of the operational outline to meet for a confirmatory briefing so deployment could occur at first light the very next day. This action would take precedence over routine debriefings.

Reporting

Recording the finer details of daily operations played a large part in the programme's success. With all field staff maintaining daily activity diaries, as well as teams completing the 'daily activity log', no detail went unrecorded. This ensured that weekly and monthly reports sent to the Regional Manager and Programme Operations Manager were accurate.

Numerous activities at many levels were reported and this information was a routine function for management. Stocktaking, product reconciliation and budget reports had significant importance for the MoH. Stocktaking had to be thorough and our technician Tash Symonds, an ex Navy Storeman, was a great asset and proved invaluable. The most expensive items on site were the treatment products S-methoprene and *Bti*, so product reconciliation was undertaken by Shaun Maclaren and Steve Crarer.

Reporting to the media was the responsibility of Bryn Gradwell, who proactively produced operational updates using the MoH sanctioned communications strategy (see Chapter 12). An established communication link between the Auckland and the Kaipara community newspapers and the Operations Manager ensured media reports were accurate, key messages were retained and the trust remained.

Reporting to the stakeholders was a shared responsibility by management and senior staff who produced newsletters for the landowners and interested organisations/agencies, and more formal reports to the consenting authorities and affected councils.

Exception reporting occurred as and when required, and covered larval/adult finds in new areas, as well as any injury or near miss events as required by health and safety legislation.

Equipment and vehicles

The procurement of good equipment and sound, reliable vehicles to undertake the activities and the challenges ahead for the Kaipara MRC was an expensive and thought-provoking exercise. Lessons had been learnt from the east coast operations, so the right equipment and best available vehicles for the job were obtained. Work was routinely conducted in a hostile, aggressive saline environment and this was known to shorten the life of vehicles and equipment considerably. A constant programme of cleaning, maintenance and servicing was the answer. Each field team had:

- a twin-cab four-wheel-drive utility, with trailer and quad bike
- a track-driven Argo for swamp access
- tandem trailers for product carriage or multiple all-terrain vehicle transport
- numerous backpack sprayers and granule spreading blowers
- all-terrain-vehicle-mounted sprayers and spreaders
- a pellet blower for mounting on the helicopter.

Equipment and vehicle maintenance

Routine checks, including running maintenance, were carried out by field staff. Designated staff members with additional mechanical knowledge were tasked to conduct minor repairs and keep an eye on servicing requirements. Daily checks or 'first parades' for all vehicles, as set out in the operating procedures, were carried out by a team member before departure every morning. Regular servicing and major repairs were undertaken by local mechanics. As mentioned previously, part of achieving a successful outcome for the programme was by gaining local support and involvement. Although the majority of vehicles and equipment originated or was purchased through Napier saleyards, it was considered an advantage to have them serviced and maintained by local agents and their mechanics in Helensville. As the programme progressed, purchases for replacement equipment were done locally.

Wash point design and implementation

To avoid any transmission of *Campto* eggs or larvae from the infested locations within the eradication zone, there was a requirement and standard operating procedure for all vehicles and trailers to be cleared and cleaned of any mud or residue.

The 'brains trust' consisting of TO Dean Singe and his sampling team were given 72 h in which to research and produce operational plans for a drive-through wash point. Alas, much of the equipment would have to be imported from overseas and the time and costs involved would have been impractical. Back to square one and there was still no effective way of cleaning/clearing the vehicles. A local drainage and pump mechanic recommended that we adopt the same wash down operation as seen on any dairy farm. It would be effective, easy to build and maintain, and a fraction of the cost of the original proposal. All water and dirt would run off into a holding pond that could be treated with S-methoprene and included in the treatment schedule. This extremely proficient method of thoroughly cleaning the vehicles was additionally beneficial for maintaining the carriage and body work, which was often corroded by the hostile saline terrain.

Visitors who came to observe the activities in the field were also required to use this wash point facility upon their return to base and before departure to their home locations.

Landowner liaison

From the outset, it was all too clear that successful eradication could only be achieved if landowners and farmers occupying the land were supportive of the programme and its goals, and allowed unhindered access by foot, vehicle and helicopter to coastline areas of their properties. We encouraged staff to establish good working relationships with landowners and occupiers at the earliest opportunity, and many developed ongoing friendships allowing access to land for fishing and hunting. An adopted telecommunications protocol warned the farmers 72 h in advance of intended aerial treatments or visits from the field teams, allowing them to move stock should they wish. Adam Mason, TA recalls:

> *'We had had a lot of rain lately so it was very wet, we got through the low alright, then we found ourselves at the base of a hill with a very sorry looking raceway to get up. We got about half way up then, and what do you know, the wheels started spinning and we were stuck. We locked up the truck and started to walk out; we knew it was going to be a long walk to anywhere because of how high the hill was. The track was slippery as! We got out after what felt like a lifetime in the sticky mud. We found a farmer to pull us out; it just cost a box of the good old universal currency – BEER!'*

Troubleshooting complaints

Complaints were few and infrequent, and those that occurred were in the early months of the programme when relationships, trust and respect were still to be earned and the activities in the area were new and uncertainty existed.

When they did occur, it was important to respond by meeting with those affected at the earliest and most suitable time for them. Dealing with an issue caused by the programme could never be dealt with over the telephone; neither was shying away an option because it only delayed the inevitable confrontation and possible media attention. A plan was devised and adhered to, which met with approval from the MoH. Complaints were always handled personally and resolved by either Bryn or Shaun. The complaints we did receive were either made by visitors or newcomers to the Kaipara who had no knowledge of the eradication programme in progress.

Weather patterns

The Kaipara Harbour was unique in that high tidal events affected not just one coastline but the four coastlines demarcating the harbour. This meant that a mixture of experience and local knowledge gained from farmers and landowners, as well as preparedness, allowed for surveillance operations to be organised in a way that allowed us react in a controlled manner. Tide charts, local meteorological service reports and internet weather forecasts such as 'Buoyweather' were all used when preparing operational plans.

Much of the surveillance undertaken was able to be planned in advance due to the fact that mosquito hatches around the majority of the Kaipara would occur as a result of expected extra high tides, such as spring and king tides. Routine high tides with a strong onshore wind would frequently have the same result as a spring or king tide.

A surprising factor was that, due to the size of the harbour, it would frequently result in different weather events, such as heavy rain that would cross the harbour or, in some cases, differing wind directions. This meant that operations such as aerial treatments were planned to be flexible, with a recognised back up plan. Swift communications between pilot and operational control allowed all parties to be made aware of any changes to the treatment plan.

Kaipara treatment plan

Ongoing intensive surveillance (Fig. 8.4) was used to identify the extent of the treatment blocks. The plan was the same as for the east coast of the North Island; that is: (1) maintaining an effective lethal concentration of S-methoprene in all identified positive habitats for at least two treatments; and (2) then maintaining an effective lethal concentration of S-methoprene in all inundated habitats for two summers and for three water events. This involved aerial applications of XR-G granules every 21 days starting in September 2002, once sufficient stocks of S-methoprene had been obtained.

The treatment plan was an ongoing, rolling, zone-based approach, as was used in the successful east coast programmes (see Chapter 5). The objective of the treatment programme was to maintain a lethal concentration of S-methoprene in all inundated habitat where *Campto* had been detected, so some flexibility was inherent in programme design and the programme would often move to event-based treatment of wet habitat, depending on the results of surveillance. S-methoprene was applied as pellets to drains and deep habitat on a 30-day cycle or XR-G granules for other habitat on a 21-day cycle. Delivery

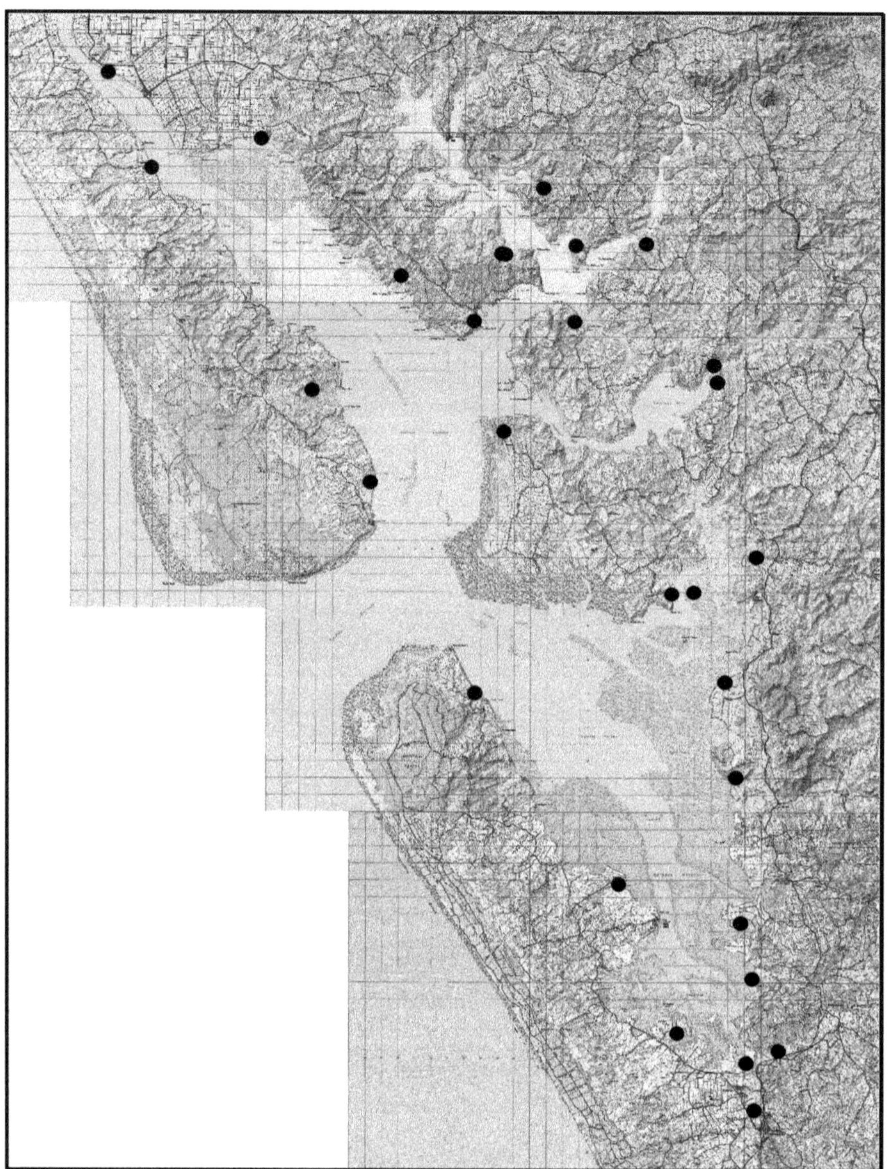

Figure 8.4 Sentinel site locations around the Kaipara Harbour.

platforms included aerial (helicopter) and ground-based equipment. A unique aspect was the artificial flooding of farm drains with salt water to control weed growth. This would induce a hatch of *Campto* larvae, which were then treated with S-methoprene. This process was considered useful in speeding up the elimination of *Campto* from these drains.

Bti was to be used to reduce the population of *Campto* further if sufficient stocks of S-methoprene were not available and to cope with extensive water events to supplement the use of S-methoprene in known positive sites.

Although 2710 ha of potential habitat had been identified around the Kaipara Harbour, it did not all become inundated at the same time. The actual size depended on whether it was low or high tide, or somewhere in between. Size was also influenced by varying topography and climatic conditions at different points in the harbour. Usually 1250 ha per month required treatment with S-methoprene.

This treatment regimen should have, over time, prevented new adult emergence from occurring by killing larvae and/or preventing successful pupation. It was anticipated that, within this time period, all viable eggs that had become wet would have hatched and those that had remained dry would no longer be viable. Extending this over two summers was consistent with the Napier eradication programme and significantly increased the likelihood of successfully eradicating the species.

Treatment efficacy was assessed through the surveillance programme (see Chapter 9), but also through additional field collections of pupae to monitor successful adult emergence (or, in our case, unsuccessful adult emergence).

Aerial applications of S-methoprene commenced in September 2002 and were planned to continue until March or April 2004, but were not completed until June 2004.

Sentinel monitoring

The Kaipara sentinel programme evolved out of the Hawke's Bay sentinel programme. Sites were selected in January and February 2003, based on site visits, regional location, positive and negative sites and overall habitat suitability. The team needed a better understanding of the effect of rainfall with respect to whether water pooled or had little impact, and how much it varied across the Kaipara region. Numerous sentinel sites contained larvae and thus were being regularly treated. Twice weekly surveillance helped build up a record of treatment effects and whether there was a gradual reduction in immature biomass, and presumably *Campto* eggs lying dormant in the mud and saltmarsh vegetation

The sentinel data were entered onto their own Excel spreadsheets to enable tables and charts to be generated showing potential influences at each sentinel.

There were 28 sites initially, with the maximum being 34 (Fig. 8.4). Each sentinel site had a permanent light trap and these sites generally continued to hold water during the summer months. A procedure for sampling and assessing each sentinel site twice a week was documented under the developing NZ BioSecure Procedures Manual. Each site was marked with a numbered stake and visited twice weekly to read rainfall and temperature. Salinity was recorded from the nearest tidal station. Photographs were taken at the sentinels on a regular basis and surveillance teams were able to discuss changes and observations routinely during daily debriefings.

Tide data from the Auckland Regional Council and National Institute of Water and Atmosphere station based in Pouto gave us an actual and a predicted tide height for the northern Kaipara regions. From experience, applying the predicted tide data with the actuals showed very little visible differences. Storm surge indicators from meteorological services were also monitored. The ongoing surveillance review for localities included salinity and water temperature at sentinel sites.

Trigger event surveillance was carried out in response to rain events and storm surge events that combined rain and tide. Surveillance in areas following trigger events was routine and standard practice. How to tie these environmental data into the day-to-day operations and make them readily accessible to field staff and, in relation to a site and property, was one of the aspects of the People Data database that was initially developed in the Kaipara programme (see Chapter 9).

Killing *Campto*

Prior to the S-methoprene treatments commencing in October 2002, there had been 6 months of surveillance, monitoring and treatment with *Bti* as per the control and containment protocol. From April to September 2002, Bryn and Regan, the two NZ BioSecure full-time staff, carried out 11 571 dips – finding 293 *Campto* larvae and catching 46 *Campto* adults from 800 trap-nights. From October, a fully trained team was operating from the new base at Parakai. In the 3 months October to December, the team carried out 23 341 dips, finding 854 *Campto* larvae, and caught 37 adults in 460 trap-nights.

During January–June 2003, surveillance around the Kaipara intensified and 93 470 dips found 2433 *Campto* larvae and 1383 adults were caught in 4627 trap-nights. But, over the next 6 months, the effect of the programmed S-methoprene treatments was evident. From July to December 2003, the team conducted 110 568 dips, finding 731 *Campto* larvae, and undertook 6893 trap-nights, catching only 2 *Campto* adults, with the last one found in September 2003.

During the first 3 months of 2004, 42 530 dips were carried out, with only five *Campto* larvae found and no adults were caught in 3122 trap-nights. For the next 9 months of 2004, there were 168 653 dips, with no *Campto* larvae and no adults found in 7679 trap-nights. The same intensity of surveillance continued until June 2005, with no *Campto* larvae or adults found. In July, the TAG had approved a recommendation to move from sentinel surveillance to post-inundation surveillance in areas of the Kaipara that had not been positive for *Campto* for at least 2 years. Sentinel surveillance continued at 14 sites and these were checked every 4 days. Additional traps were deployed and enhanced larval surveillance undertaken post-inundation.

In summary, the last adult was detected at Kaipara Harbour in September 2003 and no larvae were found after February 2004. Treatment of habitat was completed in June 2004 and, if no further *Campto* were detected, the Kaipara eradication programme would have been completed in February 2006. This was not to be ... after 20 months with no *Campto*, it again appeared in November 2005.

Campto raises its ugly head

On hearing the news, Bryn Gradwell said:

> *'Most people can recall where they were or what they were doing and how they felt when they heard news of the 9/11 attack on the World Trade Center, of the assassination of JFK or death of Princess Diana; I remember the gut wrenching feeling when my cellphone relayed a message that an adult* Campto *had been found in a trap at Kaipara. I was on leave watching my granddaughter competing in the school cross-country race.'*

No larvae or adults had been detected since February 2004 when [what appeared to be] the last larva had been found. Now an adult was detected in a trap at the Carter's Bach sentinel site. Immediately, delimiting was undertaken in a 5 km perimeter of the trap and was then extended throughout the southern Kaipara, but no larvae were found. The surveillance programme (including the sentinel traps) operated continuously but no further detections (trap catches or larval finds) were ever made at Carter's Bach.

Then, on 9 December 2005, two fourth instars were found at a site at Jordan's Drains during enhanced surveillance following the Carters Bach finding and heavy rainfall in the area, which would have prompted any surviving mosquito eggs to hatch. The site, along with adjoining drains and drains leading to the northern end of Jordan's Lake, were treated with S-methoprene. Delimiting was undertaken for a range of 3 km and additional adult traps were installed.

On 19 December 2005, a female adult *Campto* was detected in one of the sentinel light traps in this area. Further treatment of all drains and surrounding properties to a distance of 3 km was undertaken at Jordan's Drains. This treatment regime covered 160 ha of potential habitat and continued every 21 days for a further 6 months, requiring nine treatments in total.

Both Carter's Bach and Jordan's Drains were areas that had previously been heavily colonised and Jordan's had been an area of the most recent findings before these latest results. We do not think that these few *Campto* came from undetected cryptic sites because surveillance continued following treatment. Rather, they may have been the tail end of instalment hatching, suggesting possibly that drought-resistant eggs may remain viable for up to 36 months because, in this case, the new detection of larvae had occurred as long as 32 months after the last adult detection in the Kaipara.

Enhanced surveillance was implemented in what was considered the 10 sites of greatest risk, in addition to the existing ongoing surveillance activities in the Kaipara. Throughout the first quarter of 2006, no *Campto* larvae or adults were detected but, in April 2006, 19 larvae were found and four adults were trapped. On May 8, more larvae were found and in July, a further five larvae found.

No further *Campto* adults were trapped after the April catch and no further larvae found after July 2006. The 160 ha of habitat around Jordan's Drains received treatment until June 2006, but the MoH had applied for additional funding for ongoing treatments to a wider area of the southern Kaipara (Gilbert 2006). On 1 July 2006, the eradication programme was transferred to the MAF and additional funding provided for ongoing treatments to an 850 ha area of the southern Kaipara.

In August 2007, NZ BioSecure reported to a newly constituted TAG that 16 sentinel light traps had operated continuously for 6 months previously around the southern Kaipara and no *Campto* adults had been trapped. All sentinel habitat water levels had been low until July 2007 when the Kaipara had been subjected to two storm surges causing an estimated 75% of all habitat to be inundated. In July, the area treated was 977 ha, compared with 683 ha for the treatment in May 2007. Between March and July, there had been 19 675 dips and 5455 mosquito larvae found, but no *Campto*!

The death knell

In October 2007, the MAF changed contractors, replacing NZ BioSecure with Flybusters Consultants. Treatments in the southern Kaipara continued until the final application was made on 17 April 2008. It had been recommended, and agreed by the TAG, that intensive surveillance of the entire eradication zone should follow until August 2008. The criteria for eradication were met. The last *Campto* larvae were detected in July 2006 – two summers had passed and 20 months since the last life stage was found, signalling the completion of the Kaipara eradication programme and the elimination of *Campto* from the Kaipara Harbour.

Figure 8.5 'Happy chappies': (from left) Adrian Brocas (TA), Rebecca Brokenshire (TA), Chris Gaskill (Admin. Officer), Monica Singe (Technical Manager), Adam Mason (TA), Alice Petuha (TA), Regan Courtney (TO), Bryn Gradwell (Operations Manager), Andrew Macdonald (TO Information Systems), James Wakeford (TO), Dean Singe (TO), Shaun Maclaren (Logistics Coordinator), Mark Dellow (TA), Frank (HHB Pilot), Paul Green (TA), Eliane Lagnaz (TA) Daniel Rowntree (TA), Gareth Southcombe (TO), Adam Kellian (TA) and Joelle McCormick (TA). (Image: Tash Symonds)

Bryn Gradwell, Operations Manager 2002–2007, said:

> *'I know that all the staff (Fig. 8.5) who had been involved in this programme felt a huge sense of pride and satisfaction when the removal of all* Campto *from the Kaipara had been declared.'*

After August 2008, the whole area of the Kaipara Harbour was placed under the watchful eye of the National Surveillance Programme.

Acknowledgements

We are very grateful to Mr Hally Toia of Te Runanga O Ngati Whatua, Mr Waata Richards of Huranui Marae and Mr Wikiriwhi Hetaraka of Te Uri O Hau Environment Operations, all of whom were instrumental in providing assistance to the MoH and NZ BioSecure during the resource consent consultative process.

References

Duffy J (1977) *Ventures in World Health. The Memoirs of Fred Lowe Soper.* Pan American Health Organization, Washington DC.

Gilbert S (2001) Government Decision on the Long-term Response to the Southern Saltmarsh Mosquito Incursions. Ministry of Health, Wellington, NZ.
Gilbert S (2006) 'Kaipara Eradication Plan Revised, December 2005'. Chief Technical Officer, Ministry of Health, Wellington, NZ.
McGinn D, Garner S (2001) 'Briefing paper on the Kaipara Harbour *Ae. camptorhynchus* situation'. Ministry of Health, Wellington, NZ.

9

Developing detection and surveillance

Graham Mackereth, Monica Singe, Mark Disbury, Sally Gilbert, Noel Watson, Craig Williams, Rachel Cane and Scott Ritchie

This is not New Zealand!

Arriving at the edge of the Ahuriri lagoon on a warm summer's day in the weeks following the detection of *Campto*, Graham Mackereth took in the view – the green hills and pine plantations running down to the estuary, sheep, cattle and a few horses grazing, across the estuary a small plane landing at the airport, the estuary continuing out to sea under the rail and road bridges that mark the edge of Napier city. All classical NZ, until he got out of the car, when a vicious aerial assault started and he was forced to move quickly. His first thoughts were 'This is not New Zealand! We don't have mosquitoes like this!'

Graham Mackereth was collecting blood from possums that had been trapped along the edge of the lagoon as part of a survey for Ross River virus. The possum blood, along with blood from horses and cattle, would be sent to Australia for testing. We wanted to be sure the virus had not arrived with the mosquito. Each trapped possum was anaesthetised among buzzing invaders, the blood drawn and then the possum was dispatched because they are a noxious pest in NZ. As he drew the blood, the mosquitoes would line up on the possum's eyelids and feed, so it was a race! Graham took a photograph (Fig. 9.1), proving that possums were hosts for this mosquito and could be an important part of the Ross River virus cycle, should it establish.

Back at the car, he packaged the blood samples for the first stage of their journey to Australia. It did not take long, but it was time enough for a few mosquitoes to invade the vehicle, bringing with them the first forebodings as to how far they may have spread.

Initial concerns that Ross River virus and other arboviruses may have established at Napier were allayed by negative surveys in animals and people. Figure 9.2 was taken during the possum survey. The MoH surveyed 53 people from the Hawke's Bay region who had reported being bitten. Twenty people were re-tested to detect sero-conversion to Ross River virus, but all samples tested negative for antibodies.

The only effect of the incursion was the vicious biting. It was unprecedented. Horses near the lagoon wore their covers in the heat of summer; their owners were trying to protect them from the nuisance biting. There were thousands of animals and people in the flight range of this mosquito. Ross River virus was not introduced with the mosquito, but we all became infected with the resolve to get rid of this exotic nuisance.

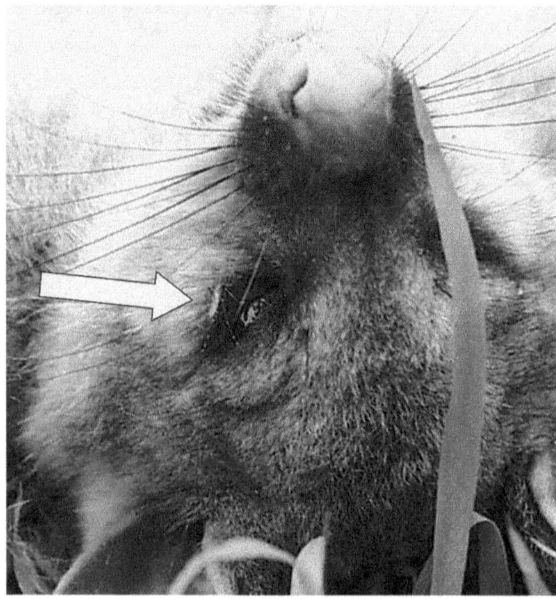

Figure 9.1 *Campto* feeding on the eyelid of an anaesthetised possum during a survey for Ross River virus, Napier 1999. (Image: Graham Mackereth)

Figure 9.2 Livestock Officer Tony Harris holds a possum in preparation for anaesthesia and blood sampling, Napier 1999. (Image: Graham Mackereth)

Napier infested

The Ahuriri Estuary at Napier was well and truly infested at the time of first detection. Vicious daytime mosquito biting is rare in NZ and unusual enough for people to report it. In December 1998, there were a lot of mosquitoes at Westshore Primary school biting the children and staff during the day. On 21 December, the caretaker caught some and took them to the Napier City Council who in turn took them to Noel Watson, a Health Protection Officer at the Hawke's Bay Public Health Unit. Noel oversaw the exotic mosquito surveillance programme at the port of Napier and sent them by courier to the MoH-contracted entomologist Gene Browne in Auckland. In the congested pre-Christmas courier system, the package was lost to science. On 23 December, Noel collected more samples from the Napier Aero Club: it did not take long, because they chased him back to his car!

The mosquito was identified as *Aedes camptorhynchus* by Gene on Christmas Eve.

On 28 December, the initial delimiting survey started and the team assembled immediately identified that the mosquito was well established on the strip of land raised by the 1931 earthquake on the northern edge of Napier city. Subsequent surveys found the mosquito further up and down the Hawke's Bay coast. At the time of first detection, *Campto* was well established and was already causing nuisance. How did *Campto* reach such high numbers without earlier detection?

In this chapter, we describe the surveillance for exotic saltmarsh mosquitoes before and after the *Campto* incursion. The art and science of surveillance for saltmarsh mosquitoes in saltmarsh habitat took many years to reach maturity on a national scale. It is a story of a rise from patchy regional initiatives to consistent systematic scientific coordinated national surveillance.

Surveillance before 1999

John Handiside and Mark Bullians conducted a national surveillance assessment for the Health Funding Authority (Handiside and Bullians 1999) to determine if *Campto* was more widespread than Napier – important information for the decision to eradicate at Napier. The assessment included some sampling as well as review of surveillance activity in each region. They concluded that, with the exception of Northland and Waikato, there was an acceptable level of confidence that *Campto* was not present across the country. They noted that, at the time of their report, Northland, despite being considered most at risk of incursion, had not had surveillance since Dr Marshall Laird's survey during the 1980s–1990s (see Chapter 1).

To bring surveillance into line with MoH guidelines, they estimated that an additional \$630 000 per annum was needed, with a one-off investment of \$150 000 for equipment, software and training. They noted the need for salinity meters, light traps, training and staff resources, pointing out that some regions had not undertaken any surveillance due to other workloads.

Although the report indicated that there was much to be done in surveillance, NZ authorities were aware of the risk of exotic mosquitoes and were working to exclude their establishment and develop plans to manage incursions. Along with its biosecurity partners, the MoH had prepared for an eradication response. The 1997 Kay Review, commissioned by the MoH, informed the training for local public health staff on mosquito exclusion and surveillance techniques. However, new risk-based national surveillance arrangements had only been in place for a short time before the discovery of *Campto* in Napier. The identification and confirmation of *Campto* in NZ resulted from the recognition of the risk of establishment by Henry Dowler and the MoH, and the subsequent training of the Health Protection Officer and funding of Gene Browne to provide the mosquito identification service.

The early days at Napier

Noel Watson's description of surveillance in the early days highlighted how much there was to learn. He described the surveillance learning process as being from 'headless chickens to seasoned swamp bunnies':

> *'We started without the knowledge, technology and resources that have since been developed and refined through our early experiences. Each week in January 1999, a new group of Health Protection Officers (HPOs) from around the country arrived at Napier to help with surveillance. They were enthusiastic but, like us, lacked experience in finding mosquitoes in coastal swamps and drains. It was a steep learning curve. Most of the HPOs only stayed 1 week, by Wednesday they were getting into the routine, by Friday they had gone. Fridays were spent trying to arrange more HPOs for the following week.*
>
> *'I split the surveillance area into zones and made up larval sampling forms with a copy of the topographical map on the back. When the samplers found larvae, they would mark it down on the map. This was not ideal and some had real difficulty recording where they got the sample from.*
>
> *'We were going across a lot of private land at time looking for larvae and although we sent information out to landowners and there was publicity in the newspaper, we were never sure always exactly where one person's boundary ended and another's started. We used to tell the guys to always wear hi-viz gear and have ID on them. In that way, they would not look like someone casing the place out, up to no good. I also used to advise them that if the owner did decide to shoot them, he would get a good target to aim at so hopefully they would get a quick, efficient and painless death ... and the police could quickly identify them from their ID.*
>
> *'The other trick was I told them was to always wear a hat. This was not only being sun-smart, but some of the swamps were deep and you could sink down into them. If they wore a hat, that would help us tell where they went under.*
>
> *'We had a master map on the wall of the main debrief room and pins were used to show where* Campto *had been found. Eventually the map became so covered in pins that you could not see the map! The MoH provided a copy of the GIS system ArcView 3.*
>
> *'Another issue we had was a lack of 4 wheel drive vehicles and people experienced with driving them. We used the 2 wheel drive hospital pool vehicles, and once you got them off road, you were in trouble. One day, a helicopter ground crew were running out of* Bti*, so we had to deliver some by a car pool vehicle. The landing zone was right at the back of the lagoon farm and there had been significant rain, so the track was muddy. To prevent getting stuck, we went at fast as we safely could. This was working well*

until the car slid off the track into a fence post and bent it over. We pushed the car out and delivered the Bti. *On the way back, we slid into the same post, pushing it back into the correct vertical position!*

'It was easy to create Campto *habitat with vehicle tyre track ruts. We had to learn to be careful in some areas and either walk it or use our four-wheel quad bikes that did not make deep wheel ruts.*

'We did not have any adult traps in NZ at the time of the discovery of the Campto *mosquito. Sometime in early 1999, we got traps from Mike Lindsay in Perth. The power source was two D-sized batteries to run the fan motors. The fans and motors were converted aeroplane model motors and propellers. If you put the batteries in the wrong way around, the fan would run the other way, and actually keep the mosquitoes out instead of sucking them into the collection bags. The container holding the dry ice was a paint-tin with moulded insulation material inside it. I had to check the traps every day to change the D-sized batteries and replace the dry ice. It took the majority of the day to drive around the 12 traps, in a big arc to the west of the main area just north of Napier.*

'Later we got better adult traps that were powered by 12 volt car batteries and had computer cooling fans, and CO_2 gas cylinders instead of dry ice. This meant we did not need to check them daily.'

The rise of entomology laboratory services

Following the *Campto* find at Napier, the MoH's consultant entomologist, Gene Browne, provided an identification service based at the Napier MRC. Jessica Taylor was hired as a research assistant. Among other things, Gene and Jess tested the effects of S-methoprene on various non-target aquatic fauna to see what impact the product may have on the environment. Eventually Jess took over the identifications for the saltmarsh eradication programme. Jess then trained Mark Disbury in identification of NZ species. The laboratory was for a short time shifted to the spare bedroom of Mark Disbury's house in Napier.

In 2001, a new building was found and a laboratory set up with nice white benches, quarantine area and double doors. From this laboratory, Jess and Mark provided an identification service for mosquitoes intercepted at the border and from the *Campto* programme. Any suspected exotic species were referred to Richard Russell in Australia. For confirmatory identifications, specimens were photographed and the photos emailed to Richard for immediate confirmation. The actual specimen would be sent by overnight courier to Sydney for final confirmation.

As *Campto* was found in different parts of the country, the laboratory grew and became recognised as the NZ Biosecure Entomology Laboratory or NZBEL. Specimens were being collected from Gisborne, Porangahau and Mahia. When the Kaipara population was discovered, it became apparent that more field staff would be required. In 2002, Jess left the laboratory and NZ BioSecure employed three people to meet the increasing demand for identification services. Joanne Christie was also employed to produce a quality management system, incorporating standard operating procedures as well as training and orientation guidelines.

A purpose-built laboratory was established by NZ BioSecure in Napier in 2003. The new laboratory accommodated all four staff at separate work stations simultaneously, as well as housing the pinned and slide-mounted preserved collections. Additionally, new offices were provided, making the electronic and paperwork components of the work much easier.

The laboratory orientation and training procedures were enhanced by the establishment of an annual training course provided by Darryl McGinn. Each year, the course was tailored to fit the explicit needs, such as permanent mounting of important specimens – larvae and adults were the main focus one year and the use of new Australian identification keys and specimens to aid in the identification of exotic species was the focus the next year.

In October 2005, NZBEL services were provided from Lower Hutt. The new laboratory was bigger than previous laboratories and had a separate invertebrate-containment facility. Through the *Campto* programme, real laboratory capability in NZ was developed – moving from a spare bedroom to a modern laboratory with appropriate equipment and procedures.

From 2007 to 2010, identification services for eradication zones were provided by FBA Consulting, a technical division of Flybusters Insect Control Ltd, from their Glenfield laboratory.

The rise of a national database

More than 10 000 individual sample forms per year were incorporated into a database at the laboratory. The data were used to provide mapping for surveillance activities, as well as to report on surveillance effort by region. From the outset, data submitted with samples from all around NZ were not standardised and often incomplete. There was the problem of the manual entry of 10 000 14-digit GPS coordinates already transcribed from the GPS units. The time required to enter data, and the associated recording and entry errors, prompted improvements to the system. What was needed was a web-based database that enabled field staff to enter their own data, standardised the data entered and enforced completion of mandatory fields. Also desirable was an electronic device to record sampling information in the field and automatically record information such as date, time and GPS coordinates.

In 2003, the MoH agreed on the need for a database and hand-held data capture devices. One aim of the system was that it would automatically flag delays in receipt and identification of specimens. The need for this was underlined by an embarrassing delay in the identification of the index *Campto* larvae from Murawai near Gisborne collected in July 2000 and, due to unfortunate confusion over delivery arrangements, not examined until 4 October 2000! In July 2004, the database went live and is still in operation today. The hand-held data capture device was not successful at the time, because this technology was not at that time advanced enough for successful deployment.

Surveillance at Napier: is eradication working?

In eradication zones, surveillance was used to determine where to apply treatments and monitor the success of treatments. The effectiveness of eradication at Napier was evident by the steady reduction of adult female *Campto* found in light traps during 1999 and up to February 2000, when, of adult female mosquitoes collected, only one of nine egg-bearing females displayed normal ovarian development. Figure 9.3 shows monthly data for the

Month/ Number	Jan-99	Feb-99	Mar-99	Apr-99	May-99	Jun-99	Jul-99	Aug-99	Sep-99	Oct-99	Nov-99	Dec-99	Jan-00	Feb-00
Adults	40	1715	767	444	10	1	7	2	5	11	11	30	12	1
Light traps	9	10	10	9	8	6	5	7	7	33	36	36	36	36
Larvae	123	253	501	769	526	258	139	4	7	32	38	259	16	27
BTI														
S Methoprene														

Month/ Number	Mar-00	Apr-00	May-00	Jun-00	Jul-00	Aug-00	Sep-00	Oct-00	Nov-00	Dec-00	Jan-01	Feb-01	Mar-01	Apr-01
Adults	0	4	0	0	0	0	0	0	0	0	0	0	0	0
Light traps	36	36	36	40	40	28	28	28	28	28	12	12	12	12
Larvae	63	99	0	8	4	0	0	0	0	0	0	0	0	0
BTI														
S Methoprene														

Figure 9.3 Timeline showing adult and larval *Campto* numbers collected in relation to sampling effort, and months during which S-methoprene and *Bti* were used at Napier.

numbers of adults and larvae found, and the control and surveillance methods being applied. An expert review committee, which met in Napier to review eradication progress in February 2000, concluded that, since the start of the programme, there had been a 99.96% reduction in the adult saltmarsh mosquito population. Larval populations showed a similar decline in both numbers and positive sites. Adult and larval numbers dropped rapidly in the first 5 months of the programme. Flooding 10 months into the programme resulted in the hatching of eggs (presumably laid on higher ground than previously flooded) and a small increase in trapped adults and larvae that month. The last adults were detected in April 2000 and the last larvae in July 2000. This was an astounding result for a response team that was learning as it went.

Noel Watson's early experience of surveillance at Napier was a far cry from the sophisticated surveillance used in the latter eradication zones, such as the Kaipara and Wairau. The events relating to the find at Whitford 2 years later (in March 2002) are illustrative of the progress made in surveillance and response capability.

Diana Court from the Auckland Public Health Unit called Mark Disbury, an entomologist at NZ BioSecure's laboratory, for advice on the best time to conduct a survey in the Whitford area. Based on rainfall and tides, she ended up going out on a Monday. Specimens were collected and sent to the laboratory, and they arrived early on the Wednesday morning and were examined straight away. They were *Campto* larvae!

By Wednesday evening, the plan to conduct delimit sampling at the site was ready. On Thursday morning, response staff conducted delimiting surveillance from the town of Miranda through to the Waitemata Harbour, while Brian Kay and several Auckland public health unit staff conducted a survey of the index site and surrounding area.

By the end of Thursday, delimiting surveillance was completed, with the exception of offshore islands, with all specimens identified. Permission to treat the site had been organised and media releases made. On Friday morning, the treatment was conducted and the offshore islands were surveyed. The whole job was completed by early afternoon and ongoing surveillance and treatments were scheduled.

National saltmarsh surveillance: a paradigm shift

Surveillance in eradication zones was so intricately linked with treatments that it was managed by the eradication teams. Surveillance outside of eradication zones was carried out on a regional basis by public health unit staff, overseen by the MoH. However, there were issues and difficulties.

Saltmarsh surveillance was added to existing port-based container-breeding surveillance responsibilities in early 1999, but no new resources were made available to undertake this additional work. There was a lack of time and resources available to survey in remote and often inhospitable areas, and, to be carried out effectively, the programmes required a good understanding of the biology of the mosquitoes.

Health protection officers from public health units around the country were accustomed to undertaking port-based surveillance for exotic container-breeding mosquitoes; this formed a small part of their many duties. Surveillance for mosquitoes in saltmarsh habitat required a paradigm shift that was difficult to make and accommodate with other work.

Staff needed to know what suitable habitat was and where it was. Sampling had to occur an appropriate time after rainfall and/or tidal events had induced a hatch of eggs. Surveillance had to occur in places that were difficult to get to. Four-wheel drive vehicles and bikes, and sometimes helicopters, were needed. Larvae had to be hunted down and ladled up without too much disturbance, often from difficult-to-reach places. Light trapping of adults was also a logistical challenge requiring constant nursing and maintenance of the equipment.

Public health unit surveillance success

Despite the extent of the habitat to be surveyed and the lack of resources, public health units are credited with detecting infestations at Murawai near Gisborne, Porangahau south of Napier, Mahia, Kaipara Harbour, Whitford and Whangaparaoa.

They did not, however, find the heavy infestation in the Vernon lagoons at Wairau near Blenheim. This was reported by duck shooters who noticed the aggressive day-biting mosquitoes in May 2004. NZ BioSecure staff member, Steve Crarer, and wife Hope carried out surveillance by helicopter, to determine the habitat size and potential. They were pretty dumbfounded by what they saw.

The Wairau Lagoons cover ~1500 ha, with 500–600 ha saltmarsh habitat comprised mostly of *Sarcocornia* with areas of *Juncus*. It was obvious that most of the farmland bordering the lagoons had been reclaimed because there were extensive drain and runnel networks.

They quickly found *Campto*, ranging from early third to late fourth instar larvae. Adults were also present in significant numbers and aggressively attacking them. *Campto* was found south of the Wairau Lagoons, with the southernmost established population being around Lake Grassmere. All life stages of *Campto* were present, despite the fact that surveillance had been undertaken over the previous 3 years with negative results.

Growing concerns

The responsiveness of the population at Napier and subsequent sites to the control tools *Bti* and S-methoprene, and improved operational experience in their application, provided those involved in the field and in the TAG with increasing confidence in the robustness of the eradication methodology.

In fact, all sites responded well to treatment, perhaps with the exception of Wairau and Kaipara, with the cryptic second finding of *Campto* in South Kaipara. This confidence was essential for recommending eradication at new sites that were subsequently detected. However, with new sites being detected, it was increasingly challenging to communicate confidence to those who were not as close to the programme that *Campto* was not widespread.

There were many officials at the MoH working to gather technical advice and implement the government's decisions. Sally Gilbert was the one to break the news to the TAG each time a new site was found. As a member of the TAG, Graham Mackereth was privy to the emails from Sally announcing each new site and calling for a special meeting of the TAG. In his view, her steadying contribution to the government response was no small part of the reason we succeeded. The most disturbing emails were those announcing the finds in Kaipara in February 2001 and Wairau in May 2004. With these finds came questions about the effectiveness of surveillance and advisability of eradication, and some of these are recounted in Chapters 3 and 4.

In early March 2002, after the Kaipara find and the MoH recommendation for a $30 million dollar eradication proposal, Dr Barry O'Neill, Group Director, Biosecurity Authority, MAF, contacted Graham Mackereth. Based on the new finds in 'totally unrelated geographical areas', Barry considered that *Campto* must have been in NZ for some considerable time, and raised the obvious implication that populations were going to exist in other areas. Barry's concerns were widely felt. Was the mosquito already endemic? He also outlined the difficulty in understanding the national delimiting work that has been done and the sensitivity of the surveillance activity to determine the existing location of *Campto*, and questioned whether the MOH had been able to delimit the existing distribution of *Campto* with any degree of confidence.

Barry had put his finger on the nub of the problem. Although we had confidence that, once detected, we could eliminate *Campto* from a site, we understood that at the time of detection adults (and possibly eggs in translocated soil) could have been transferred elsewhere, and it could be years before new populations derived from these transfers would be detected. Continuous intensive inundation dependent surveillance was needed for at least 3 years in any habitat that may have received adults or eggs from known and unknown infested sites.

The most difficult technical question to answer in the programme was: What was the sensitivity of surveillance? Was surveillance sensitive enough to detect new sites before they themselves had transferred mosquitoes to still other sites? In our view, the development of the national surveillance programme and the associated capabilities was the second most significant achievement of the eradication effort (the first being eradication itself). The story of its development follows.

Review of national surveillance 2002

In their review, Ritchie and Russell (2002) found that the tasking of public health units with both 'container-breeder surveillance' and 'saltmarsh surveillance', each with a different rationale and methodology, was resulting in less than optimum surveillance programmes. They recognised that mosquito surveillance and control programmes were in their infancy in NZ and that many public health staff had no entomological skills before the recent invasion of *Campto*.

Thus, the review was critical of some *ad hoc* and naïve methods employed. They provided a set of guidelines based on best practice in Australia and the USA that could be adopted for use in NZ. Because *Campto* was newly established, populations were very low and finding them was literally a 'needle in the haystack' exercise. The reviewers recommended several steps be taken.

The first step was to look in the right areas. Public health units were working with a vegetation-based map of possible habitat created by Landcare Research. The map identified 619 zones with habitat suitable for the saltmarsh mosquito, involving a total area of 86 900 ha. A further 356 zones were identified as having evidence of some suitable habitat. This map delineated mangrove and saltmarsh areas throughout NZ. Although it highlighted areas of potential tidal inundation, most of the area highlighted was low marsh subject to daily tidal inundation and not suitable for saltmarsh mosquito production (Ritchie and Addison 1992). The real habitat was in saline-influenced areas just inland and included disturbed pastureland with saline drains, such as at Kaipara, and even reclaimed saltmarsh habitat with saline soil used for grazing, as seen at Wairau. The review advocated that NZ mosquito workers investigate areas adjacent to saltmarsh areas. This was done by road- and helicopter-based surveys, and mapped with GPS and GIS systems.

The next step was to look at the right times. Potential *Campto* habitat was targeted for larval surveillance after flooding events (trigger events), typically tidal flooding and/or heavy rain. Many field staff relied upon routine surveys of saltmarsh areas without careful monitoring of flooding events that could trigger a hatch. The reviewers pointed out the importance of identifying trigger events by closely monitoring the rainfall and tides, and resulting flooding, in suspected areas. This could be done using a series of rain gauges and sentinel sites where residents could be contacted to provide data and confirm flooding of habitat.

Some staff had submitted first or second instar larvae that were too young for accurate morphological identification. Sampling 2–3 days after flooding was recommended so larvae had grown large enough to be seen readily, and to take large numbers of dips along grassy margins of pools, drains and saline habitat where larvae aggregated. The reviewers strongly recommended that standardised surveillance methods be developed and implemented and called for sufficient numbers of adequately trained staff.

The third step was to establish appropriate adult trapping. Surveillance for adult *Campto* was an integral part of most regional surveillance programmes. However, trapping methods and schemes were inadequate, especially early in the *Campto* eradication programme. Various different traps were deployed at the time. The mosquito-collecting efficacy of the traps varied considerably and some traps were only suited for specific types of mosquitoes. Provision of batteries to power traps and carbon dioxide (CO_2) to attract mosquitoes was a serious logistic hurdle for many public health units.

The review also suggested use of human-bait sentinels for detecting adults. The aggressive day-biting nature of saltmarsh mosquitoes made the use of human sentinels a powerful method for detection of adults. Because NZ is not connected with mosquito-borne pathogen transmission, this was an option. Staff can cover large areas, collecting attracted mosquitoes with aspirators or sweep-nets at many sites.

Towards a national surveillance programme

The 2002 Ritchie and Russell report recommended that saltmarsh surveillance be undertaken 'towards a national uniformity of approach'. By 2004, the MoH had recognised that one of the major risks to NZ public health was from the establishment of exotic mosquitoes capable of being vectors for infectious diseases. However, border mosquito surveillance programmes did not provide sufficient confidence that saltmarsh mosquitoes were being detected and/or excluded.

As a result, the MoH sought tenders for a national surveillance programme for saltmarsh mosquito species from 1 July 2004 that would provide consistent best practice surveillance activities, and maximise opportunities for early intervention in response to saltmarsh mosquito incursions.

The MoH anticipated that the first year would focus on planning the programmes, deciding on priority areas (e.g. high-risk estuarine habitats), developing training aids and guidelines, and so on, as well as rolling out the programme. The programme would follow the MoH's guidelines for best practice for saltmarsh mosquito surveillance including:

1. identification of all habitat
2. comprehensive surveillance regimens based on, as a minimum:
   - biological cycles
   - meteorological/hydrological events
   - seasonal cycles
   - emergency triggers risk assessment
3. identification services to support the surveillance activities including, as a minimum:
   - best practice mosquito identification
   - relationship with international experts for confirmation work
   - agreed turn-around times
   - timely and clear reporting
   - contribution to the national mosquito database
   - quality assurance measures
4. an externally peer-reviewed regular audit programme.

The service provider would need to work with public health units who provided border mosquito surveillance, and with the *Campto* eradication programme, to ensure synergies of mosquito surveillance activities nationally.

The MoH's evaluation panel selected NZ BioSecure because of their skilled staff, excellent quality processes and expertise in delivering *Campto* surveillance and response services. Their proposal demonstrated an excellent understanding of services required and added innovation through the use of several modelling programmes and other software. The proposal demonstrated excellent linkages with international experts, internal audit and external review. A unique feature of the proposal was the use of Bayesian statistics to provide a statistical measure of success or failure of surveillance. The current national programme is described in Chapter 10.

The National Saltmarsh Mosquito Surveillance Programme

NZ BioSecure successfully tendered for a contract with the MoH to develop and implement the National Saltmarsh Mosquito Surveillance Programme (NSP). Bryn Gradwell oversaw operations while technical development was managed by Monica Singe. The NSP began in 2005, with modelling of potential habitat, aerial surveillance and ground surveys. Once habitat had been identified and categorised, surveillance effort was prioritised according to the assigned category. Monthly surveillance reports included total time in the field, total distance covered, numbers of larvae and adults collected from saltmarsh habitats outside the eradication zones, and the species identified.

NSP surveillance based on prior probabilities

The surveillance programme needed to recognise that saltmarsh mosquito populations do not occur at random; there are particular habitat conditions under which saltmarsh mosquitoes are more likely to exist. So, rather than random sampling of habitats, a risk-based approach was taken. In addition, it was not possible to carry out the same level and frequency of surveillance within all areas of saltmarsh habitat, particularly in those areas that are very remote or located on small offshore islands.

NSP habitat identification

At the onset of the NSP, there was no existing central database of the saltmarsh habitat present in NZ. All saltmarsh habitat needed to be identified and mapped before any effective national surveillance effort could commence. Satellite-based land-cover databases could not provide the resolution and accuracy to delineate many of the smaller habitats.

A complete national ground survey of all NZ coastline to locate all potential saltmarsh habitats would have been extremely time consuming, expensive and possibly inaccurate, given it was hard to see everything from the ground alone. Aerial surveillance was used in order to ensure all suitable habitat was identified and classified.

Between May 2005 and January 2006, we flew around the entire NZ coastline and the location of all potential habitat was recorded and photographed (Fig. 9.4) using a Hughes 500D helicopter and a Cessna 172 fixed wing aircraft. This series of photographs represented the total potential saltmarsh mosquito habitat nationally and was used as a base for an 'on the ground' habitat assessment and confirmation of habitat, a process generally referred to as 'ground-truthing'. The combination of aerial visual habitat identification, followed by ground surveillance, was the preferred method of surveillance throughout the Kaipara eradication programme.

Figure 9.4 Flight path taken during habitat identification for the national surveillance programme, May 2005 to January 2006. (Image: courtesy of Southern Monitoring Services Ltd)

Table 9.1. Likelihood categories for habitat based on their suitability score

Score	Likelihood category
13–15	Highest
10–12	Middle upper
7–9	Middle lower
4–6	Lowest
< 4	Archived

NSP habitat classification

Although field staff in eradication programmes became very astute at classifying habitat on a scale of poor through to prime or pristine habitat, habitat classification for the NSP was based on a scoring system. During ground-truthing, important information was collected regarding suitability of habitat for *Campto* infestation, including indicating vegetation types, topography, land use, accessibility, water influences and salinity values. The degree to which these parameters indicated suitable habitat was parameterised during eradication programmes.

The NSP partitioned resources preferentially to high-quality habitat, because it had a higher prior probability of being positive, and often such habitat had greater consequences if it remained undetected. Each site was scored using fixed parameters: habitat quality, proximity to port, population density, previously *Campto* positive habitat and climate suitability. Habitat size was also assessed. Each site score was totalled and placed within a likelihood category, as shown in Table 9.1.

NSP measures of surveillance

A statistical measure of success or failure of surveillance based on factors that influence the life cycle of saltmarsh mosquitoes was required. NSP adopted time spent sampling and assessing habitat as the measure of surveillance effort. This was based on track data (GPS tracks of where the surveillance personnel walked) and the habitat classification. Graham McBride of the National Institute of Water and Atmospheric Research (NIWA) provided statistical expertise. McBride and Smith (2005) applied statistical principles and developed an estimate of the number of sampling hours required per hectare. This was given by the following formula:

$$h = \frac{1}{q} \frac{\log(\beta)}{\log(1-r)}$$

where:

- h = number of sampling hours required per hectare
- β = beta; the risk of failing to detect an incursion. This risk was varied depending on the likelihood category.
- q = sampling rate – observations (actual and observed dips) per hour
- r = non-zero probability that larvae are actually present in a habitat on a particular occasion (even if only one larva is present).

The formula was used to determine the minimum surveillance hours required to be at least 95% confident that there were no exotic mosquitoes.

Surveillance operations

The country was divided into five zones. Each zone had surveillance field staff living locally and carrying out surveillance on a quarterly basis at each of the categorised sites. The annual work plan required the prescribed annual hours, derived from the formula, to be spread throughout the year and particularly at optimum surveillance times and environmental conditions.

Andrew Macdonald (Information Systems Coordinator, NZ BioSecure) developed and implemented tracking surveillance in the Kaipara in 2005 using the software Trackmaker. GPS units were used in the field to record the data, providing a real time record of surveillance or observation effort at any given site. These data were used as the measure of surveillance effort when reviewing the annual programme. In addition to track and waypoint data, field staff recorded weather conditions and took photographs and voice recordings at surveillance sites to allow the control centre staff to review the surveillance (Fig. 9.5).

The ideal technology for surveillance allows prospective data collection in the field while providing relevant historical data such as tracks, photos, voice recordings, sampling and sample results (adults and larvae), as well as environmental data and landowner details. There should also be an easy transition of field data to central data. The same view of the data should immediately be available in the office as in the field.

To integrate the data collected with operations, Andrew reviewed software options and chose a combination of three pieces of software: GPS Trackmaker, Manifold and Media Mapper. Manifold was used to overlay tracks on land maps to provide evidence of where field staff had been at each site. Media Mapper was used to take a time stamp of a photograph and voice annotation and link it to the track, allowing all data to be reviewed together.

The software was used as the basis of the NSP. Media files were processed and linked to maps, track data was applied to individual sites and time surveyed calculated and compared with the annual hours required to meet the formula requirements. Monthly and quarterly updates were supplied routinely and in response to trigger events where required.

Figure 9.5 Essential surveillance equipment included GPS and digital cameras. (Image: Andrew Macdonald)

Table 9.2. Zone and site surveillance summary 2006 and 2007

Zone	2006			2007		
	Number of categorised surveillance sites	Total hours required	Total hours spent	Number of categorised surveillance sites	Total hours required	Total hours spent
01	376	579.02	505.76	354	585.39	877.04
02	95	199.08	194.65	118	215.66	389.91
03	35	74.71	87.08	28	42.3	76.03
04	148	209.16	318.92	148	200.67	363.65
05	58	69.04	73.80	61	85.82	116.73
Total	711	1131.00	1180.21	709	1129.83	1823.35

NSP results

In 2006 and 2007, the data collected were reviewed in accordance with the statistical measure of compliance. The saltmarsh habitat sites were categorised and the surveillance hours required for the given calendar year calculated for each. The 2006 and 2007 surveillance output is summarised in Table 9.2.

During 2006, over 7000 larvae and almost 3000 adults were collected by dipping. One incursion was detected in the Coromandel in May 2006, with 117 *Campto* larvae identified. The subsequent Coromandel eradication programme delimit indicated that the population biomass was relatively low when compared with previous *Campto* eradication programme biomass numbers at the time of detection. This supported the success of the NSP objective for 'early detection' of exotic saltmarsh mosquito species. All other specimens collected were not exotic saltmarsh species.

During 2007, over 13 000 larvae and 3000 adults were captured and identified. There were no exotic saltmarsh mosquito species collected.

Adult surveillance

Adult surveillance was never considered a priority for the NSP: well-executed larval surveillance was considered to provide a higher likelihood of early detection of mosquito populations. However, the programme did carry out some adult trapping in conjunction with the larval surveillance. For example, adult traps were often deployed during summer and autumn periods when there are fewer water bodies for conducting larval sampling.

Light trapping was used in eradication zones. In 2008, Scott Ritchie was contracted to review adult mosquito trapping in the eradication zones at Kaipara, Coromandel Peninsula and Wairau. Non-standard trap designs and CO_2 delivery remained a problem (Ritchie 2008). Furthermore, many of the traps were set in exposed, windy areas, rather than in shaded, protected sites that could serve as mosquito harbourage sites, minimising trapping efficiency (Fig. 9.6).

Surveillance in the end game: are they really gone?

The decision to cease treatment in an eradication zone relied upon the 'absence' of *Campto* as indicated by surveillance and was based on assumptions concerning egg survival and

Figure 9.6 ABC light trap at Wairau. While offering a great view of the famous sauvignon blanc vineyards of Marlborough, the site is open and exposed to the sun and wind, reducing mosquito catching efficacy. (Image: Scott Ritchie)

the sensitivity of surveillance methods. The Mosquito Research Group at University of South Australia led by Craig Williams was commissioned to quantify egg survival and the sensitivity of different surveillance methods.

Campto eggs hatch in instalments (like other *Aedes* mosquitoes), and a study showed no more than 56% hatching at the first inundation and up to six inundations required to exhaust the egg bank. To quantify egg survival, *Camptos* from South Australia were captured, fed blood in the laboratory and the eggs laid were incubated at a range of temperatures and photoperiods matching those at Kaipara Harbour. Embryos were regularly examined for viability. Egg survivorship was 13% at 15 months (455 days) when the project ceased. A few eggs may last up to 2 years (Bader and Williams 2011). When combining the natural embryonic attrition rate and the probable inundation hatching events, it was calculated that three inundations over 11 months should be sufficient to exhaust an egg bank (Bader and Williams 2011). This finding was consistent with the eradication programme assumptions made back in 1999 at Napier and used at the various treatment locations thereafter. The long outlier egg survival times could be managed by extended surveillance after treatments had ceased.

The sensitivity of surveillance for very low numbers of *Campto* depended on the methods used and on the human factors associated with field operators. Three field operators from NZ went to Adelaide and were judged by the University Mosquito Research Group to be expert at *Campto* larval surveillance, although there were differences in effectiveness among operators (Bader and Williams 2009). It seemed that if *Campto* were there to be found, the operators would have found them. However, in an experiment in coastal

Figure 9.7 Research Officer Christie Bader conducting census of a pool in coastal saltmarsh near Adelaide, South Australia by exhaustive pumping. Negative *Campto* dipping results were proven to be false negatives when small numbers of larvae were pumped out immediately after. (Image: Craig Williams)

swamps near Adelaide, larval surveillance of isolated pools was conducted until a team of several operators could not detect a single larva. The pools were then pumped dry through a sieve (Fig. 9.7), which revealed a small number of *Campto* larvae still remaining! The finding confirmed the need for repeated sampling in eradication zones.

To understand the sensitivity of adult mosquito traps, *Campto* were marked and released and recaptured using a matrix of CO_2-baited EVS traps (Rohe and Fall 1979). It was found that the use of multiple traps at a site significantly increased the probability of detection. With a single trap, there was only 30% probability of detecting a cohort of 56 marked *Campto* adult females, which increased to 100% when seven traps were used (Williams *et al.* 2012). The addition of octenol (1-octen-3-ol) to such traps has been shown to increase catches of *Aedes* in Australia (Van Essen *et al.* 1994) and significantly improved *Campto* collections with EVS traps in Adelaide (Bader *et al.* 2009).

The research confirmed the need for repeated rigorous larval surveillance coupled with targeted use of CO_2- and octenol-baited traps spanning multiple inundations over at least two summers.

References

Bader CA, Williams CR (2009) A novel method of assessing larval mosquito surveillance techniques: results from a very brief field study. *Mosquito Bites in the Asia Pacific Region* **4**, 16–21.

Bader CA, Williams CR (2011) Eggs of the Australian saltmarsh mosquito, *Aedes camptorhynchus*, survive for long periods and hatch in instalments: implications for biosecurity in New Zealand. *Medical and Veterinary Entomology* **25**, 70–76. doi:10.1111/j.1365-2915.2010.00908.x

Bader CA, Williams SR, Williams CR (2009) 'Sensitivity of surveillance for southern saltmarsh mosquitoes in New Zealand: Final Report, Project 10550/2008'. Report to Biosecurity New Zealand. University of South Australia, Adelaide.

Handiside J, Bullians M (1999) '*Aedes camptorhynchus* National Surveillance Assessment'. Report to the Health Funding Authority, Wellington, NZ.

McBride G, Smith B (2005) 'Statistical aspects of Saltmarsh Mosquito Surveillance Programme'. Unpublished report for SMS Ltd. National Institute of Water and Atmospheric Research Limited, Hamilton, NZ.

Ritchie SA (2008) 'Assessment of current Southern saltmarsh mosquito adult trapping programme, 2008'. Report to Ministry of Agriculture and Forestry, Wellington, NZ.

Ritchie SA, Addison DS (1992) Oviposition preferences of *Aedes taeniorhynchus* (Diptera: Culicidae) in Florida mangrove forests. *Environmental Entomology* **21**, 737–744.

Ritchie SA, Russell RC (2002) 'A review of the New Zealand Mosquito Surveillance Programme'. Report for the New Zealand Ministry of Health, Wellington, NZ.

Rohe DL, Fall RP (1979) A miniature battery powered CO2 baited light trap for mosquito borne encephalitis surveillance. *Bulletin of the Society of Vector Ecology* **4**, 24–27.

Van Essen PA, Kemme JA, Ritchie SA, Kay BH (1994) Differential responses of *Aedes* and *Culex* mosquitoes to octenol or light in combination with carbon dioxide in Queensland, Australia. *Medical and Veterinary Entomology* **8**, 63–67. doi:10.1111/j.1365-2915.1994.tb00387.x

Williams CR, Bader CA, Williams SR, Whelan PI (2012) Adult mosquito trap sensitivity for detecting exotic mosquito incursions and eradication: a study using EVS traps and the Australian southern saltmarsh mosquito, *Aedes camptorhynchus*. *Journal of Vector Ecology* **37**, 110–116. doi:10.1111/j.1948-7134.2012.00207.x

10

The National Saltmarsh Mosquito Surveillance Programme

Monica Singe, Mark Disbury and Darryl McGinn

Background

The National Saltmarsh Mosquito Surveillance Programme (NSP) aims to minimise the risk of establishment of exotic mosquitoes of public health importance that breed in saline saltmarsh habitat. This programme had its genesis in the NZ Government's eradication response for the exotic Southern Saltmarsh Mosquito (*Campto*) discovered in Napier in 1998 and finally declared eradicated from numerous subsequent locations across both islands on 1 July 2010. In 2005, the MoH established the NSP and administered operational contracts for its delivery until July 2010, when responsibility for the programme transferred to the MAF.

The programme is risk management driven and uses a saltmarsh-mosquito-specific surveillance protocol that is statistically based to provide confidence in its results. All known saltmarsh habitat within NZ has been mapped. Allocation of resources (annual surveillance hours/site) for sampling saltmarsh mosquito immature stages and adults is prioritised using a statistical algorithm accounting for habitat size, its tidal influences, presence of residual surface water, habitat quality (for mosquito breeding), prior *Campto* positivity, proximity to port of entry, climate and relevant human activity. This statistical method provides 95% confidence of correctly identifying an exotic mosquito incursion in saltmarsh habitat within 12 months of arrival (McBride and Smith 2005).

There are several native mosquitoes of NZ and other exotic, but established, low-risk species with breeding habitat intersecting with saltmarsh. The NSP routinely collects these within the course of site sampling. Species determination of collected mosquitoes is completed in the NSP laboratory in Gracefield, Lower Hutt (near Wellington). All mosquitoes received from the field are identified to species and reported within 24 h of receipt.

The sensitivity of the NSP sampling methodology was confirmed in 2006 by the successful detection of a small, and apparently new, local population of *Campto* in the Coromandel Peninsula. This detection allowed for a timely response and successful eradication at this location in much less time than other *Campto* sites discovered with already established and dispersed populations.

NSP operations: NZ BioSecure 2005–2010

From 1999, surveillance of known saltmarsh habitat outside the eradication zones was carried out on a regional basis by public health unit staff under the direction of the MoH. The review by Ritchie and Russell (2002) had recommended that a more targeted approach to surveillance would be appropriate and a team of experienced mosquito surveillance specialists would allow for more widespread, standardised surveillance and provide opportunity for development and improvement of existing surveillance methodologies. The TAG was also commenting on the need for improved national surveillance as a critical element in the *Campto* eradication protocol. To complement the first ports of entry surveillance programmes, the MoH obtained funding for the production of a NSP to be developed and implemented in 2005.

At the time, NZ BioSecure represented the core expertise for mosquito surveillance in NZ, and methodologies and procedures being trialled in the Kaipara region were being developed with a view to capturing saltmarsh habitat not just locally but nationally.

Surveillance by NZ BioSecure had long since moved on from the number of dips and larvae counts; they had gadget geeks and IT gurus who all wanted to try new technologies and techniques for collecting and analysing surveillance information, and applying it to enhance surveillance and capture information that would provide a historical baseline dataset of saltmarsh habitat.

The primary aims of the NSP during NZ BioSecure's contract period (March 2005–June 2010) were to produce a comprehensive database of all saltmarsh habitats nationally (assessed and categorised according to their relative likelihood of incursion and suitability), establish a statistical measure of the surveillance effort, and capture and record all NSP surveillance data electronically. A further aim was to provide assurance there were no undetected pre-existing *Campto* populations within NZ by being able to detect *Campto* at low population densities, if present. During 2005 and early 2006, the NZ coastline was aerially mapped (see Fig. 9.4) and each identified potential site followed up with ground surveillance confirmation of habitat. Each habitat site was categorised through ground surveillance and the statistical measures (see Chapter 9).

Field staff were positioned throughout the country to provide surveillance in each of the five zones. Technical and field personnel were primarily working full time in Auckland (Monica Singe, Andrew MacDonald, Gareth Southcombe, Dean Singe, Jess Roberts, Adam Kellian, Mark Dellow, Adam Mason, Rueben Gerbick, Janelle Underwood and Mark Leach), Gisborne (Rob Waihi), Blenheim (Tash Symmonds and Bruce Hammond), Christchurch (Regan Courtney) and part time in Wairarapa (Mark Disbury and Cliff Dawson) and Invercargill (Mary Anne Timpany).

Later in the programme, a Keri Keri office (James Wakeford) was opened and the east coast office relocated to Napier (Dan Edwards).

The team was managed through the control centre based in Auckland and the NZ BioSecure Entomology Laboratory in Wellington, which provided support for operational and laboratory aspects.

Following the initial aerial mapping and ground-truthing, the number of sites across the zones was established as Zone01 with 376 sites (including 24 sites from the Chatham Islands), Zone02 with 95 sites, Zone03 with 35 sites, Zone04 with 148 sites and Zone05 with 57 sites. There were an additional 366 sites, which had been archived based on the opinion of staff that the habitat was very poor or not saltmarsh habitat at all. The number of sites changed throughout the year(s) as new sites were identified or poor sites archived. Archived sites were also occasionally visited to ensure no changes had occurred, especially following wet events.

An alternative means to reviewing the surveillance data collected nationally was of primary consideration when the methodologies for the NSP were being developed. Track logging using wrist-mounted GPS units (Fig. 9.5) was already being used in the Kaipara and capturing excellent surveillance and reporting data, providing the ability to see areas surveyed and, just as importantly, providing visual confirmation of areas not yet surveyed (so to be included in future surveillance planning).

The use of GPS and track-recording software allowed for statistical analysis and confidence of detection of saltmarsh exotics by highly skilled observations, and from the sampling experience of field staff hunting in receptive habitats. This was the basis of the NSP methodologies – providing operational data on areas visited and not visited, and distinct locations for positive and negative samples. Such detailed and accurate operational data provided a strong basis for specific site analysis for future site management. But NZ BioSecure staff believed additional information from actual on-the-ground surveillance observations was also of significant value and needed to be routinely captured for historical and real-time operations.

One of the key aspects of the successes of the eradication programmes in east coast and Kaipara was the practice of daily debriefing. This daily meeting, whereby staff reported aspects of their surveillance and eradication activities back to the team members, enabled operational information to be discussed with all members of the team. In addition, with the large size of the eradication zones, this enabled the majority of team members to become very familiar with the entire habitat (negative and positive) under each programme. On a national scale, this daily debriefing process was not practical for a NSP. Andrew MacDonald introduced voice-recording cameras and field staff were required to record observation snippets on their cameras when taking an image.

These voice annotations (time stamped and linked to a digital photograph using Red Hen Media Mapper software) became the primary tool for collecting daily site observations, much like a daily debriefing. Technical staff, including Andrew MacDonald, Gareth Southcombe, Janelle Underwood and Monica Singe, routinely reviewed voice annotations linked to site photos and provided feedback to field staff regarding priorities for routine, but especially trigger event, surveillance. The significance of the voice annotations was particularly valuable for the Coromandel Peninsula find – being able to go back and review the ground-truthing observations made in the previous year at the time of the visit was extremely helpful for delimit planning and response, and subsequent eradication programme operational aspects.

Track data provided the evidence of surveillance effort visually once downloaded; also, being time stamped, it was able to be linked to the voice annotations and photographs, and thus captured for a specific site. The track data was compiled and summarised as kilometres travelled (actual ground surveillance hours, not including vehicle time) and time in the field for a given site. These data accumulated against each site and were assessed against the annual calculated required hours.

Manifold (GIS) software allowed each site to have a defined, yet movable, boundary to capture all data recorded for a given site. This movable boundary was excellent – saltmarsh habitat can change from week to week and season to season, and being able to push out the sites' boundaries enabled a more accurate visually mapped representation of what was actually occurring on the ground over time.

Weekly technical review of these data at the control centre allowed all aspects of each site's management to be recorded accurately, and provided not only site-specific information but also an operational overview of single sites. This phase of the NSP produced a comprehensive baseline dataset with 5½ years of site-specific data readily reviewable through Manifold mapping software, overlaid with localised environmental data, including tide, rainfall, and temperature.

The surveillance year was divided into quarters and each site was visited at least once each quarter to ensure surveillance captured a variety of habitat conditions throughout the seasons. This method also provided assurance that all sites should have completed their required annual hours to meet the methodologies statistical measures. An adult trapping programme was also carried out throughout the period and additional surveillance hours were accumulated and applied for adult light-trap surveillance. We did this by applying a boundary around each trap location and each site within captured 5 min surveillance for each 24 h trap period. With the focus being on early larval detection, and the known variability in adult trapping efficiencies, the adult traps were not considered a primary surveillance method and we didn't want trapping hours to overly influence the sites' annual hour requirements.

Field staff provided monthly programmes of likely surveillance locations and weekly reports including track and digital data (voice annotations, photographs and environmental data) were collected summarising activities completed. The weekly reports, in conjunction with the track and digital data, represented daily debriefing material that the Information Systems Coordinator and Technical Manager reviewed weekly, providing feedback on surveillance priorities and any data management and environmental factors that may have arisen.

Monthly reports were compiled to provide the MoH with ongoing status reports regarding surveillance activities, including sample results, hours of surveillance, areas visited, trap-nights and general information about the programme's development and implementation. We also produced an annual summary report of surveillance undertaken and completed.

A further tool under development, during what should have only been the first 2 years, was the NSP database or 'PeopleData' developed by Syfir and Generis. The various sources of data were amalgamated in the NSP database, providing a central location to review and assess all data for each site. Behind the operational surveillance deployments, and media collection and review, was the development of this database. In the early version of PeopleData, NSP staff had regular and reasonably good access to data, but then the programme changed to an updated version and, although data were entered and maintained, we couldn't access the promised capability.

Operationally, we were recording landowner information, environmental conditions, including temperature, tide and rainfall, and uploading site track and media data and applying it to relevant individual site locations. This provided a comprehensive site-by-site history of surveillance activities undertaken (see Table 10.1 on page 190). There were reporting capabilities built into the system that enabled the generation of single site history reports, sub-district reports and zone reports, all of which when working were valuable to inform field staff and management.

Throughout the NZ BioSecure period of the NSP, there were annual audits, both internal (conducted by Southern Monitoring Services Quality Manager, Cliff Dawson) and external (conducted by Darryl McGinn, Mosquito Consulting Services Pty Ltd Australia). On all audit occasions, the feedback and compliance were generally positive, and any recommendations or suggestions were readily adopted where possible.

In addition to audits, operations were supported by the NSP Standard Operating Procedures, performance appraisals and reviews, monthly reporting, and landowner and stakeholder reporting on a regular basis. During 2005 and 2006, management conducted a roadshow to all the public health services to provide information on the programme and how we had used the surveillance information collected by these officers previously in the

Box 10.1: National Surveillance Programme Staff story – Dan Edwards

One of the best things about the NSP was the wide range of areas and habitats you found yourself in, with most being either a long way off the beaten track or through private property, limiting access. From idyllic private beaches near Young Nicks Head to areas of Kaikoura coast where seals bask metres away from grazing sheep, the scenery would equally be at home on any postcard. As well as the amazing views, there was an incredible array of wildlife to be seen: pods of dolphins off Karikari Peninsula, pheasants and quails around Ohiwa Harbour, gathering legal paua (NZ black abalone) from rock pools without getting your feet wet along the Wairarapa coast, and spectacular fishing spots along the beaches and rocky outcrops of the Bay of Plenty where snapper and kingfish were plentiful.

The other side of the remoteness of these sites reared its ugly head when things went wrong. I had a couple of misadventures along Whakaki beach, east of Wairoa, where the location exacerbated what should have been only minor problems. On my first quad bike trip with a more locally experienced colleague, returning from the furthest sites we ran into a rather large inconvenience: what had been a minor trickle across the beach from a lagoon was now a raging torrent that had carved a large and impassable channel through the beach, no doubt due to the help from the regional council directed digger parked on the far side. What should have been a 5 km straight run down the beach turned into a 15-km scenic route through a couple of farms and a short run down State Highway 2. Although the regional council played the unwitting bad guy on this occasion, at another time I was also towed a significant distance by a council employee from the same spot after a major quad bike meltdown; it was just pure chance that he happened upon a broken quad bike with an extremely stroppy rider, because traffic along this stretch of beach is extremely low.

Likewise, getting stuck in these places was more than just embarrassing; more than a few farmers and locals benefitted to the tune of boxes of beer after rescuing a hapless mossie chaser stuck in sand, mud, ruts or even perched precariously on the edge of a stop bank. What made it worse is that in the majority of cases, our vehicles seemed to get stuck within 100 m of the desired destination.

system. We also provided datasets to each unit with information on sites in the area now under the NSP. The roadshow was successful and feedback from colleagues was very positive.

Success – Coromandel *Campto* detection

By far the signal achievement of the NSP was the detection of the *Campto* population in the Coromandel peninsula. Surveillance was planned in the Coromandel for 8–12 May 2006, in response to a period of high rainfall and high tides that occurred during the weeks just prior. Surveillance had been originally planned for 26–28 April, but the wet weather and storm events that week meant surveillance was delayed to ensure better sampling conditions, which would be perfect after the flooding event.

On Tuesday 9 May 2006, Jess Roberts and Mark Dellow discovered what they believed was *Campto* at sites Boll01 (a single second instar) and Waim01 (numerous larvae of all instars). Adam Kellian was sent over on the Wednesday morning, joining Mark and Jess (NZ BioSecure musketeers) to further sample the area, deploy a light trap (with *Campto* adults being trapped that night) and apply an emergency treatment response. Shaun Maclaren also joined the team later that day with a second vehicle.

Surveillance finds that week confirmed six initial positive NSP sites located within the Coromandel Peninsula. In all, 19 of the 28 sites in the area – with assigned surveillance hours using the detection probability formula method following the ground-truthing in September 2005 – were visited that week. With the exception of one site (Dent01), all initial detections occurred in known NSP sites that had been assigned likelihood categories, and which had been allocated surveillance hours using the statistical measure. Two archived sites (i.e. with no surveillance hours assigned) were also subsequently identified as positive; during the delimit, these were visited in response to the positive finds nearby and the large flood event in the area.

Particularly amusing for staff at the time was a review of Adam Kellian's voice annotation for site Dent01, which he suggested was poor habitat, and it turned out the positive area of that site was located behind the woolshed. Using this as an example, we adjusted the procedures with field staff requiring all archived sites to be visited when in the area, and observations recorded to ensure any habitat changes were noted, and sites could be upgraded into higher priority sites if required.

The initial ground-truthing of Coromandel habitats occurred during September 2005 and many of the sites were dry or had little water present. However, several were wet and

Figure 10.1 Survey site Boll01. (Image: Adam Kellian)

larval samples were collected. Voice recordings and tracks were recorded and site names established for the area. The ground-truthing was carried out by Adam Kellian and Shaun Maclaren. The following are the voice recording transcripts recorded by Adam during ground-truthing.

> Site Boll01 *'Sampling up by Colville there's a house being built here, they told us the landowners name is Bolland, so the site ID is Boll01. It's some pretty good habitat, it's only a small site about as big as ... a tennis court ... lots of Bachelors Button, a few hoof prints, little bit of water there, definitely get saltwater influence, a few salt water drains around the perimeter, but they are pretty crap, 2 or 3 ppt salinity, this little site here is probably medium – high priority'.* (See Figure 10.1)

> Site Colv02 *'Looks pretty good, a few drains, lots of Bachelors Button, lots of rushes, might be a little bit tidal though these drains, have a lot of crab holes and stuff through them. Of the edges of the drains there's a lot more Bachelors Button and a few hoof prints that's where the better habitat would be. The three sites we've had around Colville have all been pretty good; I'd definitely come back and have it as part of a surveillance plan. All three of these sites are medium–medium high. It's quite a big area, about two rugby fields. At the foot of the hill there's a bit of runoff and its fresher, but down the bottom here it's a lot more tidal and salty. Probably worth keeping an eye on for sure.'* (See Figure 10.2)

Figure 10.2 Survey site ColV02. (Image: Adam Kellian)

The subsequent eradication programme delimiting survey results indicated that the population numbers at the Coromandel sites were relatively low when compared with previous *Campto* population numbers at the time of first detection.

The team was delighted – although that expression was probably somewhat inappropriate (Monica Singe can still recall Sally Gilbert's face when the word 'delighted' was uttered at the TAG meeting) but we all felt 'YES! The system worked and our efforts had paid off, by not only detecting an exotic, but detecting it in numbers lower than those of the previous populations when first detected' It strongly supported the success of the NSP objectives for 'early detection' of exotic saltmarsh mosquito species, in this case a previously undetected *Campto* population. The developing programme routinely faced critique and questions of accountability, so it was a bitter-sweet find for the team and we were all hugely proud of the programme's 'success'.

One of the biggest questions the find raised was how did it get to Coromandel? Jess Roberts and Mark Dellow spoke to the landowner at Boll01 at the time and it turned out he lived in the Kaipara and had been commuting weekly to the Coromandel to build a house for the last 2 years. We all felt it was a big ask to ignore that coincidence.

Laboratory identification

Although the field staff were pretty sure they had found *Campto*, we really needed to be absolutely positive because a finding prompted initiation of a 2-year plus response programme with all the associated costs.

The Coromandel *Campto* index specimens were forwarded through to the NZ entomology laboratory, as per all specimens collected, but flagged for urgent attention. The larvae were mounted on glass slides and inspected using a compound microscope. Identification was determined to be *Campto* at the laboratory but, again considering the amount of working hours and money potentially being committed, further confirmation was required.

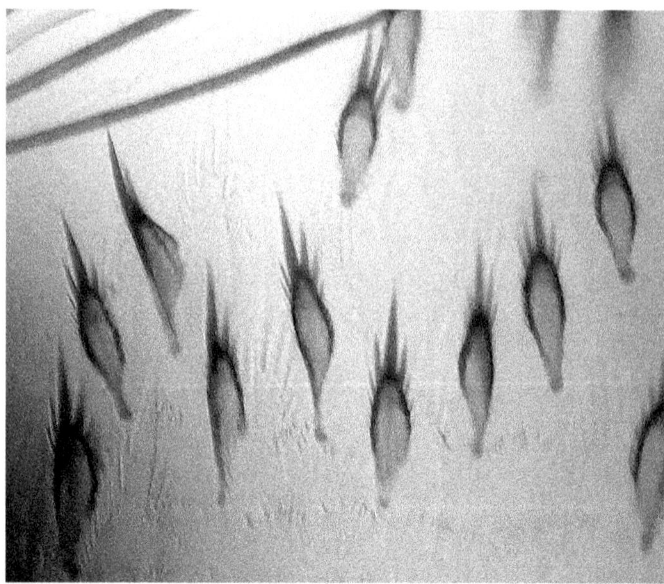

Figure 10.3 Looking down a microscope at *Campto* larval comb scales. (Image: Mark Disbury)

Figure 10.4 Why send actual larvae when the images are good for identification – photomicrograph of *Campto* larval ventral brush. (Image: Mark Disbury)

The arrangement was for specimens to be sent to Richard Russell at the University of Sydney at Westmead Hospital in Australia for confirmation, but, even using 24 h couriers, specimens could take up to a week to arrive due to customs and quarantine delays. To circumvent these delays, the NZ BioSecure laboratory generated image portfolios of specimen features used in the identification process. By simply using a digital camera to take photographs down the eyepiece of the microscope, a series of up to 20 images (see Figs 10.3 and 10.4 as examples) formed a portfolio file that would then be emailed to Richard for him to look at; if more detail was needed, he could simply email back his requirements.

This technique meant that his confirmation of the identification would be received much faster and usually on the same day (Disbury *et al.* 2008). Using this method meant that all the identification confirmations of *Campto* in the Coromandel were received within 2 days of the find, allowing confidence for all that the response that was already underway was not misguided.

A history of NSP surveillance outcomes

During the first full year (2006), not all sites achieved their required surveillance hours, mainly due to problems with landowner access permissions and site access difficulties, but also due to the teams coming to terms with operational requirements.

The statistical analysis and interpretation of the actual hours completed versus the required hours uses a negative predicted value, calculated upon the conditional information that we did '**not**' find any exotic larvae in a given year. The more we looked and did not find any exotics, the larger the negative predicted value became, and 85% of all sites

Table 10.1. National Surveillance Programme collection statistics, 2006–2010

Year	No. sites	No. larvae collected	No. adults collected	Surveillance hours required	Surveillance hours completed
2006	711	7056	2904	1131	1180
2007	709	12 897	3076	1130	1823
2008	743	12 651	6205	1266	1901
2009	766	7601	4278	1360	2075
2010	777	8989	15 556	1344	1571

surveyed during 2006 met or exceeded this value. Only 12 sites did not meet the 95% threshold applied, and over 7000 larvae and almost 3000 adults were collected during 2006.

Despite being disappointed that we didn't meet all the measures applied, we were pretty happy with the first year's efforts. The single exotic incursion detected in the Coromandel in May 2006, with 117 *Campto* larvae identified, was the icing on the cake (bad for you, but still sweet), and the view that we now had a better understanding of our operational requirements and commitments meant our surveillance performance could only get better for the coming years (Table 10.1).

2007 surveillance

A total of 680 categorised sites was initially identified for surveillance for the 2007 calendar year (Singe and Southcombe 2007). This increased to 699 and then 709 throughout the report period, as more sites and surveillance information were gathered from the field. The majority of the sites and surveillance hours remained in Zone01.

During 2007, a total of 1823 surveillance hours was undertaken; 1474 of these hours were assigned to categorised sites. This represented a significant increase in surveillance hours compared with 2006 (1180 h).

To assist with understanding field surveillance staff observations of the habitat conditions, they were asked to report and record their tracks using a specific waypoint icon representing 'poor', 'average' and 'optimum'. These icons, once in the tracking software, were able to be grouped to provide additional reporting information on the quality of the surveillance habitat for the month period. A further initiative, implemented during 2007, was direct contact with regional councils and local authorities.

The Technical Manager sought information from the councils regarding any potential new resource consents that may have an impact on coastal sites, resulting in the generation of saltmarsh habitat. Many of the councils provided regular consent notification and, where applicable, site locations were reviewed by field staff.

Of those sites where surveillance was conducted, 19 did not fall within the calculated negative predicted value percentage applicable to the sites category during 2007; however, only eight were less than negative predicted value. Overall, 96% of sites met or exceeded a 95% negative predicted value. This was an 11% increase on the 2006 surveillance effort. Almost 13 000 larvae and over 3000 adults were collected during 2007 (Singe *et al.* 2008).

Since the inception of the NSP, the NZ Government wanted to maintain an independent oversight of the programme by requiring an annual review and audit. Because of his field experience and formal role as a member of the TAG, Darryl McGinn was engaged by the NSP to be its independent external auditor. The audits were generally conducted about the time of the MoH Border Health Advanced Surveillance Workshop attended by Darryl as part of directing staff: delivering lecture-based information and practical training of health protection officers in field sampling techniques.

The audits generally consisted of matching actual practices observed in the field and laboratory with written NSP operating procedures, and reviewing surveillance outcomes in terms of quantity and quality of the surveillance effort. Darryl McGinn would spend time in the field with NSP staff observing and coaching, but also learning. This was a great time to note changes in field practices and techniques that had evolved with the NSP, and evaluate them in terms of programme efficiency and effectiveness. Improved practices would be written into the NSP operating procedures. It also provided the opportunity to make sure field sampling was as sensitive as possible within the reality of time and cost.

The issue of maintaining surveillance sensitivity was a growing one by mid-2006. Because *Campto* in NZ was diminishing towards zero as a result of the eradication programme, surveillance teams were encountering the insect less and less. How does the NSP maintain high sensitivity for detecting *Campto* when it has been eradicated in several zones already and was becoming increasing rare?

The answer was to deploy NSP surveillance officers to active saltmarsh habitat in Australia, where they became reacquainted with the appearance, behaviour and diversity of habitat utilisation of saltmarsh mosquito species there, and bring back that experience and apply it to exotic species detection in NZ saltmarsh. In February 2007, the first rotation of

Figure 10.5 2007 NSP Staff Training: (from left) Dan Edwards, Mark Dellow, Jess Roberts and Monica Singe. (Image: Darryl McGinn)

Figure 10.6 2007 NSP Staff Training: (from left) Shaun Maclaren, Darryl McGinn, Mark Leach, James Wakeford and Dean Singe. (Image: Darryl McGinn)

NSP staff visited south-east Queensland to undertake saltmarsh mosquito surveillance refresher training with Darryl (Fig. 10.5). A second rotation was in February 2009 (Fig. 10.6). NSP staff who attended were Monica Singe, Shaun McLaren, Dan Edwards, James Wakeford, Mark Leach, Jess Roberts, Mark Dellow and Dean Singe.

The south-east Queensland training was generally over 6 days of field contact in a wide variety of habitats where saltmarsh species could be found. The timing of the visits coincided with spring tides to enable NSP staff to witness tidal inundation of dry saltmarsh, observe *Aedes* hatching rapidly *en masse*, and observe larval development and behaviour. Adult trapping was also undertaken, but the emphasis was on developing sensitivity to detection of larvae in saltmarsh habitats.

The 'saltmarsh' habitat was extended to include adjacent freshwater habitats with slight marine influence, because saltmarsh mosquito activity can sometimes be found there also, and non-saltmarsh species such as the high-risk *Culex annulirostris* are regularly found in the fresher margins of saltmarsh.

One of the common themes that always came out of the international training was the high potential for saltmarsh mosquito species to use atypical habitat: the far margins of classic 'saltmarsh'; the edges of adjoining 'fresh water'. This was one of the most important teachings. Effective field surveillance should not be restricted to any specific definition of what is 'saltmarsh', nor governed too much by lines on a map. It is difficult to characterise in strict technical or scientific terms where to look for saltmarsh mosquitoes; however, the word 'intuitive' is sometimes used to describe competent sampling. It is not surprising that many NSP field staff were also hunters and fishermen. It is the same approach: understand your 'prey'.

Another challenge was the potential boredom of negativity. Most staff had a fair amount of experience and were used to not finding *Campto*, but they were finding reasonable numbers of other mosquitoes, some comparatively rare, except in very dry periods where the challenge was just finding adequate habitat to sample; these were the most challenging times. All in all though, the large areas covered by each NSP field officer saw them somewhere different each day, interacting with different people and keeping them less likely to slip into a rut.

2008 surveillance

A total of 690 categorised sites was initially identified for surveillance for the 2008 calendar year (Singe and Underwood 2009). This increased to 743 throughout the report period when the Kaipara and Grassmere eradication zone sites were incorporated into the programme from 1 September, and subsequent to a few other sites being reassessed and their categorisation altered. The majority of the sites and surveillance hours remained in Zone01.

A total of 1901 surveillance hours was undertaken during 2008, with 1530 of these hours achieved through on-the-ground tracked surveillance. Additional liaison was undertaken in the Kaipara area when the saltmarsh sites in the eradication zone were handed over to the NSP. This included updating ownership and occupancy details because no such information was made available by the eradication contractors.

Freshwater sites were also a hot topic with field staff: we wanted to expand surveillance into fresher water bodies and, as a precursor to understanding the degree of freshwater at the saltmarsh margins, field staff recorded freshwater waypoints with specific icon markers including little fish and water symbols. Once downloaded, these were able to be separated out and the data stored for subsequent use. There were 416 archived sites identified in 2008 (compared with 366 in 2006 and 381 in 2007) and an additional 65.5 track surveillance hours and 50.2 light-trap-surveillance hours were also completed within archived sites during this period. This low level of surveillance in the archived sites aimed to assist with the detection of environmental and/or habitat changes in these unlikely sites.

Total zone surveillance was incomplete but better than previous years, with only three sites (87 in 2006 and 18 in 2007) not being visited at all. The primary reasons for being unable to visit these sites remained consistent with previous years – unavailability of landowner permission and physical access difficulties. However, NZ BioSecure made significant efforts to gain the confidence of all landowners and this was reflected in the reduction of sites unable to be visited nationally. Sites not visited were always incorporated into the adult light-trap programme where possible to capture a degree of surveillance for these sites. Overall, 98.5% of all sites met or exceeded a 95% negative predicted value; this was a 2.5% increase on the 2007 surveillance effort, and over 12 000 larvae and 6000 adults were collected during 2008.

In 2008, the MAF had contracted Massey University to update the exotic mosquito risk assessment and the MoH was also interested in revisiting its border health programme of surveillance in first call sea/air ports previously reviewed by Kay (1997) and Ritchie and Russell (2002). The new review was largely in response to NZ's ratification of the revised World Health Organization *International Health Regulations 2005*, which fundamentally changed the philosophy of managing public health risks across international borders. The change was from one of following a prescription of standards (e.g. surveillance only within

400 m of the port), originally aimed at restricting *Aedes aegypti* vectored Yellow Fever, the focus of the *International Health Regulations 1969*. The *International Health Regulations 2005* were an outcome-based approach across a broad range of biological, chemical and radiological public health risks.

Given the coincidence of timing between the MAF's risk assessment update and the MoH's interest in border health compliance with the *International Health Regulations 2005*, there was a decision by the NZ Government to combine the two exercises. In 2008, Darryl McGinn was contracted by the MoH to undertake the port surveillance review and use the findings to produce a joint report updating a wider exotic mosquito risk assessment to be included with the Massey University work (McGinn 2008). He would be accompanied by JR Gardner (MoH Senior Advisor) to visit all of the NZ sea- and airports regarded as potential pathways for exotic mosquitoes.

National Saltmarsh Mosquito Surveillance review visits conducted at the same time included:

- Silverdale HQ for annual NSP systems audit and programme performance. Field visits to selected saltmarsh sites with NSP field staff to audit field processes.
- Tauranga for field visits to several active and archived saltmarsh sites. NSP field staff identified previously archived sites that they recommended should be reactivated into regular surveillance due to changed site conditions. McGinn concurred and noted the utility and dynamism of the site classification processes in accommodating changing environmental determinants of risk.
- Christchurch for field visits to Marshlands and Ellesmere with NSP staff. Several constructed fresh and saline wetlands have been featured within new residential developments. These were considered in terms of changing risk for *Campto* and potentially freshwater risk mosquito species. It was considered that constructed wetlands adjacent to international ports and airports should be included within Border Health mosquito surveillance. As with Auckland International Airport, it seemed logical to include freshwater wetland (constructed or natural) within the scope of the NSP.

The 2008 surveillance review detailed a large number of observations and recommendations for improvements, which are too numerous to be included in this chapter, but the major observations are summarised below.

In 2008, almost all border surveillance was aimed at container-breeding *Aedes* species (e.g. *Aedes aegypti* and *Aedes albopictus*) as per the out-dated *International Health Regulations 1969*. Surveillance for other high-risk mosquito species had not been adequately designed or implemented at ports of entry.

High-risk mosquito species (exotic to NZ), including *Culex annulirostris*, would likely arrive in NZ differently to *Aedes* species and have great potential to establish in fresh/brackish ground pools well outside the classic (and out-dated) 400 m port surveillance zone.

The NSP was designed for detection of exotic saltmarsh species only. It does that job very well and, because of the knowledge of saltmarsh species utilisation of marginal saltmarsh habitat, some adjoining freshwater sites are subject to NSP surveillance. There are significant freshwater habitats adjacent to ports of entry that are receptive to high-risk exotic species but are, however, not subject to surveillance by either Border Health or NSP. There should be an extension to surveillance into these near-port freshwater sites to manage the risk of high-risk mosquito species establishing in NZ.

2009 surveillance

With 766 sites and an additional 433 archived sites, 2009 was the largest full year of surveillance carried out by NZ BioSecure (Singe 2010), and it was the last, with the programme transferring to an alternative provider on 30 June 2010 following the declaration of eradication of *Campto*.

The Coromandel Eradication Zone sites were incorporated into the NSP in May 2009, adding several active and archived sites, and surveillance hour requirements, to the programme, and a complete reassessment of the area to ensure landowner and site information was up to date.

As in previous years, habitat quality assessment was provided by field staff while conducting surveillance: this incorporated field staff's experienced opinion regarding the habitat. This had been an important ongoing feature of the programme and reflected an accumulation of 4 years observations, enabling greater surveillance at sites where staff identified an increased likelihood that exotic mosquitoes would be found.

Since 2006, questions regarding the appropriateness of the negative predicted value for this type of surveillance had been discussed at various levels and been revisited by Graham McBride.

A total of 2075 surveillance hours was undertaken during 2009, with 1574 of these hours achieved through on-the-ground tracked surveillance. An additional 58 surveillance hours, including 36 light-trap-surveillance hours, were also completed within archived sites during this period. Total zone surveillance was incomplete, with 26 sites not being visited at all (although 24 of these sites were located on the Chatham Islands where no surveillance was planned for the period).

Overall, 96.6% of all sites met or exceeded a 95% negative predicted value (but if the Chatham Island sites were excluded, the percentage of sites that met or exceeded a 95% negative predicted value rose to 99.7%). Over 7000 larvae and 4000 adults were collected during 2009. These numbers were lower than previous years and believed to be as a result of relatively drier climate throughout 2009.

2010 surveillance

Site reviews and surveillance hours were calculated for the 2010 surveillance year, using the same review-and-assess process of previous years. Only three additional sites were archived and a few minor site scores altered. The surveillance sites and associated hours for 765 sites, and 436 archived sites, represented 5 years of site visits by experienced field staff, and we firmly believe that these figures and assessments reflected an accurate representation of NZ saltmarsh habitat.

With NZ BioSecure exiting the NSP, our aim was to ensure that a minimum of 50% of the hours would be completed before 30 June 2010. As we drew towards exiting the programme, the control centre reviewed and audited all historical data, finalised datasets and assisted staff where possible with the handover phase. The final landowner and stakeholder update was distributed on 28 June 2010. We completed 828 surveillance hours up to 30 June 2010, being more than half the required 1344 h. NSP management prepared for closure with the passing over of assets, data, methodologies and standard operating procedures.

Following on from the successful eradication of *Campto* from the last site in NZ, the NSP has added those areas to its site list and continues to function. NZ BioSecure passed the baton to Mosquito Consulting Services Ltd (Darryl McGinn as Principal) and all the field staff joined the new team and carried on with the work as if nothing had changed. There have been a few developments in technologies, such as the move to use smart phones in the field to replace several pieces of equipment, cameras, GPS units, pens and paper, while saving on manual processing of data by linking directly into the database. Other than this, the programme continues with the same review processes as it has since 2005 and, as of the time of writing (mid-2013), there have been no other NSP *Campto* finds since the Coromandel, and in fact no other discoveries of any mosquito species not expected to be found.

The NSP grew out of the government's need to be assured that *Campto* distribution in NZ was not outrunning the eradication programme. Surveillance techniques developed in the programme from the time of the Napier discovery, and refined to meet the mammoth task in the Kaipara, were an integral part of the eradication programme and applied in each eradication zone separately – but what about the rest of the country? These surveillance techniques were packaged into a specification for a stand-alone programme that allowed a new organisation to operate and apply surveillance to areas outside active eradication sites.

For the first time, government could rely on a uniform and technically competent approach to surveillance in known and potential saltmarsh mosquito habitat throughout the country. The development of the NSP and its success in the Coromandel Peninsula played no small part in strengthening government's resolve to maintain support for the eradication programme after Coromandel and particularly during the critical transition of eradication responsibilities from the MoH to the MAF.

Acknowledgements

Thanks to Dan Edwards for providing the text box account.

References

Disbury M, Cane R, Russell RC (2008) Remote identification of exotic mosquito specimens using digital photography. *Australian Journal of Entomology* **47**, 128–130. doi:10.1111/j.1440-6055.2008.00638.x

Kay BH (1997) 'Review of the New Zealand Programme for the Exclusion and Surveillance of Exotic Mosquitoes of Public Health Significance'. Report for the New Zealand Ministry of Health, Wellington, NZ.

McBride G, Smith B (2005) 'Statistical aspects of Saltmarsh Mosquito Surveillance Programme'. Unpublished report for SMS Ltd. National Institute of Water and Atmospheric Research Limited, Hamilton, NZ.

McGinn DR (2008) 'Mosquito Surveillance Review 2008'. Report for the Ministry of Agriculture and Forestry and the Ministry of Health, Wellington, NZ.

Ritchie SA, Russell RC (2002) 'A review of the New Zealand Mosquito Surveillance Programme'. Report for the New Zealand Ministry of Health, Wellington, NZ.

Singe M (2010) 'NSP Surveillance Report 2009'. Report for the Ministry of Health, Wellington, NZ.

Singe M, Southcombe G (2007) 'NSP Surveillance Report 2006'. Report for the Ministry of Health, Wellington, NZ.

Singe M, Underwood J (2009) 'NSP Surveillance Report 2008'. Report for the Ministry of Health, Wellington, NZ.

Singe M, Cane RP, Disbury M (2008) Development of the New Zealand National Exotic Saltmarsh Mosquito Surveillance Programme (NSP). *Mosquito Bites in the Asia Pacific Region* **3**(1), 31–34.

11

Maintaining environmental integrity in invasion areas

Helen Gear, Travis R. Glare and Maureen O'Callaghan

Introduction

The MoH's guiding principle in all of its work is to 'first do no harm' and it is this philosophy that informed the Chief Technical Officer's approach to all decisions that would affect the environment. In this respect, the environment was treated with the same care and respect that the MoH affords individuals and communities. In fact, in this programme, the only thing that was not treated according to this mantra was the mosquito itself.

The importance of maintaining NZ's status of being free of disease-spreading mosquitoes has already been covered in Chapter 1, and the surveillance programme and other steps taken to keep NZ free of these pests have been covered in Chapters 9 and 10. As part of the MoH's preparation for such an incursion, it commissioned a report (Glare and O'Callaghan 1998) to assess the potential environmental and health impacts of the bacterial larvicide *Bacillus thuringiensis israelensis* (commonly known as *Bti*). So, even before an actual incursion, the MoH had in its tool box a potential control agent that would be effective against mosquitoes and could be used with minimal impact on other organisms in the environment.

Little did it know that, within 12 months of the report being delivered, it would need the findings of the report to inform the development of a full-scale mosquito eradication programme.

Once *Aedes camptorhynchus* (*Campto*) was identified in Napier in December 1998, there followed an intense period of activity through to April 1999, where information was collected about this particular mosquito, its ecology, the environment in which it had established and the potential effects of the available control agents. This showed that the only effective options available for eradication of *Campto* were to change the mosquito's habitat (to make it unfavourable) or to use mosquito control agents. Obviously, changing the habitat over large areas was not only going to be impractical (how could you remove or modify all potential breeding areas and how could it be guaranteed that every last small puddle/hoof print of water was removed?), but it would also dramatically change conditions in that small part of the ecosystem in which the mosquito had set up residence.

Internationally, the use of chemical-based control agents has been well accepted. For those agents most commonly used, there was a wealth of literature available on their effects on other (non-target) species. Ensuring that the active ingredients did not cause damage to areas in which the mosquitoes were found, and that they did not further endanger any species that were already struggling to survive, was a high priority. It was already known that *Bti* could be used with minimal side-effects, but it was not persistent in the aquatic habitat and so there were concerns that it would not be able to do the job on its own. Early on, the juvenile hormone mimic S-methoprene was identified as a possible candidate (Chapter 2) to be an alternative chemical, and a thorough literature review on methoprene was commissioned in early 1999 to ensure that the science supported its claims.

Based on these findings, the Chief Technical Officer (Health) agreed to the use of *Bti* and S-methoprene, along with local habitat modification (where appropriate), to undertake the eradication. As the programme extended from months to years and the number of sites increased, both literature reviews were updated to satisfy the Chief Technical Officer (Health) that there was no new evidence to suggest that these control agents caused harm to other species.

In essence, this chapter follows the same process used by the Chief Technical Officer (Health) to make a decision on the shape of the final control programme by:

- describing the environments in which *Campto* established
- looking at potential non-target effects of the two control agents that were chosen for eradication
- covering the habitat modification approaches that were used
- reflecting on the effects (or lack) of the control programme on the natural environments in which it was used.

Environments suitable for *Campto*

Campto is a cool-tolerant species that is widely distributed on the southern coastal areas of Australia, from southern New South Wales and Victoria, to Tasmania (including King and Flinders Islands), South Australia and south-western Western Australia. Although it is generally regarded as a coastal saltmarsh mosquito, it is also found in riverine flooded marshes where brackish water occurs, and occasionally in inland regions where rain- or irrigation-water persists on saline soils.

In NZ, potential *Campto* habitats were limited to areas of shallow brackish water found in estuaries and lagoons (coastal lakes found around selected river mouths). NZ has a total of 15 134 km of coastline and over 300 estuaries. There was plenty of potential habitat, but we were grateful that we did not have to deal with freshwater habitats as well.

Estuaries exist where the fresh water from a river mixes with salt water. They are formed where the topography constrains the complete mixing of fresh and sea water during the tidal cycle. This most often occurs behind barriers such as sand spits and coastal embayments, at river mouths and in drowned river valleys with gently sloping substrates. Salinity then tends to decrease with distance from the sea. The inland limit of estuaries is formally deemed to end where salinity reaches a dilution of 5% of the marine concentration (Clarkson *et al.* 2003).

Coastal lagoons have salinity ranges from almost fresh water, through brackish to almost pure sea water and they can change over time. The depth and salinity of many lagoons are influenced by tidal movement. The barrier bars behind which they form are often not permanent, but are formed and re-formed by wave action, flooding rivers or a combination, and changes are often associated with storm events. Lagoons range in extent from less than a hectare to hundreds of hectares.

NZ estuaries and lagoons with brackish water typically contain a complex mix of species, comprising vegetation, invertebrates, fish and birds.

Vegetation – whether it is microscopic in the form of phytoplankton or smaller herbs, or shrubs or trees such as mangroves – is the base from which ultimately all species in an ecosystem derive their food. Vegetation provides the nourishment at the bottom of each food chain, but not always directly.

Estuaries and coastal lagoons contain high levels of phytoplankton because of their nutrient-rich status. Plant matter in estuaries has been shown to be four times more productive than a rye grass pasture and 20 times more productive than the open sea. This results in these areas being extremely rich in organic matter, creating a range of diverse habitats that are rich in crustaceans and invertebrates; these, in turn, create large quantities of food that attract large populations of fish and birds.

Communities of higher plants in NZ's estuaries and lagoons are dominated by sea grass and mangroves (*Avicennia resinifera*), which are found in the more sheltered harbours in the top half of NZ's North Island and, where water is more transient or where soil is saturated by the high water table, by saltmarsh. The dominant plants on saltmarsh are sea rush (*Juncus kraussii*) and jointed rush or oioi (*Apodasmia similis*). Often present on slightly higher land are the marsh ribbonwood shrub or makamaka (*Plagianthus divaricatus*) – this scrubby vegetation is found especially along the banks of tidal streams.

Beyond the saltmarsh, where the land becomes drier, turf-forming plants survive and a salt meadow develops. Saltwort (or samphire) (*Sarcocornia quinqueflora*), a native succulent, forms distinctive red, grey or green colonies alongside or sometimes intermixed with mats of creeping herbs such as remuremu (*Selliera radicans*) and shore primrose or mākoako (*Samolus repens*). Saltmarsh vegetation gives way to coastal scrub on dry land or to freshwater swamp in wetlands beyond the influence of salty water.

The primary consumers of vegetative matter, microbial products and plankton in estuaries and lagoons are the burrowing invertebrates, such as crabs, shellfish, snails and marine worms, and the filter feeders, such as cockles and pipis.

A large number of insect species also feed on vegetative matter and their diversity is often used as a measure of environmental health for estuaries. Many studies, both in NZ and overseas, record the number of species and individuals to calculate ecosystem health. Numerous studies exist on NZ areas, with a good indicative list (based on a study in Ahuriri) including the following range of invertebrates: water bugs (various Hemiptera); assassin bugs (Reduviidae); damsel flies (Zygoptera); dragonflies (Anisoptera); water beetles (Dytiscidae); moth flies (Psychodidae); marsh fly (Sciomyzidae); non-biting midges (Chironomidae); spiders (Arachnida); a range of mosquitoes (Culicidae); Crustacea, such as false shrimp/mysid shrimp (Mysidacea) and seed shrimps (Cypridopsinae); water fleas (Daphniidae); and water snails (Browne 2005).

The high productivity of estuaries and lagoons has resulted in them being particularly important breeding grounds for several fish species in NZ, both freshwater and marine. At least 30 types of marine fish use estuaries at significant times in their life. Some enter the estuaries daily with the tide, such as sand flounder (*Rhombosolea plebeian*), yellow-bellied flounder (*Rhombosolea leporina*), common sole (*Peltorhamphus novaezeelandiae*), kahawai (*Arripis trutta*), grey mullet (*Mugil cephalus*) and yellow-eyed mullet (*Aldrichetta forsteri*). They graze the tidal flats on the incoming tide feeding on shellfish, crabs and worms. Others – snapper (*Pagrus auratus*), red cod (*Pseudophycis bachus*) and gurnard (*Chelidonichthys kumu*) – are seasonal visitors. They enter the estuary as immature fish and spend some months feeding in the rich, sheltered waters before heading back out to sea.

Similarly, estuaries are particularly important to several the native freshwaters species. Adult whitebait or īnanga (*Galaxias* species) come down rivers to lay their eggs among the plants of the upper estuaries in late summer and autumn, and then die. On hatching, the young are swept out to sea, where they spend 5–6 months. They then enter estuary mouths in spring and swim upstream to fresh water further inland. In contrast, adult eels (*Anguilla dieffenbachia*) come down the rivers and through the estuaries to spawn at sea somewhere in the tropical Pacific. Their larvae then make their way back to NZ's coastal waters in spring. They transform into a juvenile stage, known as glass eels, just before they enter estuaries, where they settle and feed for a while before swimming upstream into fresh water.

In lagoons, fish populations vary according to the salinity, with both fresh water and marine fish being present according to the location. Lagoons will often support populations of native fish, such as eels and *Galaxias* spp., including the threatened short-jawed kokopu (*Galaxias postvectis*). The Galaxids are mainly responsible for the revered whitebait delicacy in NZ and are therefore of economic importance.

Estuaries and lagoons are also vital feeding grounds for a large number of bird species. Some birds are resident in the area and nest on the estuary margins, such as the pukeko (*Porphyrio porphyrio*), black swan (*Cygnus atratus*), mallard (*Anas platyrhynchos*), grey duck (*Anas superciliosa*) and 'sparse' banded rail (*Rallis philippensis*). Others nest mainly, or exclusively, outside of estuaries elsewhere in NZ, such as on riverbeds, and visit estuaries only during the winter, such as wrybill (*Anarhynchus frontalis*), variable oystercatcher (*Haematopus unicolor*), pied stilt (*Himantopus leucocephalus*) and the white heron or great egret (*Egretta alba* subsp. *modesta*). In addition, large numbers of migratory winter birds from the Northern Hemisphere use NZ's estuaries and lagoons to recover and rebuild body condition in preparation for the next round of nesting and chick raising in the Northern hemisphere. Some, such as the bar-tailed godwit (*Limosa lapponica* from the Arctic), depend on NZ for their continued survival. Other migratory birds that arrive in NZ include the lesser knot (*Calidris canutus*) and golden plover (*Pluvialis apricaria*).

As can be seen from the section above, estuaries and lagoons are ecologically important, containing a wide range of highly productive habitats that support comparatively high resident populations of a large number of species. This makes the ongoing health of these ecosystems particularly important for the survival of many species. Furthermore, any change in the health of an estuary or lagoon will generally have an influence far beyond its physical boundaries.

Estuaries and lagoons around the world are under threat from human activity and pollution. Compared with pre-European times, the majority of estuaries and lagoons in NZ have been degraded. Removal of large areas of NZ's forests and the rise of pastoral agriculture on higher ground has resulted in increased sedimentation rates and higher nutrient loads in many NZ rivers. Industry often introduces pollutants into the waters and activity such as sand dredging and fishing has caused physical damage to many estuaries and lagoons. This has placed many of these ecosystems under pressure and already limits their ability to support the range of species that they did in the past. This knowledge further reinforced the Chief Technical Officer (Health)'s desire to run an environmentally responsible eradication programme that would minimise the programme's effect on these vulnerable ecosystems.

Areas in NZ infested with *Campto*

At the peak of the incursion response, *Campto* was present in four main geographical areas, as described below. All the areas involved in the incursion were coastal, and included both rural and urban settings. In all cases, the infected sites represented a small percentage of the possible ecosystems in the area that were suitable habitat for *Campto*. Importantly, none of these areas was home to particularly rare or seriously endangered species. All the areas where *Campto* set up residence shared the sorts of complexes described above, but each also had its own notable characteristics, and several contained threatened, but not endangered, species.

Hawke's Bay and Poverty Bay

Campto was first identified in NZ in Hawke's Bay in the Napier District in December 1998. As previously described, the Napier City Council was alerted when it received complaints about mosquitoes from Napier's northern Westshore and Bayview suburbs. Specimens were found in December of that year in an area bounded on the southern side by Ahuriri and to the east by the Main Outfall Channel (known as the 'landcorp farm'). The ~10–20 km of drainage channels through the landcorp farm provided an ideal habitat for *Campto*. Near to Napier, it was also found in several sites south of Ahuriri, including the mouth of the Clyde River and the Porangahau lagoon. In late 2000, *Campto* was then identified in the northern Hawke's Bay near Maungawhio lagoon near Mahia (95 km north of Napier) and at Muriwai, south of Gisborne City.

Hawke's Bay and Poverty Bay have a Mediterranean-like climate, with long hot summers and mild winters. Typical temperatures range from 10°C to 30°C (maximum daytime), and sunshine hours average ~2200. Napier has an average annual rainfall of ~800 mm, with Gisborne receiving a slightly higher 1000 mm per annum. Westerly winds predominate, but afternoon sea breezes from the east are a common feature during the warmer months.

Hawke's Bay has a population of 155 000 (June 2012 estimate). Of these, 57 800 live in the Napier district and 74 300 live in the Hastings district (Napier and Hastings are 21 km apart). Hawke's Bay is renowned for its horticulture, with large orchards and vineyards on the plains. In the hilly parts of the region, sheep and cattle farming predominates, with forestry blocks in the roughest areas. Hawke's Bay is a popular tourist destination, with a reputation for good weather, good culture and a unique concentration of 1930s Art Deco architecture. Tourism makes a significant contribution to the regional economy.

Gisborne is the largest district council in the North Island, covering 8355 km^2 of land, with 1855 km of local roads. The population of Gisborne district is ~44 499 (2006 census), with 41 922 living in Gisborne city. Local industries are principally primary based and, like Hawke's Bay, include horticulture and wine growing on the flats, with farming and forestry further from towns and on the hills. Tourism also provides significant input into the local economy, especially over the summer, with the main attractions being Gisborne's good weather, high sunshine hours and numerous beaches.

Ahuriri Estuary is a modified estuarine wetland complex immediately north of Napier and is the most significant habitat of its type located on the east coast of the North Island between Wellington and the Bay of Plenty. Of the original 3840 ha of estuary that existed before the 1931 Napier earthquake, only 470 ha remain today, which makes this protected area especially important for wildlife.

Landcorp farm, the area of the initial infestation, borders the Ahuriri lagoon. The area was predominantly pasture, with two distinct regions of saltmarsh, dominated by

Sarcocornia, that totalled ~1–2 km^2. The area contained ~10–20 km of drainage channels of variable width and several large shallow depressions, all of which were proving very attractive to *Campto*.

The farm region also included the Department of Conservation-controlled wild life refuge adjacent to the suburb of Westshore. This estuary provides important feeding and resting areas for over 70 species of migratory and domestic waders, gulls (Laridae), terns (Sternidae), herons (Ardeidae) and shags (*Phalacrocorax* spp.). Migratory birds that feed here through the summer include the bar-tailed godwit, lesser knot and golden plover. NZ birds that set up home in Ahuriri over the winter include the wrybill, pied stilt (*Himantopus himantopus*) and royal spoonbill (*Platalea regia*) from the South Island. Porangahau beach at the southern most point of this incursion area is a breeding site for the nationally endangered banded dotterel *(Charadrius bicinctus*) and variable oyster catcher.

The northern-most site to be treated in Poverty Bay was the Te Wherowhero lagoon – a 160 ha tidal lagoon at the base of the peninsula of Young Nicks head near Gisborne. This lagoon borders the Murawai swamp system, which is the biggest freshwater wetland remaining in this ecological district. Te Whereowhero lagoon is also home to the banded dotterel.

Mangawhio lagoon is a small lagoon at the base of the Mahia peninsula and close to the Mahia township. It had no special distinguishing characteristics other than being a wetland of reasonable size and health.

Kaipara

Campto was first found in the southern Kaipara district in February 2001 . Kaipara is situated in the North Island of NZ, north of Auckland. The district includes the low hills around the northern shores of the Kaipara Harbour – a large natural harbour open to the Tasman Sea, which is fed at its northern end by the Wairoa River and its tributaries. Kaipara Harbour covers 947 km^2 (366 sq. miles) at high tide, with 409 km^2 (158 sq. miles) exposed as mudflats and sand flats at low tide. The harbour shoreline is convoluted by the entry of many rivers and streams and is ~800 km (500 miles) long, being the drainage catchment for ~640 000 ha of land. Large numbers of rivers and drainage ditches provide ideal habitat for *Campto*.

The Kaipara Harbour includes sites of high ecological importance, including the wetlands and saltmarsh in areas such as Manukapua (Big Sand Island), Papakanui Spit, Waionui Inlet and Omokoiti. *Campto* was found at several sites around the Kaipara Harbour, at Whitford inland to the east and 39 km to the east on the Whangaparaoa Peninsula (one of the northern most suburbs in Auckland).

Kaipara has a warm-temperate climate, with warm, humid summers and mild, damp winters. Nearby Auckland is the warmest main centre of NZ and is also one of the sunniest, with an average of 2060 sunshine hours per annum. The average daily maximum temperature is 23.7°C in February and 14.5°C in July. High levels of rainfall occur almost year-round, with an average of 1240 mm per year spread over 137 'rain days'. Westerly winds predominate.

The Kaipara district consists of the towns of Dargaville, Ruawai, Paparoa, Maungaturoto, Kaiwaka and Mangawhai, as well as the rural area that surrounds them. It has a population of 19 100 (June 2012 estimate), of whom ~4450 live in Dargaville, the seat of the district council. Auckland city (the largest city in NZ with a population of over 1 million) lies just south and east of the southern end of the Kaipara Harbour.

Human activity in the area is diverse. Urban areas are broken up by rural areas that are primarily involved in dairy farming, exotic forestry and horticulture (in particular sweet potatoes). The Kaipara Harbour is well used by both commercial and recreational fishermen, while the Whangaparaoa peninsula is now predominately covered with housing. It is also home to several attractive beaches, which are well frequented by locals and tourists. The painted apple moth eradication response, which involved the aerial spraying of several of Auckland's central suburbs, was underway during the early 2000s and sensitised the Auckland population to the use of insect-control agents.

Kaipara Harbour is the largest harbour in NZ and contains a wide range of habitats, ranging from tidal reaches to freshwater swamps and coastal scrublands. The area includes 125 km^2 of mangrove forest, with sub-tidal fringes of sea grass. The Kaipara is a migratory bird habitat of international significance. It is one of NZ's five most important areas for wading birds and during the summer it is used by 16% of all of NZ's waders (some 35 000 birds). Forty-two coastal species are known to frequent the harbour, with it being critical to the survival of the black stilt (*Himantopus novaezelandiae*), banded dotterel, wrybill and NZ fairy tern (*Sterna nereis davisae*). Rare migratory species that visit the harbour include the bar-tailed godwit, lesser knot and turnstone (*Arenaria interpres*). Significant local resident populations of black swan, pukeko and grey duck also breed in the area. Areas adjacent to the harbour also support some rare botanical species, including native orchids, the king fern (*Ptisana salicina*), and the endangered kaka beak (*Clianthus puniceus*).

The marine and estuarine areas in the Kaipara Harbour are noted breeding grounds for snapper, grey mullet, sand flounder, common sole, kahawai, white trevally (*Pseudocaranx georgianus*), gurnard, yellow-eyed mullet and skates (Rajidae), rays (Batoidea) and sharks (Selachimorpha). Of particular importance is the significance of Kaipara for snapper populations. In 2009, National Institute of Water and Atmosphere (NIWA) scientists discovered that 98% of snapper on the west coast of the North Island were originally juveniles from nurseries in the Kaipara. Snapper is NZ's largest recreational fishery and is also a commercial fishery with an annual export value of $32 million.

Conservation groups (including 'Forest and Bird' and the 'National Wetland Trust') consider that this area is so ecologically important that they are seeking Ramsar status for the Kaipara Harbour. This will give international recognition to the importance of these wetlands.

Marlborough – Wairau and Grassmere

Campto was first identified in 2003 in the Wairau/Grassmere region on an estimated 1200 ha of potential habitat (Wairau Lagoons: 1200 ha and Lake Grassmere: 30–80 ha). Marlborough is located in the upper north-east of the South Island. The infested area lies 20 km south-east of Blenheim, the largest town in the region. Around 1800 ha of the lagoons are owned under Crown tenure as conservation estate, with the balance owned by Marlborough District Council and Mount Vernon estate. Lake Grassmere is a shallow saline lake 38 km south-east of Blenheim that is used as a commercial salt evaporation operation. *Campto* was found in the runnels in fields surrounding the lagoons, brackish streams that traverse surrounding farmland and more-or-less permanent saline ponds (Stark 2005).

The climate is even with low rainfall (655 mm annually) and average temperatures of 8 to 18°C. The prevailing wind is north-westerly, especially in the evenings. It is drier than other incursion sites and ~2°C cooler over winter.

Much of the Marlborough's population is found around the coastal plains around, and to the south of, the mouth of the Wairau, and in smaller settlements along the coast of the Marlborough sounds. Blenheim is the main town in the region with a population of 28 200.

Marlborough is predominately a sheep farming and wine growing area, with some horticultural production close to Blenheim and forestry on the hills. Marlborough is best known for its wine and today represents 62% of total vineyard area in the country. Marlborough also has a significant amount of tourist activity, with most visiting to access the Marlborough sounds, sample wine or gain access to the Southern Alps.

Close to the incursion area, there is a significant wetland and a highly significant Maori archaeological site with a strong *Iwi* (tribal group) presence. Tourism activities are being planned for this area.

The Wairau/Vernon lagoons are significant as a reserve for wildlife and are also known as a significant recreational and archaeological area. Approximately 1800 ha are held as Conservation Area, and managed and administered by the Department of Conservation. The western 40% of this area has an overlying status of Wildlife Refuge. The area attracts around 90 species of birds and the lagoons are especially important for shags, swans, geese and ducks, and gulls and terns. They also support a great diversity of waders and shorebirds. Most notable is the largest breeding colony of royal spoonbill. Winter visitors include the endangered black stilt, white heron, Australasian bittern (*Botaurus poiciloptilus*), Pacific reef egret (*Egretta sacra*), and wrybill. Other notable species include the resident banded rail and a small breeding population of the banded dotterel. Summer visitors from the northern hemisphere include the far eastern curlew (*Numenius madagascariensis*), Asiatic whimbrel *(N. phaeopus variegatus*), red-necked stint (*Calidris ruficollis*), sharp-tailed sandpiper (*C. acuminata*) and curlew sandpiper (*C. ferruginea*). Ramsar status is being sought for this area, which also hosts 22 of NZ's 42 native fish species. The lagoons are notable as a nursery area for various fish, especially for yellow-belly (*Rhombosolea leporine)* and sand flounder.

Coromandel Peninsula

Campto was first identified in the Coromandel in May 2006. The Coromandel Peninsula is situated in the North Island, ~65 km east of Auckland and 130 km north-east of Hamilton. It extends 85 km north from the western end of the Bay of Plenty, forming a natural barrier to protect the Hauraki Gulf and the Firth of Thames in the west from the Pacific Ocean to the east.

The east coast of the Coromandel has mainly medium-sized estuaries and wide sandy beaches separated by rocky headlands and cliffs. Some open coast beaches are backed by extensive areas of sand dunes. The west coast of the Coromandel Peninsula and the Firth of Thames is a mixture of rocky coastal fringe and narrow sand and gravel beaches.

Although the peninsula is close to large centres of population, such as Auckland 115 km by road to the west, and Tauranga 100 km to the south-east, its rugged nature means that much of it is relatively isolated, and the interior and northern tip are both largely undeveloped and sparsely inhabited. A Forest Park covers much of the peninsula's interior. The peninsula is steep and hilly, and is largely covered in temperate rainforest.

Campto was discovered on 300 ha in the northern Coromandel near the Waiaro River, Colville, Waikawau Bay, Kennedy Bay and in Whangapoua/Matarangi.

Coromandel has a subtropical climate with warm humid summers and mild damp winters. Rainfall can be heavier than the other incursion sites, with an average of 1400 mm per annum (up to 3500 mm per annum in some of the higher areas) and it has average temperatures of 13°C in the winter and 22°C in the summer. The prevailing wind is westerly.

Owing to the nature of the land, much of the Coromandel's population is concentrated in a small number of towns and communities along the south-eastern and south-western coasts. Only five towns on the peninsula have populations of over 1000 (Coromandel, Whitianga, Thames, Tairua and Whangamata) and, of these, only Thames has a population of over 5000. The population of several of these centres is highly seasonal, with many Aucklanders having holiday homes in the Coromandel. During the Christmas and New Year holiday period, activity in the area is significantly increased by families and travellers from around the North Island, particularly in Whangamata, and Whitianga. Pastoral farming is the major land use apart from forestry (exotic and native). The other economic driver is tourism, especially ecotourism, with the majority of tourists visiting over the warmer months.

Colville is the most northerly settlement on the Coromandel Peninsula and the most populous area within the treated zones.

Incursions of *Campto* on the Coromandel were limited to estuaries at the four sites named above. There were no outstanding features recorded for any of these 4 sites. However, such areas on the Coromandel are known to be home to good populations of waders and waterfowl. The most common of which are white-faced heron (*Ardea novaehollandiae*), black swan, mallard, grey duck and, sometimes, New Zealand shoveler (*Anas rhynchotis*). Rarer visitors are white heron and cattle egret (*Bubulcus ibis*), while fernbird (*Bowdleria punctata*) can also be found in suitable cover. Because of Coromandel's equable climate, kingfishers (*Halcyon sancta*) often flock here in winter for the insects they need to sustain themselves during the cold weather.

Environmental effects of the control agents

As mentioned above, the MoH's philosophy of 'first do no harm' drove the selection of the control agents to be used in the eradication programme. As also previously mentioned, the MoH had commissioned a literature review (Glare and O'Callaghan 1998) of *Bti* even before *Campto* was first detected in Napier, and S-methoprene was identified in early 1999 as a complementary low-impact chemical worthy of consideration. Accordingly, a literature review of methoprene was commissioned from the same authors in 1999 to identify whether in fact it did have low non-target impacts. Several other chemicals were available in NZ and registered for use against mosquitoes at this time, but their use was never considered because they were less specific to mosquitoes and were known to have greater environmental impacts than the two being reviewed.

The following section looks at the primary findings of these two reports and the potential impacts the reports identified in relation to the environment.

The use of S-methoprene has been well discussed in Chapters 2 and 6, and so this chapter will limit itself to comments on non-target species. *Bti*, however, has not been as extensively covered and more background is provided below before discussing its potential environment impacts.

Bacillus thuringiensis israelensis (*Bti*)

In 1975–76, the World Health Organization sponsored a project in Israel examining mosquitoes for the presence of pathogens or parasites. During this survey, a new *Bacillus thuringiensis* (*Bt*) strain was discovered from the Negev Desert (Goldberg and Margalit 1977), which was significantly more toxic to mosquitoes than other known bacterial strains available at that time. Although other dipteran-active *Bts* were known, *Bti* was found to be relatively specific to Diptera and was quickly shown to be toxic to a range of mosquito species.

Bacillus thuringiensis occurs naturally in NZ, especially in soil and insects. Aquatic environments have not been extensively sampled, but strains similar to *Bti* have been found. The natural role of *Bt* is not clear, however, and several hypotheses exist. It could be a natural insect pathogen, although it rarely recycles in insect hosts, or it could be a soil bacterium with unassociated insecticidal properties.

Bacillus thuringiensis is a Gram-positive, rod-shaped, spore-forming bacterium. *Bti* culture products contain the spores and parasporal crystals of *Bti* H-14 serotype, which when ingested by the larval stage of the mosquito cause death. They act after ingestion, when the parasporal crystals are solubilised in the alkaline larval midgut and proteolytic activation of the soluble insecticidal crystal proteins occurs. The toxin binds to a receptor on the midgut cell wall, resulting in pore formation in the cell and leading to death of the larva.

Bti-treated mosquito larvae generally cease feeding within 1 h, show reduced activity by 2 h, extreme sluggishness by 4 h, and general paralysis by 6 h after ingestion (Chilcott *et al.* 1990). *Bti* is highly pathogenic against Culicidae (the mosquitoes) and Simuliidae (blackflies – known as 'sandflies' in NZ), and has some virulence against certain other Diptera, especially Chironomidae (the non-biting midges). There are few records of susceptibility outside these dipteran groups, and most other hosts within Diptera require high doses to kill them.

Products based on *Bti* have been used in many countries, with extensive mosquito- and blackfly-control programmes based on *Bti* occurring in West Africa, USA and Europe. In particular, *Bti* has been used in areas considered environmentally sensitive. With respect to mosquito control, one disadvantage of *Bti* is its relatively low persistence in the environment – thus limiting the time that it is active. For this reason, a complementary and more persistent chemical tool was needed to provide control over a longer period. Hence the selection of S-methoprene.

The literature review by Glare and O'Callaghan (1998) assessed studies that had been conducted on the effects of *Bti* on a range of species and for a range of secondary effects. These were considered under the following headings: insects; fish and amphibians; mammalian toxicity; persistence and activity in the environment; gene transfer; and anticipated effect on NZ flora and fauna, as described below.

Numerous studies on the effects of *Bti* on invertebrates, vegetation, fish, birds and mammals have established that *Bti* causes little non-target mortality. Invertebrates as a group are considered to have the greatest potential for an adverse reaction to *Bti*. The 1998 report reviewed studies of effects on nine orders of insects including 148 species. Of the invertebrates tested, only Chironomids (non-biting midges) showed some susceptibility, but only at high rates of application (at least 10 times the rates needed for mosquitoes) and so it was considered that any reduction in natural populations would have been minimal.

The 1998 review looked at studies on 17 fish and two amphibian species, and found no evidence of detrimental effects. It recognised that the World Health Organization had also reviewed several laboratory and field studies on frogs, newts, salamanders and toads, in which no adverse effects were recorded (WHO 1992).

Mammalian safety tests showed a very low safety risk from direct exposure. One cautionary note with *Bti* was a trial that showed that mice were affected by the solubilised endotoxin when it was administered by injection. The solubilised toxin had also been found to be cytolytic to human erythrocytes. However, solubilisation occurs at high pH (such as in insect guts) and does not occur in mammalian guts. No effect had ever been found after ingestion by mammals and the report concluded that *Bti* had no significant effect on mammals. Humans are, of course, mammals and it was noted that *Bt*s have been used for many years in agriculture and forestry situations with no reported problems. Most workers in the field of safety of *Bti* had found no reason to discontinue the use of *Bti* on grounds of risk to human health (e.g. Siegel and Shadduck 1990; Becker and Margalit 1993; Drobniewski 1994; Becker 1992).

Concerns about mammalian susceptibility to naturally occurring *Bacillus* species have arisen because one species, *B. anthracis*, is a virulent mammalian pathogen and others, including *B. cereus*, have been isolated from wounds (Warren *et al.* 1984). However, extensive trialling across several species (including immune-suppressed mice), has found no toxicity to *Bti*. Proven cases of *Bt* causing clinical disease in mammals remain extremely rare and, after reviewing available data, Drobniewski (1994) concluded that the risk to public health from *Bti* was extremely small.

The safety of *Bt*, when used over human populations, had been further supported by a study commissioned on the health of Aucklanders following the control of tussock moth using *Bacillus thuringiensis konkukian* over urban Auckland in 'Operation Evergreen' in 1996–97. *Btk* is similar to *Bti* but is a different *Bacillus* serotype that is often used to control Lepidopteran pests. The study concluded that in the 2 years following the operation, there were:

- no miscarriages or premature deliveries associated with *Btk* spraying
- no allergies caused by the bacteria
- no increased attendance at Baycare health centre as a result of the spraying
- no increased incidence of measles or meningococcal disease.

Persistence of an insecticidal agent has a strong influence on the risk to non-target organisms and the environment. A persistent agent is more likely to control the target pest, but is also more likely to have unintended effects. *Bti* does not usually persist very long after application; generally, toxicity to mosquitoes persists for only days and efficacy can be reduced within 24 h. Formulation and application techniques can extend the persistence of activity for over 1 month in some situations, but activity remains sensitive to factors such as UV degradation, and persistence cannot be extended any further.

Due to the ubiquitous nature of *Bacillus thuringiensis*, concerns were initially expressed about the potential for gene transfer between the *Bti* and natural populations. Most toxin-encoding genes in *Bti* are based on plasmids, extra-chromosomal DNA that can be transferred from cell to cell. Studies on transfer of genetic elements in the environment are in their infancy, but, although it was demonstrated in the laboratory that *Bt* strains can transfer toxin-encoding plasmids to other bacterial species, there is no evidence that such transfer has ever resulted in a dangerous new combination in nature.

The authors of the review concluded that *Bti*'s effect is restricted in host range to the insect subclass Nematocera, with relatively few effects recorded outside this group. As a consequence, non-target mortality would be expected in only the Culicidae, Chironomidae and Simuliidae, of which there are representatives in NZ.

It was considered, however, that if effects did occur they would only occur at the area of application, because *Bti* does not recycle readily in populations. Further, the Simuliidae, in particular, are also a pest species in some areas of NZ, their control with *Bti* has previously been investigated (Chilcott *et al.* 1983) and actually there may be some benefit in that respect.

Methoprene

S-methoprene is a true analogue and synthetic mimic of a naturally occurring insect endocrine system product called juvenile hormone (JH). Methoprene is not directly toxic to insects but prevents the emergence of adults from the pupal stage, which eventually die. Sublethal effects from methoprene application include reduced fecundity, abnormal morphologies/development, alterations in pheromone production and altered behaviour. Methoprene may also be toxic to mosquito eggs under some circumstances.

Methoprene had previously been used extensively for pest control but the 'general substance' contained both R- and S-isomers and was not as clean as the S-isomer based product. In 1998, David Sullivan (*pers. comm.)* reviewed the environmental and health impacts of S-methoprene for the Chief Technical Officer (Health). This review showed that when used at label approved rates, S-methoprene would not:

- kill or harm fish
- cause frogs to develop abnormally
- prevent normal development of crustaceans, such as crabs and lobsters, or prevent the development of chitin (the outer shell). There are chitin inhibitors developed for insect control, but S-methoprene is not one of them.
- cause the food chain to be reduced for predatory animals. Mosquito larvae are still present as a food source and, when eaten, S-methoprene will not accumulate in the predator.

Because S-methoprene was relatively unproven in NZ at least, it was decided that a literature review should also be conducted by Glare and O Callaghan (1999). This review was more detailed than the 1998 survey of *Bti* because the use of methoprene in mosquito programmes had not been as widespread as *Bti*. Many laboratory studies showed some non-target effects but the authors noted that subsequent field studies at control level application rates resulted in little or no effect. The authors looked carefully at the full range of life forms that could be affected and noted the potential effects on plants, insects, aquatic organisms, frogs and mammals as described below.

With regards to plants, there was no effect observed except for one study that showed slight phytotoxicity through the slightly delayed appearance of female flowers in *Cucurbita pepo* (spaghetti squash).

Insects studies showed greater effects, with mortality in 12 orders of insects and in mites when they were treated at a species specific dose of methoprene. Susceptibility to methoprene, however, varied greatly, and was dose dependent. Lethal doses for mosquitoes were much lower than for other classes of insects, with Lepidoptera requiring dose levels around 100 times higher for equivalent mortality. This resulted in methoprene at field rates being effective against mosquitoes without affecting non-target insects. It was noted that, internationally, methoprene is used selectively to control insect pests, especially mosquitoes, horn flies, ants and fleas.

When the reviewers considered field applications, they noted that methoprene resulted in few negative impacts on non-target organisms, with the exception of some aquatic invertebrates, particularly benthic communities; however, such communities appeared to recover quickly (Majori *et al.* 1977; Bircher and Ruber 1988; Yasuno and Satake 1990; Hershey *et al.* 1995; Retnakaran *et al.* 1974).

Honey bees are often singled out in studies because of their importance in pollination for both domestic and wild plants. There were some contradictions in the reported effects of methoprene on honey bees. Barker and Waller (1978) concluded that methoprene was relatively safe for honey bees, while other work suggested that methoprene may be harmful because it had been shown to significantly reduce wax secretion, indicating methoprene may be sublethal and poisonous to worker honey bees (Muller and Hepburn 1994).

Studies on rotifers, molluscs and microcrustaceans found little effect, but there was some indication that methoprene may have an effect on macrocrustaceans, such as shrimps and crayfish, and that these effects were probably a result of interrupted larval development. The US Environmental Protection Agency (1991) summarised available fish studies, concluding that methoprene at levels higher than were used for mosquito control is moderately toxic to warm water, freshwater fish and slightly toxic to cold water, freshwater fish (Anon 1973), but the review concluded that acute fish toxicity would not be expected during control programmes because the concentration of methoprene in water at any one time is unlikely to exceed 2 ppb (Ross *et al.* 1994).

The review identified only one area of concern and that was its possible role in the development of deformities in some frog species in North America. Methoprene was suggested as one of the possible causes of the deformities in the early 1990s, generating a lot of publicity and speculation. Possible causes included a parasite of frogs, ultraviolet radiation or chemical contamination (Kaiser 1997), or some combination of the three (Manuel 1997). However, there was no definitive evidence that S-methoprene, or its breakdown products, caused the types of malformations being observed at the concentrations being applied. Although NZ is home to several indigenous species of frogs, none of these species live in the saltmarsh habitat and so this particular concern was not sufficient to prevent its use.

Studies have shown that methoprene has no effect on mammals. Garg and Donahue (1989) reviewed the activity and safety of methoprene used to control insect pests of cattle, dogs and cats, and reported that methoprene could be considered safe by insecticide standards, with no recordable effects. Further, toxicological evaluations conducted in the laboratory on pigs, sheep, cattle, dogs, rats, rabbits, hamsters and guinea-pigs have revealed no clinical signs of toxicosis. Based on these and similar studies, the WHO had concluded that it could approve methoprene for use in drinking water for control of container mosquitoes because it was regarded as posing minimal or no risk to humans, animals or the environment.

Following application against mosquito larvae, methoprene is more persistent in the habitat than *Bti*, but methoprene is rapidly degraded in the environment and in animals. Methoprene undergoes hydrolysis, demethylation and oxidative scission in microbes, plants and insects. In animals, another pathway is used to convert it into natural biochemicals. Metabolic bioproducts identified are biologically innocuous compounds. Extremes of temperature can influence persistence and activity, but methoprene is relatively unaffected by temperatures between 10 and 25°C and ultraviolet light rapidly degrades methoprene.

Anticipated effect on NZ flora and fauna

After these two reviews, S-methoprene was regarded as suitable for use. The only species that could be identified that would potentially be adversely impacted were NZ's mosquito species and some non-target benthic organisms.

The Chief Technical Officer (Health) recommended that S-methoprene and *Bti* should be used because they were the least environmentally damaging pesticides available for mosquito control or eradication. Both chemicals had low toxicity when used at field rates against mosquitoes, and posed low risk to humans and other non-target species. At these rates, repeat applications may reduce benthic insect biomass and, in particular chironomid midges, which make up a large part of the diet of many organisms that inhabit wetlands. However, field studies undertaken in North America had shown that this had not resulted in flow-on effects to other species, because organisms that fed on chironomids were able to switch to other food sources. Further, because pesticide applications in the eradication programme would be made to relatively small areas, recolonisation by any affected non-target organisms would rapidly occur from surrounding untreated areas.

It is notable that, although the Department of Conservation (NZ's manager and protector of its natural environments) had had some reservation about the use of these chemicals, these reviews and their recommendations resulted in the Department giving their blessing for their use. At the end of the day, the Departmental staff considered that the long-term adverse effects of the mosquito would outweigh any temporary effects of the pesticides.

Habitat management and steps taken to minimise the effect of the treatment programme

Once the decisions had been made by the Chief Technical Officer (Health) to proceed with the eradication programme using *Bti* and S-methoprene, the operational managers applied these chemicals to maximise success and minimise the adverse impacts on ecosystems and humans. This meant they needed to look at each area, use the science to help them apply the chemicals at the right time and make habitat changes to prevent *Campto* surviving in hard to get to or in cryptic spots. Because of the capacity of *Campto* to reproduce in large numbers, survive adverse conditions and disperse widely, an objective of killing less than 100% of the mosquitoes would not have resulted in successful eradication.

Habitat elimination and modification were also an important part of the eradication programme within defined areas. Habitat modifications were always restricted in size and to quite defined areas so that, as for the pesticide use, any negative effects were expected to be limited and short lived.

Any modifications were made to remove areas where *Campto* could live and breed, remove vegetation that may prevent spray penetration or to flood soils and so stimulate hatching before spray treatments. These activities were undertaken by landowners or by staff employed as part of the incursion response. No formal analysis of the impact of these activities was undertaken by either the TAG or the Chief Technical Officer (Health). However, every action was carefully considered, most were temporary, and only those that promised a positive result were used. Activities included:

- clearing drains to remove weeds and ensure water flows increased, and preventing ponding that would be suitable for *Campto*, This was particularly important in areas such as Ahuriri where the neighbouring Landcorp farm had 20 km of drainage ditch on land with minimal fall.

- artificial flooding of farm drains with saline water to promote egg hatch so that treatments could be timed most effectively. This was used extensively in the Kaipara to minimise the number of broad-scale sprays that were needed in a season to kill all of the larvae.
- allowing other drains to dry out completely during drier periods of the year so that they did not have to be treated.
- filling depressions to eliminate ponding – an activity that was used in Ahuriri where the large shallow depressions were of concern.
- removal of dense/thatched vegetation was used in several sites. In the Wairau, Department of Conservation staff were particularly appreciative when the programme paid for the removal of exotic trees and shrubs from one of their reserves, not only assisting the eradication programme but removing unwanted vegetation from a conservation area.

There were times, however, where it was decided that habitat modification should not proceed. This was particularly so at Shakespear Regional Park, west of Kaipara, where the wetland was being actively managed to encourage its regeneration and it was deemed that altering the habitat would have an adverse impact on its recovery. The area was left unmodified.

Operational managers were also careful to minimise their impact on people and wildlife. Pesticide application always involves either helicopters or other machinery. These will potentially disturb wading birds and waterfowl, scare livestock and annoy humans. As a consequence, the following protocols were put in place during treatment. To minimise the disturbance of water birds, helicopters did not fly over wading bird roosting sites at high tide or in the early morning when the majority of the birds were roosting. Farmers were either asked to remove livestock from adjacent areas or care was taken to ensure that livestock remained in the same paddock (at times this meant checking them after treatments and returning them to their paddock). In sites where there was a lot of human activity, care was taken to avoid times when visitor numbers were high (i.e. treatment was undertaken early in the day). In the Wairau, one application was delayed by a week to accommodate duck shooting!

So were there any lasting effects of the eradication programme?

It cannot be said definitively whether there were any lasting environmental effects on the treated areas, because there was no organised monitoring put in place to measure any impacts on the treated ecosystems. Not that this wasn't considered – it was. In 2005, the MoH commissioned a report to determine whether monitoring should be put in place, and the findings from this report recommended that monitoring was not required. Based on the two reviews described above, and their updates, Stark (2005) noted that the reviews had identified that 'the environmental impacts on non-target species of the eradication programme were considered to be very low. There was sufficient use of *Bti* and S-methoprene and reporting on their relative environmental safety to consider that these treatment agents were very unlikely to cause any long-term or irreversible impacts to damage to non-target species'.

Stark (2005) also noted that 'While some impact on non-target organisms (especially in aquatic communities) was predicted, the effects of methoprene application were considered to be less harmful than those caused by most mosquitocidal pesticides. While S-methoprene had longer persistence than *Bti* after application, there was no indication in the literature of permanent disruption to ecosystems after methoprene application. Thus, use of S-methoprene and/or *Bti* in an eradication programme was considered as safe and should not result in long-term negative effects to ecosystem function'.

Although there was no formal monitoring of the treated areas, some were being actively managed by either Department of Conservation or the local regional council. Two in particular (as discussed below) are regularly reported on and neither has reported significant issues arising from the eradication programmes.

Ahuriri is assessed annually by the Hawke's Bay Regional Council and, in the Kaipara, Auckland Regional Council has put in place regular monitoring from 2005. The baseline survey of the Kaipara showed that 'A 2005 broad-scale survey of habitats and ecological communities suggests that many areas of the southern Kaipara display high diversity and several organisms living in the harbor are large and long-lived. Several invertebrate species commonly associated with pristine environments (sponges, ascidians, bryozoans, hydroids, echinoderms and pipis) were also found'. This survey took place during the eradication.

Although there is therefore no definitive evidence that the eradication programme had no effect on the environment, we do know that no noticeable effects have been recorded despite quite high levels of activities in these areas. New Zealanders at large would also agree that, without *Campto*, our natural environments have been maintained nearer to their natural state and are much nicer places to visit than they would have been if this mosquito had been allowed to establish and disperse unhindered.

References

Anon (1973) *Altosid.* Technical Bulletin. Zoecon Corporation, Palo Alto, California, USA.

Barker RJ, Waller GD (1978) Sublethal effects of parathion, methyl parathion, or formulated methoprene fed to colonies of honey bees. *Environmental Entomology* 7, 569–571.

Becker N (1992) Community participation in the operational use of microbial control agents in mosquito control programmes. *Bulletin of the Society for Vector Ecology* **17**, 114–118.

Becker N, Margalit J (1993) Use of *Bacillus thuringiensis israelensis* against mosquitoes and blackflies. In: *Bacillus thuringiensis, an Environmental Biopesticide: Theory and Practice.* (Eds PF Entwistle, JS Cory, MJ Bailey and S Higgs) pp. 147–170. John Wiley & Sons, New York.

Bircher L, Ruber E (1988) Toxicity of methoprene to all stages of the salt marsh copepod, *Apocyclops spartinus* (Cyclopoida). *Journal of the American Mosquito Control Association* **4**, 520–523.

Browne G (2005) Biosecurity surveillance of mosquitoes in New Zealand with a case example of methods used for the eradication of the Australian southern saltmarsh mosquito *Ochlerotatus camptorhynchus.* PhD thesis. University of Auckland, NZ.

Chilcott CN, Pillai JS, Kalmakof J (1983) Efficacy of *Bacillus thuringiensis* var. *israelensis* as a biocontrol agent against larvae of Simuliidae (Diptera) in New Zealand. *New Zealand Journal of Zoology* **10**, 319–325. doi:10.1080/03014223.1983.10423921

Chilcott CN, Knowles BH, Ellar DJ, Drobniewski FA (1990) Mechanism of action of *Bacillus thuringiensis israelensis* parasporal body. In *Bacterial Control of Mosquitoes and Black Flies.* (Eds H de Barjac and JD Sutherland) pp. 45–65. Unwin Hyman, London.

Clarkson BR, Sorrell BK, Reeves PN, Champion PD, Partridge TR, Clarkson BD (2003) *Handbook for Monitoring Wetland Condition. Coordinated Monitoring of New Zealand Wetlands.* Ministry for the Environment SMF funded project, Wellington, NZ.

Drobniewski FA (1994) The safety of *Bacillus* species as insect vector control agents. *The Journal of Applied Bacteriology* **76**, 101–109. doi:10.1111/j.1365-2672.1994.tb01604.x

Environmental Protection Agency (1991) *R.E.D. Facts: Methoprene.* Pesticides and toxic substances no. 738-F-91–104. Environmental Protection Agency, Washington DC.

Garg RC, Donahue WA (1989) Pharmacologic profile of methoprene, an insect growth regulator, in cattle, dogs, and cats. *Journal of the American Veterinary Medical Association* **194**, 410–412.

Glare TR, O'Callaghan M (1998) 'Environmental and health impacts of *Bacillus thuringiensis israelensis*'. Report for the Ministry of Health, AgResearch, Lincoln, NZ.

Glare TR, O'Callaghan M (1999) 'Environmental and health impacts of the insect juvenile hormone analogue S-methoprene'. Report to the Ministry of Health, Lincoln, NZ.

Goldberg LH, Margalit J (1977) A bacterial spore demonstrating rapid larvicidal activity against *Anopheles sergentii, Uranotaenia unguiclata, Culex univattatus, Aedes aegypti* and *Culex pipiens. Mosquito News* **37**, 355–358.

Hershey AE, Shannon L, Axler R, Ernst C, Mickelson P (1995) Effects of methoprene and *Bti* (*Bacillus thuringiensis* var. *israelensis*) on non-target insects. *Hydrobiologia* **308**, 219–227. doi:10.1007/BF00006873

Kaiser J (1997) Deformed frogs leap into spotlight at health workshop. *Science* **278**, 2051–2052. doi:10.1126/science.278.5346.2051

Majori G, Bettini S, Pierdominici G (1977) Methoprene or Altosid for the control of *Aedes detritus* and its effects on some non-targets. *Mosquito News* **37**, 57–62.

Manuel J (1997) Frog deformities research not leaping to conclusions. *Environmental Health Perspectives* **105**, 1046–1047. doi:10.1289/ehp.971051046

Muller WJ, Hepburn HR (1994) Juvenile hormone III and wax secretion in honeybees (*Apis mellifera capensis*). *Journal of Insect Physiology* **40**, 873–881. doi:10.1016/0022-1910(94)90021-3

Retnakaran A, Jobin L, Buckner CH (1974) 'Experimental aerial application of a juvenile hormone analog against the Eastern Hemlock looper *Lambdina fiscellaria fiscellaria* (Guen.) in Anticosti Island in July 1973'. Information Report, Insect Pathology Research Institute, Saulte Ste. Marie, Ontario, Canada.

Ross DH, Judy D, Jacobson B, Howell R (1994) Methoprene concentrations in freshwater microcosms treated with sustained-release Altosid formulations. *Journal of the American Mosquito Control Association* **10**, 202–210.

Siegel JP, Shadduck JA (1990) Mammalian safety of *Bacillus thuringiensis israelensis.* In *Bacterial Control of Mosquitoes and Black Flies.* (Eds H de Barjac and DJ Sutherland) pp. 202–220. Unwin Hyman, London.

Stark JD (2005) 'Recommendations for estimating pesticide effects on nontarget organisms during mosquito eradication programmes in New Zealand'. Report for the Ministry of Health, Wellington, NZ.

Warren RE, Rubenstein D, Ellar DJ, Kramer JM, Gilbert RJ (1984) *Bacillus thuringiensis* var. *israelensis* protoxin activation and safety. *Lancet* **323**, 678–679. doi:10.1016/S0140-6736(84)92189-5

World Health Organization (1992) 'Fourteenth report of the WHO Expert Committee on Vector Biology and Control, Safe Use of Pesticides'. Technical Report Series No. 813. WHO, Geneva.

Yasuno M, Satake K (1990) Effects of diflubenzuron and methoprene on the emergence of insects and their density in an outdoor experimental stream. *Chemosphere* **21**, 1321–1335. doi:10.1016/0045-6535(90)90148-M

12

Communications and cultural issues

JR Gardner, Val Aldridge and Bryn Gradwell

Communications: an instrument of management by means of which all consciously used forms of internal and external communication are harmonized as effectively and efficiently as possible, with the overall objective of creating 'a favourable basis for relationships with the groups upon which the organisation is dependent' (Van Riel 2007).

Communicate: Share or exchange information or ideas (Concise Oxford Dictionary 11th Edn)

Health: A state of complete physical, mental and social well-being and not merely the absence of disease or infirmity (World Health Organization)

Introduction

If this chapter had been given a more whimsical approach, it might have been introduced as something akin to this:

'It was a dark and stormy night in the December of 1998 ... Little did the residents of the sleepy Hawke's Bay suburb of Westshore realise that within a week of a drought-breaking deluge they would be tormented by a plague of blood-lusting mosquitoes that descended on them and drove them from their gardens and BBQs to seek respite indoors, battling the voracious winged hordes with cans of fly spray and lashings of DEET.'

And the narrative would then have finished with:

'Sometime, possibly in early spring of 2008, a forlorn female southern saltmarsh mosquito, desperately seeking sexual congress, fluttered through the ripe Wairau habitat hunting for a lusty male to mate with, alas the Mossie Singles Bars were void of suitable dudes and she fell to earth her appetite to breed un-satiated.'

This chapter is different from the others in that the success or failure of the communications strategies was never formally measured, although we all considered it important to national effort. The test for eradication is a simple one: no more *Campto* detected. The measure of achievement or failure of the communications is more subjective. For the *Campto* incursion, the MoH did not conduct formal surveys or polling to assess the effectiveness of the communications efforts that were delivered, but used indicators such as the numbers of public complaints, the numbers and the gist of *Official Information Act 1982* requests, and the tenor of the commentary from the print and electronic media.

The purpose of this chapter is to discuss the application of 'communications' as a core component of the management of a response to a public health contingency, and then to describe how the MoH, the MAF and the many other participants in the *Campto* eradication programme shaped their communications strategy to support the operation. The chapter has been sequenced into several sections as follows:

- first there is a general discussion on 'communications' and this will outline the relevant principles and the techniques that are available and that were applied by the MoH during the eradication programme
- a brief overview of activities during the period before the detection of *Campto* in 1998 and just after the response commenced
- a commentary on the development and implementation of the communications effort at the strategic level that followed the initial detection of *Campto* in Napier
- a detailed case study of the communications strategy for the Kaipara, including the efforts to establish relationships with stakeholders at all levels, recognition of cultural issues and rigorous efforts made to sustain communications
- a 'timeline' of the various phases of the response using injections of examples of the electronic and print media comments that were made. This will illustrate the nature of the commentary that was made.

Communications – just another health crisis

It should be understood that, although the *Campto* event was a novel one for the MoH (that is, the event was the first time a response to the detection of an exotic mosquito incursion that led to its eradication had been mounted in New Zealand), responding to emerging health crises is very much business as usual for the MoH's communications staff. Health has a high profile that attracts intense scrutiny from the public. Health issues provide the media with excellent copy, and health events can be fed into the relentless maw of 24/7 news industry where there is always an insatiable appetite for 'heartbreaking' accounts of personal misfortune.

On another scale, potential health threats to the entire community also gain good traction in the media, particularly if it is a new emerging threat that has a degree of uncertainty regarding outcomes and possible 'life-threatening' health impacts. We are well used to the international response that occurs when there is the mention of a new outbreak of SARs or the discovery of Ebola virus and the subsequent flurry of media excitement that will travel around the world (WHO 2005). It can be said with some certainty that the communications staff in the health sector have had plenty of experience of being under the 'blowtorch' when such events occur.

Is communications an art or a science? In the modern context, communications should not be mistaken for what was once called 'public relations'. Legacy public relations was often characterised by churned out media releases to the Press rendered by 'hack' journalists. The content was strongly focused on the corporate view of the world, delivered in a patronising 'we know what is good for you, so do what you are told' style. There was little, if any, cognisance of contrary opinions, interests or concerns.

Such blunt techniques are unsophisticated and their usage risks the public becoming disenchanted with the authority disseminating the information. Before 1999, the MoH had endured several health crises, (the 'Unfortunate Experiment' and the Gisborne cervical screening inquiry to name just two), which were events that had put the MoH's reputation at risk. From these experiences, it had been learned that an effective communications strategy was essential if the public's anxieties were to be tempered, and that their concerns needed to be recognised and dealt with using non-patronising language. This required communications to be delivered 'sensitively' and in a manner that allowed the public to take ownership and become part of the solution to the problem – an 'informed consent' paradigm (WHO 2005).

NZ's society at the end of the millennium had been affected by substantive changes in politics following the radical reforms of 1985 and the introduction of proportional representation in 1996. The social landscape had also changed. The wider community had become more questioning of authority and were far less inclined to knuckle under and comply with dictates. The community was suspicious of any perceived attempts at obscuration by officialdom. Any apparent stone-walling or the release of unclear or incomprehensible information leads immediately to the suspicion that the truth is being supressed/perverted and that the citizens' right to know what they are being exposed to has been abrogated. The evolvement of the world-wide web has provided a powerful tool for those prepared to challenge officialdom's version of events because it offers a very deep well of information and disinformation that could be trawled to provide plausible contrary commentary. The web also provides a very economical platform for exchanging information and generating a divergent constituency that can be mobilised to challenge officialdom's dogma.

Health is 'special'

The NZ public is easily beguiled by images of cute seals, frolicking dolphins, majestic kauris, comical keas and of course the myopic ground-hugging kiwi. Threats to these icons are met with dismay and a swift reaction to mobilise resources that can be deployed to defeat an emergent hazard. On the production side, pastoralists have the spectre of foot-and-mouth disease being introduced to our domestic cattle and sheep populations. For foresters, there is an assortment of invertebrate pests that have cast a shadow over their industry (including incursions of white-spotted tussock moth (*Orgyia thyellina*), painted apple moth (*Orgyia anartoides*) and Asian gypsy moth (*Lymantria dispar dispar*) and then there has been the pain of other incursions such as the varroa mite (*Varroa destructor*), PSA kiwifruit virus (*Pseudomonas syringae* Pv *actinidiae*) and didymo (*Didymosphenia geminate*). New Zealanders are well apprised of the possible risks to our biodiversity and our economy if unwanted organisms are introduced (see the MAF's list of biosecurity threats http://www.biosecurity.govt.nz/pests/animals).

However, it is contended that the public, although enthusiastically responsive to threats to endemic biodiversity, is far more receptive to risks (real or perceived) to human health. This has both an upside and a downside.

The **upside** is that the public's attention is easily aroused (in fact their engagement might become uncomfortably intense). Cooperation can be more easily obtained if both community and individual interests can be demonstrated to be threatened by either pathological impact, or loss of 'lifestyle values' through increased nuisance exposure, or both.

The **downside** is that in an 'emergency' situation the community's scrutiny of the remedies that are being implemented to deal with the problem will be intense. There will be challenges based on science of varying quality, as well as more emotive and less qualitative objections, if not outrage. To sustain a high-quality communications strategy in such an environment takes both careful planning and careful implementation. Dramatic public health events will energise the media who will sensationalise the occurrence. The response will be under an intense spotlight with no shortage of folk ready and eager to add their own commentary. After all, when it comes to personal health, we are all experts (and often apprehensive).

When there is a requirement to mount a response to a public health event, the need for, and the content of, a communication strategy should be one of the earliest considerations. This strategy must be reviewed and revised as the process continues. There should be a continuous effort to better understand people's concerns about the risks and how to communicate in a way that will be most effective (WHO 2009).

Characteristics of a public health event

When public health events occur, they are often characterised by:

- **uncertainty and speculation**, particularly if the event has not been forecast or has been allocated a low probability of occurrence
- **setbacks and surprises** – an event may have an unpredictable course
- decision making made on **incomplete** or **unreliable evidence**
- intensified **political aspects** if the event has the potential to increase health risks, or is socially or economically disruptive, and communication decisions may be taken away from the health professionals
- **public anxiety** (and sometimes disquiet among officials)
- **becoming eminently** newsworthy.

Prior to December 1998, the MoH's communications team had learned from a succession of bad news events and had adopted a comprehensive set of procedures for providing communications that engaged the affected communities as partners. The core principle that has to be adhered to is the precautionary principle (Box 12.1).

From this principle, there flow several parameters, which were incorporated into the conduct of the eradication programme's communications strategy:

- Maintain a health culture of informed consent (i.e. giving a priority to getting **all** the information out to the public no matter how embarrassing it might seem). If a

Box 12.1: The precautionary principle (WHO 2006)

When human activities may lead to morally unacceptable harm (that is, scientifically plausible but uncertain), actions should be taken to avoid or diminish that harm.

Morally unacceptable harm refers to harm to humans or the environment that is:

- threatening to human life or health
- serious and effectively irreversible
- inequitable to present or future generations
- imposed without adequate consideration of the human rights of those affected.

mistake is made, then declare it immediately (an example was including information about the 'lost specimens' in the media statement that announced *Campto* had been found in the Gisborne region).

- Use an 'informed patient' model (i.e. assuming the public were intelligent and would make their own minds up so needed full and frank information, and not the 1960s model of 'doctor knows best' or 'trust us, it's good for you').
- Make it clear that this is a pest with a human health impact, not just affecting (remote, commercial) forestry areas.
- Enable people to take action by including information to let them make choices about how they managed any perceived or potential risks to themselves from the mosquitoes (such as how to avoid mosquito bites) or exposures to treatment agents (fully informing people when and where applications were taking place).
- Focus on very careful use of language (e.g. we 'treated' areas (we didn't 'spray' them); the target species was 'vicious', 'aggressive' and 'disease bearing').
- Ensure very full information is made available including hand-outs on the various treatment agents used, all scientific reports were placed on the website, reviews of health/environmental assessments of products to keep up to date on the latest evidence of any potential risks.
- Respond fully and carefully to any *Official Information Act* requests, queries, concerns and other correspondence from the public, media, environmental groups and other stakeholders.
- Support media statements with 'questions and answers' and other background information for journalists to use if they wished to add more detail.
- Provide proactive advice to residents about treatments being made nearby (when and where and what).
- Make sure information is available locally (i.e. the programme was run locally and local people were on hand to talk face to face to any concerned residents if need be).
- Use a highly proactive communications strategy reaching deep into the affected community, including via schools, early childhood centres, medical centres, councils and public notices – and ongoing with regular updates being provided to local media outlets.
- Define and delineate roles and responsibilities between national and local programme delivery.

Great having a plan but that's only half of it; effective delivery is essential

Having a set of core principles to follow is one thing, but for a communications strategy to succeed there needs to be considerable discipline exercised by those who deliver it. The public's confidence in the management of the event will be lost if:

- the authority's staff appear to be ill informed about technical details
- spokespersons deliver opaque or contradictory messages
- the communications initiative is lost to another organisation or personality that assumes the mantle of being the expert on the matter and they become the 'go to' person for media comment.

To ensure none of the failings listed above are allowed to occur, there must be a degree of discipline if the communications delivery is to be effective. This is very important when the event occurs over a long time line of months or even years. The MoH maintained a disciplined approach by ensuring:

- key messages were formulated and these were adhered to by all spokespersons at all times
- spokespersons were 'hands on' technical staff, who could 'talk the talk' with authority rather than bland 'communications people'
- external experts were identified who agreed to be available to talk to journalists if required.

For the *Campto* incursion, the MoH had several spokespersons available to respond to incoming media inquiries. These people were not media 'talking heads' – they were just plain ordinary folks that were easy to understand and to relate to. They were:

- Henry Dowler, the Deputy Chief Technical Officer (Health) in 1998, who had as much knowledge as anyone in the MoH at that time regarding mosquitoes and the health risks associated with them. He was also an 'operations' guy who wasn't afraid of stepping into the swamp and getting his hands dirty. As a consequence, he could talk with authority about the operational setting.
- Dr Bob Boyd who had an impressive ability to gain the audience's attention with a 'bedside manner' that Dr Findlay would have been proud of. Dr Bob would in quiet, but reassuring tones, inform the listeners that all was well, all the risks were being taken care of and you can all rest easy in your beds.
- Sally Gilbert was another who could earnestly articulate the facts in lay persons' language. She had an 'elephantine' memory of what had gone before, what had been said before and (most importantly) who had promised what – she could never be 'blindsided' by a 'curly' question.
- John Gardner had a 'meat and potatoes, hail good fellow' manner of delivery that seemed to resonate with the media and the public.

The use of designated spokespersons was replicated at the operational level where people such as Steve Garner and Bryn Gradwell became the local 'mossie men' who were always available to provide a comment on any facet of the programme. Completing the layer of legitimacy were internationally recognised technical gurus such as Brian Kay, Richard Russell and Darryl McGinn. The key messages that were hammered again and again were:

- the potential health risks that would result if the mosquito were to became established in NZ
- the high nuisance impact of the species
- the safety of the treatment products being used for containment and eradication
- the NZ Government's commitment to deal with the problem.

At the MoH, the management of the communications followed certain principles that had already been well exercised in response to other high-profile health events. The framework was applied to the *Campto* response for the phases of detection, containment and subsequent eradication. As the situation developed and the programme moved, the phases (containment, suppression and eradication) changed and so the communications strategy evolved. The succession of new detections was to provide a useful proving ground for building a model communications strategy for underwriting public health contingency operations in a rural environment. By the time the Kaipara detection occurred in February

2001, there was a bullet-proof template available for the operational staff to use. This model was then applied to subsequent eradication zones with unqualified success.

Setting the scene

By the end of the millennium, 'mainstream' Kiwis were well advised of the importance of biosecurity to NZ's economic well-being. A succession of incursions as well as 'scares' had made the public reasonably mindful of the risks and what the responses to incursions of unwanted organisms might involve. However, before 1998, NZ had never had to confront the actuality of an incursion of an invertebrate pest that was a realistic threat to human health.

That all changed when *Campto* were discovered in Napier in December of 1998. For the MoH, the question was to be how would the New Zealanders react to the detection of a 'dangerous' vector of human pathogens in the local environment and would they wholeheartedly provide their endorsement of any response to the incursion?

NZ public had become somewhat sensitised to eradication projects

There were some grounds for apprehension. Since the 1960s, public opinion had been influenced by events in other parts of the world, with commentary by writers such as Rachel Carson (*Silent Spring*), and events such as Operation Ranch Hand (defoliation in Vietnam with 2,4,5-T contaminated with dioxin). In NZ, operation Ever Green, the eradication programme for the white-spotted tussock moth, began in October 1996 when caterpillars began hatching from their over-wintering eggs. DC6 aircraft carried out the low-level aerial spraying of the 30 km^2 of infested area in the western suburbs of Auckland which, including a buffer zone, involved 80 000 households in all.

Although this operation was a success in eradicating the tussock moth, the programme was marked by mounting public resistance from the urban dwellers whose houses were overflown during the operations, to the extent that there was intensified public resistance to spraying of residential areas.

In a free society, there will always be those who wish to express a contrary view and there are plenty of platforms for them to express their displeasure. By the time the white-spotted tussock moth eradication programmes were completed, there were small, but vocal, numbers of Kiwis vehemently opposed to 'spraying' of any sort of 'chemical' into the environment and in particular in urban areas.

The MoH realised that there were significant challenges that needed to be overcome if Kiwis were to be convinced they should commit to the implementation of an extensive, prolonged and expensive eradication programme. By having a well-educated constituency who were cognisant of the issues and the benefits, the risks of consumer resistance at the national (strategic) level would be mitigated.

It was clear that the communications plan had to reinforce the 'strengths' and alleviate the 'weaknesses'. The strengths were seen to be:

- There was strong public support for eradication from the immediately affected community. In the first instance, this was the Westfield suburb adjacent to Napier Airport whose outrage was distinctly evident and well communicated.
- The treatment areas in Napier, and at all subsequent incursion sites, were in rural, not urban, environments, and in saltmarsh areas. This meant populated areas were not being overflown and product was not being released onto urban population centres and not over residential dwellings.

- Activities focused in rural areas had another advantage – the people in such areas mostly work in the primary sector and have a good understanding and considerable experience in pest control operations. They, as a group, had a more matter of fact approach and became, once they been provided with substantive information, 'cheerleaders' for the eradication.
- The products were relatively benign. *Bti,* which was used for suppression, is an organic product that is well described in the literature and its impact was specific to invertebrates. The treatment product, S-methoprene, was a solid granule which did not create 'vapours' that might drift into populated areas. S-methoprene's inert qualities and extremely low dosage rates were well accepted by the relatively few householders in the treatment zones. The MoH communications policy highlighted these characteristics and avoided the use of emotive terms such as 'spray', instead referring to 'treatment', a word that in itself has therapeutic connotations.
- The MoH's communications staff had a long history of dealing with an often frightened or aggrieved community, and were well versed in the various techniques of communicating with a potentially hostile audience.

The threats to the eradication programme could have come from several quarters, whether they were individuals, organisations or constituencies, these included:

- Well-meaning individuals with a deep-rooted repugnance to all forms of compulsion, whether that be 'spraying' of chemicals or the imposition of 'mass medication'. This group is instinctively resistant to any initiatives by the 'government' that they felt ignored them or was going to impose on them against their will unnecessary exposure to chemicals that might have a health risk.
- Special interest groups who had concerns that their recreational activities would be compromised by the programme, such as anglers or wildfowl shooters.
- *Tangata Whenua* (local Maori with historical ties to the area – people of the land), seen as a 'core' stakeholder who needed to be persuaded that operational activities could be conducted without having an impact on the local *Iwis*' (Maori tribal groups) responsibility for being guardians of their land, and their entitlement to access, management and utilisation of their *whenua* (land), *wai* (water) and *kai* (food) resources.
- Environmentalists, not limited to just birdwatchers, but including those with a passion for the environment and for the protection of indigenous fauna and flora.
- People with commercial interests that might be at risk, whether from engagement in illicit operations (drug cultivation), which would be compromised by mosquito eradication operations, or organic farming operations with concerns regarding chemical contamination.

The strategic communications response

When Noel Watson sent the first specimens off to Gene Browne for identification on 21 December 1998, he was to light a fuse that was to burn for a decade. The identification of the presence of *Campto* in the wetlands adjacent to Napier airport and the seaside suburb of Westshore led to an immediate 'communications' response from the MoH. From the moment that Gene Browne's identification was passed to Sally Gilbert on Christmas Eve, MoH officials were scrambling to start to deal with the arrival of an exotic mosquito species that had not really been on the 'surveillance' radar. The location of the index site was also unanticipated. Most of our assumptions about, and planning for, incursions of exotic mosquitoes were turned upside down.

MoH officials set about educating themselves as fast as they possibly could about the species, its life cycle, habitat, competency at becoming a vector for human (and zoonotic) pathogens, possible control methods and data on personal protection for exposed individuals. This was necessary just to keep one jump ahead of the media and public inquiries that were flowing in from all quarters.

There was 'hot' email and telephone traffic between the MoH and the Queensland Institute of Medical Research in Brisbane. MoH officials were acquiring and analysing data before preparing Health Impact Assessments for the government. The MoH communications staff were fielding inquiries from the media and assisting in the preparation of information packs for the media and for direct delivery to the public. Call centre staff had to be briefed so they could disseminate the MoH's messages about mosquitoes and what protection measures they could take in response to the incoming calls from the 'worried well'.

Once the facts about the species, the locations it infested, the health risks and the containment options were collected and analysed, communications moved from reacting to public concerns over to informing the stakeholders what the possible courses of action were available to deal with the problem. Apart from drafting and releasing media releases, there was considerable effort required to provide briefings to a range of interested parties. MoH officials briefed central government agencies, regional and local government authorities, and non-government organisations. Table 12.1 lists a representative sample of these.

Table 12.1. Strategic stakeholders*

Office/Organisation/Agency	Remarks
Minister for Biosecurity Minister of Health Minister of Finance	Scheduled routine meetings for updates on programme Specific briefings on emerging (new detections) issues Quarterly briefings to key agencies
Biosecurity Council Chief Executives Forum	Regular briefings to agency Chief Executives
Conservation Authority	Briefings and requests for approvals to conduct operations within the conservation estate
Central Government Biosecurity Group	Briefing the regular meetings of Regional Council Biosecurity Managers
Regional councils	One-off briefings of Regional Councils where *Campto* incursions had been detected
Biosecurity Institute	Attendance at Institute Conferences to present papers on eradication programme
Key stakeholder agencies	Ministry of Agriculture and Forestry (MAF), Department of Conservation (DoC), Ministry for the Environment (MfE), Ministry of Research Science and Technology (MoRST), Ministry of Fisheries (MFish), Treasury, Ministry of Foreign Affairs and Trade (MFAT), Department of Prime Minister and Cabinet
Government	Seeking funding and promulgation of regulations

*See Chapter 3 for more detail.

There were three distinct phases for communicators at head office in dealing with the response, which were:

1. **The initial reaction (reactive phase).** This was the period from initial detection until the government's response was declared. This represented a very busy time assembling the data and rendering it into a format that could be readily assimilated by the public.
2. **Introducing the Eradication Plan (proactive phase).** Once the course of action to eradicate was determined, then the public and the other stakeholders had to be advised. Briefings and media releases were drafted and delivered, which outlined the what, where, when, who and why elements of the eradication programme.
3. **Maintaining the momentum (business as usual, with the odd hiccup).** Once the treatment programme began in Napier, the interest in *Campto* by the media and other stakeholders at the strategic level diminished. The odd event in the field would ramp up media interest, such as new detections in new locations. Once the eradication effort had bedded in, most of the communications effort was expended in the operational area.

The Kaipara Communications Plan, a case study

The Kaipara detection in February of 2001 was to be a major challenge to the NZ Government's resolve to continue with the eradication programme. The previous detections at Mahia, Tairawhiti and Porangahau were explainable as being a spill over from the index site. The Kaipara Harbour was over 370 km from Mangawhai, with several physical barriers between both sites. Second, the area of saltmarsh habitat was initially reported at being 22 000 ha, an area that would require an enormous quantity of product to treat successfully. Fortunately, the initial estimates proved to be substantially overstated and, once the area of habitat was 'ground-truthed', the area that would need to be treated shrank to just over 2700 ha (see Chapter 8). After further analysis and consultation, the government of the day bit the bullet and authorised eradication at a projected cost of some NZ$30 000 000. What undoubtedly helped the decision was the evidence that the treatment at Napier had been successful, so there was confidence in the treatment programme, but the scale was daunting!

Kaipara became the main effort for NZ Biosecure. With the wealth of knowledge accumulated from east coast operations, a detailed eradication plan was developed and endorsed by the TAG (see Chapter 4 for further information). The plan recognised that communications were an essential component. Local communications were resource intensive because the operators had to reach out to the local community by whatever means possible and the range of delivery modes is described below.

- Mail drops were made to every land occupier within the treatment zone where treatment and intensive surveillance were to be conducted, and in the surrounding buffer areas where surveillance was also occurring.
- Town Hall meetings were called once an incursion had been detected and this gave an opportunity to meet the community, listen to their fears, concerns and (importantly) ideas. 'Cheerleaders' could be recognised and the community leadership identified and cultivated.
- Briefings were given to local organisations, the district council, government agencies, the Department of Conservation, the Ministry of Fisheries and the MAF,

and to non-government organisations such as conservation groups (Forest and Bird), primary production (Federated Farmers) and hunting groups (Ducks Unlimited).
- Local news outlets, the weekly 'free' papers, and the local radio station were invariably hungry for news and had limited resources to secure it. A well-crafted media release that needed minimal, if any, subediting was usually welcomed by the local media. Close contact with the Press also allowed personal relationships to be established that could provide leverage when significant events occurred that the operators wanted to be publicised.
- One-on-one contact was made with key land occupiers. Though time consuming, these encounters could be very productive. Access to either fly or drive over private property throughout all of the seasons was critical. Although there were legislative tools that could be exercised, to use them would have been an admission of the failure of the communications plan objective of generating willing stakeholder compliance.
- Face-to-face meetings with the 'apprehensive' members of the public – those who had qualms about the treatment operation, whether that related to stock disturbance from over-flights or fears that the product had a health risk attached to it.

The Kaipara Eradication Plan (Ministry of Health 2006) included a detailed communications strategy and protocol as an appendix. A perusal of this appendix section will show that no stone was left unturned when drafting the protocol. Appendix 1 runs on for some nine pages with some 12 subsections. The protocol outlined roles and responsibilities in very clear unambiguous terms. Paraphrased extracts from the Kaipara Communications Strategy are shown below.

The **communication requirements** section identified the need for … 'timely and accurate communication with stakeholders groups'. It also articulated the contractor's responsibilities regarding responses' to meet the demands of:

- parliamentary questions
- requests for official information
- reporting to Ministers and government
- provide reports on technical information that would support research projects
- provide technical advice to the agency as required.

The **objectives** of the information protocol were to:

- to ensure that all the communications requirements relating to exotic mosquito containment, control and eradication programmes were met
- to ensure that in high pressure situations, all parties were aware of their communication responsibilities.

The **principles** that the protocol ran included:

- '…timely and accurate and appropriate information dissemination'
- comment to be limited to 'descriptive, technical and non-policy information'.

The **Kaipara stakeholders** section comprised a list of all concerned parties by classification, such as contractors, landowners, *Iwi* (Maori tribal groups), regulatory authorities, environmental groups, special interest groups, media and the general public.

Stakeholder concerns were grouped according to their particular interests and/or concerns. For example, the protocol coverage included:

- *Iwi* (Maori tribal groups) – concerned about the effects on environment especially *kai moana* (seafoods from traditional gathering areas) and *kai whenua* (foods from traditional gathering areas), health effects of the programme and the mosquito
- environmental groups – concerned about acute and chronic impacts of treatment agents on the environment
- special interest groups – concerned about the effects on fisheries, farming and growing, effects on organic certification, health concerns relating to infestation, health concerns relating to treatment agents, and diagnosis and treatment information.

To ensure the **initiatives reached the stakeholders,** various media platforms that could, and were, to be used to promote the programme were listed. A wide range of tools were used: conference calls, town meetings, mail drops, email distribution lists, print media, public notices, and so on. If available, other organisations were approached to use their mailing lists and circular faxes.

There was extensive **pre-treatment consultation** – fifteen different organisations were contacted before treatment commencing. Ten of these were various *Iwi* organisations. The others were regional and district councils, and the Department of Conservation. The level of engagement ranged from telephone discussions with follow-up hard copy to formal briefings.

Key communication messages were a vital part of the protocol. Table 12.2 is taken from Appendix 1 and shows the five key messages that were created by the project team.

The media requests for information section laid out the procedures to be followed when '...developments likely to generate media interest occur'. The emphasis was on the process, with the project manager designated the lead spokesperson.

The information protocol covered all other requests for information from third parties such as parliamentary questions, select committee inquiries, oral questions in the House of Representatives, and so on. Again the focus was on the process, and timelines for responses were specified.

Media requests for information. Once more there was a documented and detailed process that clearly outlined the actions that were to be taken by the Project Manager. The delegation of authority to the Project Manager was succinct on matters that he could deal with and those that had to be referred to the agency.

Comment on the Kaipara Communications Strategy and Information Protocol

Some might question the need for such a detailed communications plan that could be described as being pedantic, if not nit picking. Given the situation of a major commitment of resources that had the potential for political fallout if there was evidence of failure in the delivery of the project, it is not unreasonable to ensure the communications plan had to be unambiguous and directive.

The success of the communications strategy could be judged by the media commentary that was generated over the 10 years of the programme. A cursory analysis of press clippings and radio spots would indicate that positive comments outweighed negative references by more than 10 to one. Gaining the confidence of the media is a cornerstone of successful communications. This entails being totally transparent and always going the 'extra yard' to assist the media with their inquiries. Providing clear unambiguous statements with supporting technical commentary makes the journalists' task easier.

Table 12.2. Kaipara Communications Plan – key communications messages

Ser	Key message	Remarks/content
1	**Eradication with S-methoprene**	• S-methoprene is an insect growth regulator (IGR) • Used extensively in Gisborne and Hawke's Bay • Not harmful to people, stock or crops • At application rates to be used no non-target effects on other aquatic organisms • *Campto* extremely sensitive to this product • Product is effective at rates of less than 0.1 parts per billion • Slow release profile products - continuous release of methoprene over 21 to 30 days • Methoprene is biodegradable – breaks down in water in the presence of UV light and microorganisms • Low application rate – 6kg of product over a hectare of treatment area ~150 granules in a square meter - minimises any risk of non target effects • Targeted approach – very specific for mosquitoes and in particular *Ochlerotatus* subgenus mosquitoes • Solid based (sand granule) product – minimises or eliminates drift • Low application release height 10–30 m • High tech spreading equipment for helicopter imported from USA • Methoprene is an Insect Growth Regulator not a toxin • Majority of treatments on private land thus minimal effects on public areas • Consultation with affected land owners before treatments start on their property
2	***Bacillus thuringiensis israelensis* (*Bti*)**	• *Bacillus thuringiensis israelensis* (*Bti*) is an organic larvicide • *Bti* was discovered in Israel in 1978. • *Bti* products have been extensively used in mosquito and biting fly control programmes, especially in Africa, USA and Germany, and also in recent times NZ in Napier and Gisborne. • There is a well documented history of environmental safety of *Bt* strains used in pest control. The environmental safety of *Bt*, coupled with the nature of the toxicity and specificity for target hosts, has led to the use of *Bt* in many pest control programmes in environmentally sensitive areas such as Napier, Gisborne. • This product has been used for control in the Kaipara Harbour since December 2001. • *Bti* has not been reported to affect a large range of invertebrate species including most aquatic fauna. • It is not toxic to bees. • Fish are not affected either in the laboratory or after field applications. • Shellfish including oysters and mussels are not affected by this product • No direct applications of *Bti* over habitat of fish or shellfish are likely to occur as these areas are tidally flushed and are not good habitat for the *Campto*. • *Bti* is considered to pose little threat to mammals. • No direct applications over stock or people are proposed. • *Bti* does not persist in the environment after application. • The liquid formulation proposed to be used has a low persistence profile.

(Continued)

Table 12.2. (Continued)

Ser	Key message	Remarks/content
3	**Programme background**	• *Campto* first discovered in Napier December 1998, eradicated from an area of 680 ha July 2002 • Populations detected in Gisborne and Mahia in October 2000, and Porangahau in November 2000, eradicated from an area of 190 ha in September 2004, August 2003 and September 2004, respectively • *Campto* discovered in Kaipara in February 2001 • Control programme implemented in December 2001 using *Bti* • Largest harbour in New Zealand • Mangawhai included in programme but no *Campto* found since April 2001 and no treatments undertaken to date; eradication completed December 2004 • *Campto* found at Whitford March 2002, eradicated November 2004 • Approximately 2700 ha of habitat to be surveyed around harbour • Cabinet decision to fund eradication July 2002 • 4 year programme – 2 years of treatment and 2 years of follow up surveillance • programme being undertaken by NZ BioSecure (a division of Southern Monitoring Services Ltd)
4	**Health concerns relating to infestation**	• *Campto* is an aggressive mosquito that will readily bite during the day as well as around dawn and dusk. • In Australia the mosquito is a competent vector of Ross River Virus Disease (RRV) • The symptoms of RRV include inflammation of the small joints, lethargy and flu like symptoms. The disease can only be diagnosed by blood testing. • There is no evidence the disease has arrived in the country with these mosquitoes. • To avoid being bitten wear long loose light-coloured clothing around dawn and dusk and avoid breeding areas where possible.
5	**Other FAQs**	*When and where were the larvae found in the Kaipara area?* Sampling was taken in the Kaipara Harbour area on 18 February, 9 days after heavy rain and high tides were reported in the area. On February 20, the MoH was told seven of the larvae found in the Rodney District of Kaipara Harbour were unconfirmed *Campto* larvae. The samples were then sent to Australia for confirmation. Since then, adult mosquitoes have also been found in the area.*Have exotic mosquitoes ever been found in this area before?* No, not before the 2002 incursion. The Kaipara Harbour has always been considered a possible breeding ground for *Campto*, but none have been found in previous surveys of the area. *How big is the area of infestation in the Kaipara Harbour area?* The infested area in the Kaipara region is the largest incursion of *Campto* in NZ. The potential habitat is ~2710 ha, which is less than was initially estimated.*How much funding did the government allocate to controlling and eradicating exotic mosquitoes earlier this year?* The total funding to date (as at 31 June 2006) is approximately $48-million nationwide over four years. The money will be used to continue to attempt to eradicate the exotic mosquito in the southern Kaipara, Whangaparaoa and Wairau (including Grassmere) as well as the Coromandel. *What do people do if they encounter exotic mosquitoes?* Mosquitoes are small frail looking insects that look similar to a midge. However the mosquito has a proboscis for sucking blood, which the midge does

(Continued)

Table 12.2. (Continued)

Ser	Key message	Remarks/content
		not. It is not possible to tell individual mosquito species apart with the naked eye. Our entomologists use microscopes to identify physical differences. Any mosquito specimens caught can be placed carefully in a small container such as a film canister or a matchbox and dropped into any Council service centre or posted to NZ BioSecure PO Box 634 Warkworth. Remember to include your contact details so we can let you know what you have caught. *What level of access required to their land?* This will be covered on a one-to-one basis *Long-term impact of treatment agents on the environment:* No chronic effects of using S-methoprene for mosquito control or eradication have been documented *Consultation process:* Individual landowners affected by treatments will be consulted with to determine any constraints on the treatment process. Consultation with other affected parties includes *Iwi*, regulatory authorities, DoC, environmental groups, special interest groups and the general public *Progress reports and effectiveness of programmes:* A quarterly report highlighting progress to date is submitted to the consent authorities and MAF. Copies of the progress section of the report will also be circulated to other interested groups. Staff will discuss progress with landowners as the opportunity arises. Staff will accept opportunities (subject to this strategy) to speak at *marae* [local Maori meeting house], schools, special meetings, public meetings etc.

Communications timeline

This section is a narrative of the actions from detection through to final eradication, using media statements to provide some context. It is not a detailed history.

1. the calm before the storm: pre-incursion – preparing for the unknown
2. damage control and preventing chaos: the immediate reaction to the initial discovery
3. tell them once, tell them again and tell them one more time: rolling out the containment and eradication effort
4. potential programme destroyer: some major headwinds have to be dealt with
5. hunting down that very last mosquito: maintenance of the communications momentum through to the end of the programme.

The calm before the storm

Some milestones: Marshall Laird Five Year Survey > Brian Kay Review > Initial training of Health Protection Officers in mosquito surveying > Production of first public health advisory for mosquitoes

Although the discovery of *Campto* was an unforeseen event, the MoH and the district health board public health units were not unprepared (see Chapter 1). The work that had been done in the previous years (e.g. Laird surveys and Kay Review), to develop a public health capacity to manage possible incursions of exotic mosquitoes, had raised the capacity of public health units to deal with exotic mosquito interceptions and incursions.

That said, it would be fallacious to say that, in 1999, the public health units were 'match fit' to meet an incursion of saltmarsh mosquitoes on NZ soil. The main effort by the public health unit staff had been concentrated on container-breeding mosquitoes such as *Ae. albopictus* and *Ae. aegypti*. These species are very different to the saltmarsh species in their habitat requirements, life cycles and potential pathways into NZ. The likely pathways for container breeders were through the first-points of entry (ports and airports), with probable interceptions occurring on or in conveyances, risk goods (in particular used tyres), in viable habitat at the point of entry or at container de-vanning sites (see Chapter 7 for further information). It would be true to say that saltmarsh mosquitoes were not on the radar as far as the public health unit monitoring effort was concerned.

Specific communications capabilities for providing information on mosquitoes of public health, in particular saltmarsh mosquitoes, were nominal. A generic Health Advisory had been drafted for distribution to the public. This document was specifically for container-breeding mosquitoes but it did include useful information on personal protection against mosquito nuisance biting. And then there was the well-tested ability of the MoH's communications staff to employ very effective communications techniques to reach the target audience within a short space of time.

Damage control and preventing chaos

Some milestones: The instalment hatching of *Campto* larvae in December 1998 and January 1999 > 'wave of nuisance biting in Westshore suburb' > Confirmation of *Campto* > Establishment of the Napier MRC > First TAG > Suppression - aerial treatment of *Bti* > Eradication proposal (S-methoprene) to government > Government signs-off eradication plan > S-methoprene treatment commences

Life for the public health fraternity in Napier involved in the *Campto* incursion during the few weeks between Christmas Eve 1998 and the end of February 1999 was to prove hectic. Not least of their worries was managing the intense interest shown by the media about the incursion. Initially, it was all about ensuring the public had all the information they needed so they understood the health risks and the measures by which they could protect themselves. Then the task became preparing the community for the implications of, first, a suppression programme (*Bti*) and then an eradication programme. Fortunately, the public support for eradication was positive, and for the well-bitten residents of Westshore, (for them) there was no alternative but action, and as soon as possible. One must concur that the media reports at that time (Fig. 12.1) would have had a powerful influence in shaping public opinion to agree to, if not demand, control and eradication.

Westshore youngster 'bitten 141 times'

Seven-year-old Raymond Ross knows how nasty the exotic mosquitoes that have infested Westshore can be — he was bitten 141 times in just two days last week.

His story has prompted health authorities to repeat warnings to residents to protect themselves against being bitten by the mosquitoes.

That meant covering exposed skin with long clothes and wearing insect repellant when outdoors, and taking steps to keep houses mosquito-free.

Raymond sustained his bites in just two days early last week, mostly while playing outside in the evenings, his mother Gaylene Ross said today.

"I counted 46 bites on the first day, and by the third day there were 141," she said.

Her son couldn't stop scratching the bites that covered his body and now had dozens of little pink scars on his legs, back and arms.

"I would give him a bath and put calamine lotion on him, but he'd still scratch."

Eventually she couldn't apply insect repellant because his skin was broken. Raymond was crying and in pain. The bites puffed up and many got infected and pussy.

Ms Ross said the mosquitoes were biting Raymond through his clothes.

"You wouldn't believe how big (the mosquitoes) are."

Since the attack her children had stopped playing outside in the evenings.

"Night-time is the worst. You can't even have the windows open."

The family had gone through three cans of fly spray to keep the house mosquito-free. The pests had been less of a problem this week, she said.

Yesterday authorities began spraying areas near the Ahuriri Estuary and Westshore in an attempt to eradicate the mosquitoes, which can carry the debilitating Ross River Virus disease.

Figure 12.1 *Napier Daily Telegraph,* 16 January 1999.

The public perception was that there was a 'dangerous' mosquito infesting Napier wetlands. The mosquito was both a 'vicious and aggressive biter' and capable of spreading a 'crippling' disease that had only been seen in NZ from the occasional imported case from Australia. In Napier, lurid tales of the vicious mosquitoes increased the public's appetite for action to deal with the intruders.

At the operational level in Napier, a Mosquito Response Team had been established and the Team Manager, Steve Garner, was very active in engaging local media, regulatory stakeholders, non-government organisations, commercial entities, *Tangata Whenua* (local Maori with historical ties to the area – 'people of the land') and community groups, while setting up and running NZ's first ever MRC team. Every effort was to be made to engage with every sector in the local community that might have been directly or indirectly affected by the event.

The next step was to prepare the public for the initial *Bti* suppression programme. Information on the possible health risks was disseminated through media releases (e.g. Box 12.2).

Box 12.2: HealthCare Hawke's Bay media release – 21 January 1999

Reaching the Community

Residents wanting information can call (06) 834 1811 to hear recorded information about the *Bti* programme and mosquito protection. People can leave a message on the hotline asking for an information fact sheet or ask to have a staff member from the public health unit return their call. The Ministry of Health also has a website www.moh.govt.nz that provides regular news updates on the mosquito operation.

The health risks were soon disseminated (Box 12.3).

Box 12.3: Radio NZ (Heugh Chappell) – 15 January 1999

***Ae camptorhynchus* spreads diseases**

In Australia, the southern saltmarsh mosquito is known to carry two serious diseases; Ross River Fever and Murray Valley encephalitis. Steve Garner, a Health Protection Officer with HealthCare Hawke's Bay says researchers from Wellington Medical School are writing a risk assessment on the bug. Mr Garner says the report is due out next week and will lead to a decision about how to deal with the pest.

Stakeholders were pulled in for consultation to ensure community support for the response (Box 12.4).

Box 12.4: HealthCare Hawke's Bay media release – 21 January 1999

Community group monitors spray operation
HealthCare Hawke's Bay has set up a community liaison group to ensure any public concerns relating to the mosquito or the treatment operation are able to be dealt with by independent community representatives. Members include representatives from the Asthma Society, Forest & Bird, Westshore and Haumoana ratepayers associations and Maori interests. Health Protection Officer Steve Garner says the Groups' role is to provide a 'sounding board' for the community. Residents wanting information can call (06) 8341811 for information on the eradication programme and mosquito protection. The Ministry of Health has a website www.moh.govt.nz that provides regular updates.

The operations were well reported in electronic and print media. And there were the inevitable distracting false positives as the public's attention was titillated (Box 12.5).

Box 12.5: Radio NZ (Heugh Chappell) – 22 January 1999

The Worried Well
Doctor says he's had several patients claiming to be suffering from Ross River virus which the exotic mosquito can carry, or Dengue Fever. However clinical tests haven't borne out the patients fears.

Then the community had to be prepared for the eradication plan. In the back of everyone's minds was the backlash to the *Bacillus* (albeit another variety) spraying for exotic moths in Auckland. The information about our *Bacillus* product, its effects on the target species and its benign characteristics was released.

Stakeholders were drawn in and advised of the plan, the product and the delivery systems (Box 12.6).

Box 12.6: MoH media release – 2 February 1999

Eradication of exotic mosquito in Napier Possible
The Ministry of Health believes 'Eradication of the exotic mosquito in Napier is possible' says Dr Gillian Durham, Chief Technical Officer. 'The expert Technical Advisory Group supports progress to date, expansion of the surveillance and the use of alternative methods which the Ministry is exploring' says Dr Durham.

Media releases led to further media interest (Box 12.7).

Box 12.7: Radio NZ (Heugh Chappell) – 15 March 1999

Eradication proposal
A Technical Advisory team is compiling a plan for government approval on eradicating the mosquito altogether. Laboratory trials are to start in Napier and Queensland to test the effectiveness of the chemical S-methoprene on Southern saltmarsh larvae from Hawke's Bay. Doctor Brian Kay from the Queensland Institute of Medical Research says S-methoprene has been used successfully in Australia. It is as environmentally as safe as *Bti* but is more effective. He is confident that if government next month gives a go ahead for eradication, that S-methoprene could wipe out *Aedes camptorhynchus* by the end of next summer.

The science support was emphasised to highlight the technical nature of the programme. The involvement of the boffins demonstrated that this was a very science-based project. And the characteristics of the eradication product and its application were carefully explained (Box 12.8).

Box 12.8: Media release HealthCare Hawkes Bay – 23 April 1999

S-Methoprene Trials to begin in Napier next week
HealthCare Hawke's Bay exotic mosquito control team is gearing up to begin field trials of the new larvicide, S-methoprene. Dr Michael Brown, from the Queensland Institute of Medical Research, will be overseeing the trials due to start on 26 April.

Media release HealthCare Hawke's Bay – 21 July 1999

Green Light for Aussie Mossie eradication with S-methoprene
Steve Garner, the control team manager said S-methoprene's slow release formulation meant a single application would provide control for several weeks. 'The product comes in granular and pellet form. The granules are about the size of poppy seeds' Steve said. 'We have calculated the dosage at ~6 kg per hectare applied every 21 days. This equates to 180 tiny granules per square metre. It will be applied aerially by helicopter and by staff working on the ground' Steve said.

The decision to attempt eradication was announced by the Minister for Biosecurity (Box 12.9).

Box 12.9: Media release, MAF – 29 April 1999

Government commits $6.5m for exotic mosquito eradication
Biosecurity Minister John Luxton today announced the government has earmarked up to $6.5 million for the next stage of the eradication programme of Southern Saltmarsh mosquito in Napier.

And the communications channels continued to deliver progress reports on activities as milestones were reached. This ensured the programme remained firmly in the public's consciousness.

The Napier MRC bent over backwards to accommodate local recreational interests (Box 12.10).

Box 12.10: Radio NZ (Heugh Chappell) – 4 May 2000

Concessions to duck hunters
HealthCare Hawke's Bay, which is confident it has killed 99 per cent of the Australian Southern Saltmarsh mosquito population in the region, is making a concession to duck hunters. Aerial treatments scheduled for this weekend have been deferred, so duck shooters who traditionally use *mai mais* [hides] on the Landcorp farm near the airport won't be put at risk.

Tell them once, tell them again and then tell them one more time

Some milestones: Treatment with XR-G begins > First signs of success > Gisborne, Mahia Peninsula and Porangahau detections > Government authorises expansion of treatment areas > NZ BioSecure become lead contractors for eradication operations.

Treatment of habitat was telegraphed to the community by frequent media releases. Progress was tracked and reported by the media (Box 12.11).

Box 12.11: Media release HealthCare Hawke's Bay – 30 August 1999

Aussie mossie update

Last week, HealthCare Hawke's Bay exotic mosquito control team treated 360 ha of mosquito habitat on the Lagoon and Landcorp farms with S-methoprene. The product was applied aerially by helicopter and by ground staff said project team leader Steve Garner.

National Radio Morning Report – 12 January 2000

Success in eradicating mosquito

The team which has been trying to eradicate the Australian Salt Marsh Mosquito is confident it has eliminated 95% of the population. Gene Browne is the team entomologist and he says 'It's been a tough job reducing the mosquito population.'

But just when it was all looking good …

Gisborne has new mozzies to swat

By PHILIP KITCHIN

JUST as Hawke's Bay seems to have all but swatted out its Aussie mozzies, Health Ministry officials are expected to open a new front in the war against the exotic mosquito in Gisborne today.

Nearly four months after suspicious larvae were found near Gisborne, the ministry is expected to confirm that the larvae are from the southern saltmarsh mosquito — which sparked a multi-million-dollar eradication war in Hawke's Bay.

Figure 12.2 *Dominion,* 12 October 2001.

Uncertainty

Mosquitoes so far contained

by John Jones

FINAL confirmation is expected today that mosquitoes found in the Wherowhero Lagoon near Muriwai are the southern saltmarsh variety which can carry the Ross River virus.

Australian experts are expected to confirm today that specimens they have been examining are Aedes camptorhyncus. That information will come before a technical advisory group in Wellington tomorrow.

Meanwhile, a team looking to see if the mosquitoes have spread beyond the lagoon is expected to finish its work by the end of this week. So far no mosquitoes have been found outside the lagoon area.

Ministry of Health deputy chief technical officer for biosecurity John Gardner said while they were waiting for final confirmation from Australia they were fairly certain it was the southern saltmarsh mosquito.

A media statement is expected after tomorrow's meeting of the TAG.

It would probably be too soon to produce a control strategy from that meeting but they would probably produce some options such as control or eradication and look at the implications of that.

The environment at the lagoon was very different from Napier. There was less water mass available which would mean less breeding sites and that might restrict the population of the mosquitoes.

The hardest part to try and determine was how the mosquitoes might have got here. Surveys had been conducted down the coast to see if there had been a "stepping stone " effect from the south. There was no evidence to indicate that.

Looking at the spot where the suspected southern saltmarsh mosquito was first discovered at the Wherowhero Lagoon are response co-ordinator Bryn Gradwell, Gisborne principal public health officer Tom Scott and Ministry of Health deputy chief technical officer for biosecurity John Gardner. Picture by Paul Rickard

"We suspect it is more likely to have been by mechanical transportation rather than by natural migration because of the very nature of the environment. There is a lot of adverse terrain along that coastline."

It could have come in the wheel well of a plane or the back of a vehicle. Mosquitoes would head to a dark place and if someone locked a door they would stay there and leave when it was opened.

Describing the response since last Thursday, he said the ministry had put together the survey team headed by Bryn Gradwell, a mosquito response co-ordinator involved in controlling the Napier outbreak.

Tairawhiti Health hosted the team which was essentially a composite one drawn from both local resources from the public health unit and Gisborne District Council, people from the eradication team and officers from other public health units such as Hutt Valley and Pacific Health.

Last Saturday they commenced a detailed survey around the lagoon. The initial survey of the inner cordon was complete and had now been extended to other possible areas. That process was ongoing and and it would be extended down the coast.

That data would be taken to the technical advisory group on Friday. This was a group of scientific people with expertise in mosquitoes and their control and would include people brought in from overseas. Some had been involved in the Napier exercise.

He said he was impressed at the way the teams had worked together, particularly the district council and public health unit staff.

It was a regular survey by Gisborne staff that discovered the outbreak.

After the Napier infestation was found Landcare did some extremely technical work which identified sites around the country where the mosquito could establish itself. From that a surveillance programme had been established with public health units. The sample was found "slap bang" in the middle of an area identified by this programme, said Mr Gardner.

Figure 12.3 *Gisborne Herald,* 7 November 2001. (Courtesy of the *Gisborne Herald*)

The MoH scrambled to respond, the TAG convened and staff were deployed to Gisborne (Fig.12.2).

Initially there were some doubts as to whether it was feasible to expand the eradication programme but the resources were found (Fig.12.3).

Potential programme destroyer, some major headwinds have to be dealt with

Some Milestones: Kaipara findings > Government bites the bullet and increases funding > Whitford and Mangawhai sideshows > Napier Eradicated > Shakespear Park > If you find it, we can kill it >

Disease-carrying mosquito defies eradication

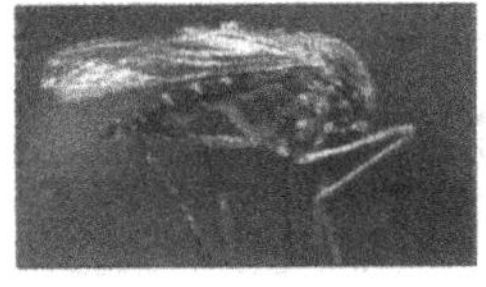

CARRIER: The southern saltmarsh mosquito can carry Ross River virus.

by Anne Beston
environment reporter

An Australian mosquito which can carry a debilitating virus has been found again in East Auckland, more than a month after biosecurity authorities hoped they had wiped it out.

Ministry of Health public health manager Graeme Gillespie said two adult southern saltmarsh mosquitoes were discovered last month in the same area where larvae were found in early March.

The ministry has hunted extensively for the mosquito since the March find and has used the insecticide s-methoprene on the 1ha infestation site at Clifton Rd, in Whitford.

Mr Gillespie said that although the latest find was "not unexpected", the ministry had hoped a lack of suitable habitat for the mosquito in the area meant it had been killed off.

The southern saltmarsh mosquito is a potential carrier of Ross River virus, which can cause dizziness, headaches and aching limbs.

In some serious cases the virus causes severe depression.

Mr Gillespie said surveillance work for the exotic mosquito was ongoing but would be stepped up during the next king tide, when the insects were more likely to breed.

Figure 12.4 *NZ Herald,* 21 May 2002.

The detection of *Campto* in the Kaipara Harbour was a major threat to the *Campto* eradication programme. The MoH was not positive that eradication was feasible (Fig. 12.4).

However, the size of the habitat, initially estimated to be 22 000 ha, was revised downward to a manageable maximum of 2710 ha. The government committed more resources and a base was set up at Parakai to manage the programme. The renewed commitment by government to continue eradication was emphasised by a high profile ministerial visit, by Minister Hobbs, to the Parakai Base opening (Fig. 12.5).

Overview of Kaipara mosquito campaign

The Parakai base for the southern saltmarsh mosquito eradication programme was officially opened this week by Minister for the Environment and Associate Minister for Biosecurity Marian Hobbs.

A powhiri by local iwi, Ngati Whatua, welcomed about 50 representatives of groups that included the Auckland Regional Council, the Ministry of Health and New Zealand BioSecure to the Green Rd control site.

NZ BioSecure was chosen to complete the programme after successfully eradicating the mosquito in Napier. The mosquito can carry the Ross River virus.

The Kaipara Harbour project is estimated to cost $30 million and will take four years to complete.

Workshops showed Parakai visitors how NZ BioSecure was tackling the saltmarsh mosquito, what the mosquito looked like, where they could be found and how they were caught.

Mrs Hobbs thanked local iwi, farmers, NZ BioSecure and everyone involved in the eradication programme.

"Biosecurity is about security of our life as we know it in Aotearoa.

"We have let a lot of things into this country, but what we are trying to do is keep our life secure," she said.

She said the $30m cost is almost the entire budget for the Ministry of the Environment.

MOSQUITO FLIGHT: Marian Hobbs about to get an aerial view of the problem area.

"My joy is there will be an end, and my other joy is this experience will supply knowledge to others."

Director of Public Health Colin Tukuitonga said that when he had just finished medical school, he experienced Ross River virus first hand when his whole Niuean community got sick.

"It wasn't going to kill people but it devastated the community. It shows how important it is to do things before they occur," Mr Tukuitonga said.

Guests were invited for lunch and helicopter trips to see the Kaipara Harbour.

Figure 12.5 *Nor-West Newsbrief* 20 February 2003.

Mozzie team sets up in Parakai

By MICHELLE HYLAND

Parakai will host the control centre for a $13 million, four-year programme to eradicate the southern saltmarsh mosquito from the Kaipara Harbour.

New Zealand BioSecure have set up camp on Greens Rd, opposite Parakai airfield.

The command centre has 17 staff, a helicopter, computers, testing devices, chemicals and a fleet of four-wheel drive vehicles.

Most of the staff are working as technical officers and assistants.

Adam Mason, 21, of Parakai, and Adrian Brocas, 22, of Muriwai, say they have enjoyed the past three months spent searching for mosquito larvae. Both are fulltime employees for the next four years.

"When we were looking for staff for the operation we did all our advertising locally. We got some really talented people and they are predominately from around the Kaipara," says Bryn Gradwell, regional manager of New Zealand BioSecure.

Employees are mainly in their 20s and from Muriwai, Parakai, Warkworth and other parts of Rodney.

Jobs include collecting mosquito larvae and setting traps, which are carried out on foot, in helicopters and in areas accessible only to four-wheel drives.

Testing is at the Parakai base and the specimens are then sent to Napier for positive identification by four New Zealand BioSecure scientists.

MOSQUITO MASSACRE: Working to combat bloodthirsty airborne pests are Adam Mason, 21, of Parakai, left, and Adrian Brocas, 22, of Muriwai. Photo: MICHELLE HYLAND

Figure 12.6 *Nor-west Newsbrief,* 23 January 2003.

The Kaipara saw the eradication programme scaled up with operations occurring in four separate regions of NZ. However, by now the operators were 'match fit'. Field staff had evolved into 'mossie hunters' with well-honed skills at detecting habitat and the pests. The Kaipara MRC that was set up just west of Helensville was well reported locally (Fig. 12.6).

Delivery systems had been refined, and the staff were very capable at identifying *Campto.* Once the 'green light' was given the biggest regional operation of the whole *Campto* eradication programme swung into action (Fig. 12.7).

■ BIOSECURITY

Kaipara mosquito blitz begins today

The Health Ministry will start spreading a growth regulator around Kaipara Harbour from today as it seeks to eradicate disease-carrying exotic mosquitoes.

The mosquito growth regulator S-methoprene, used in the southern saltmarsh mosquito eradication programmes in Hawkes Bay and Tairawhiti, is to be applied in pellet form around the harbour.

"There has never been an outbreak of a mosquito-borne disease in New Zealand and we want to keep it that way," said the ministry chief technical officer for health Sally Gilbert.

The southern saltmarsh mosquito is a potential carrier of the Ross River virus disease.

Symptoms of the disease include pain and tenderness in muscles and joints, fever, chills, sweating, a headache and tiredness. A rash may also briefly appear on the torso and limbs.

Southern saltmarsh mosquito larvae were found in the Rodney District of Kaipara Harbour in February last year. Since then, the mosquitoes have been found in about 2700ha of habitat around the harbour.

Extensive public consultation work and ground and aerial mapping of the area have taken place in the lead-up to the full eradication programme.

Ms Gilbert said the Government approved the use of a full-scale eradication programme around the area in June.

This means a two-pronged attack with biological spray Bti, which isbeing used at present, and S-methoprene, which has not yet been used in the area.

"We're targeting the lifecycle of the southern saltmarsh mosquito over two summers to ensure we treat over two breeding seasons," she said.

Ms Gilbert said Bti and S-methoprene had been given full health-impact assessments and left no long-term residue.

Disruption to the public would be limited as most sites being treated were in remote areas.

All affected landowners had been contacted and the control agents would be applied from helicopters, quad bikes and people on foot.

People wanting updated information on the application programme and sites can call 0800 MOZZIE (0800 669-943).

— *NZPA*

Figure 12.7 *NZ Herald,* 25 October 2002.

The Kaipara programme did have several outlier sites that were detected in Whitford and Mangawhai, but these were readily dealt with. There was the success in Napier to celebrate (Box 12.12).

Box 12.12: Minister Marian Hobbs, Associate Minister for Biosecurity – Speech Notes, 27 November 2003

Eradication of Southern Saltmarsh Mosquito, Aircraft Hangar, Napier Airport

> *'Today we are celebrating a significant achievement for New Zealand biosecurity, the eradication of* Aedes (Ochlerotatus) camptorhynchus *from Napier and Mahia. We've done what John Mitchell and the All Blacks (Rugby team) couldn't – wiped out a team of aggressive, invading Aussies. It's too early to gloat just yet. We have won this battle but are yet to win the war'.*

Every time a new detection occurred, the media continued to highlight the potential public health risks. This ensured that the NZ public became 'mossie' conscious and had the benefit of maintaining public awareness about the other exotic mosquito threats (i.e. the interceptions of exotic mosquitoes at ports of entry). The long-suffering laboratory staff knew that every time there was a significant 'mossie' story reported in the media (Fig. 12.8), well-meaning citizens would collect all sorts of arthropods and would then be sending these off to be identified.

>> Ross River virus

- People infected may suffer pain and tenderness in muscles and joints, particularly the wrists, knees and ankles. Flu-like symptoms, including fever, chills, sweating, headache and tiredness are common.
- Symptoms occur 3 to 21 days (an average of 9 days) after being bitten and can last months or years. They subside with few or no side-effects.
- Painkillers such as aspirin or paracetamol may relieve pain and swelling. Sometimes stronger medication is needed.
- The disease, which can only be diagnosed with a special blood test, is usually milder and runs a shorter course in children.
- The virus is not contagious and if people avoid being bitten by mosquitoes they cannot get infected.

(Source: The Ministry of Health)

Figure 12.8 *NZ Herald,* 14 May 2004.

There were further detection events and some predictable reactions (Fig. 12.9).

There was a concern that the continuing new detections could lead to reinforcing failure in that the scale of the response required would become unsustainable. At various times, such disquiets were raised by both government and the media (Fig. 12.10).

Disease-risk mosquito at new site

BIOSECURITY: Ministry's eradication programme extended to regional park

Southern saltmarsh mosquito larvae have been found on the Whangaparaoa Peninsula north of Auckland, prompting the Ministry of Health to extend its eradication programme.

The larvae were found near Shakespear Regional Park on Tuesday during routine checks. Yesterday, the Associate Minister of Biosecurity, Marian Hobbs, authorised treatment to prevent the mosquito spreading.

The saltmarsh mosquito can carry the Ross River virus. People who contract the disease suffer symptoms similar to flu which may last for months or even years. Treatment is available to relieve the symptoms.

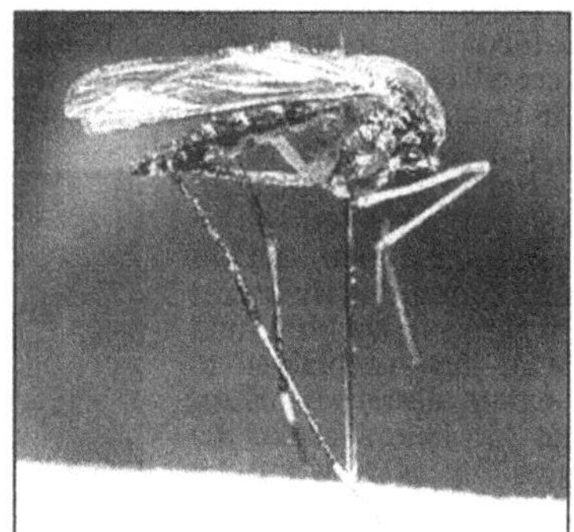

VIRUS CARRIER: The southern saltmarsh mosquito.

Figure 12.9 *NZ Herald,* 29 January 2004.

Sometimes a detection of *Campto* could be linked with interceptions of other unwanted exotic mosquito species such as occurred in February 2004 (Fig. 12.11). This allowed the public health services to leverage off the *Campto* eradication campaign to raise public awareness regarding the ongoing threat generated by interceptions of other species of exotic mosquitoes at points of entry which were also competent vectors of human diseases.

By this time, it became obvious that the NZ *Campto* eradication programme was technically very sound. Therefore, when *Campto* mosquitoes were first detected in the South Island, there was no lack of confidence that they could be dealt with. The response followed a well trodden path: delimit, set up a base, and sort the software, hardware and wetware, start treatment and conduct intensive surveillance. In a few months observe the inevitable collapse of the biomass. Once eradication was in full swing, local media turned out to report on their activities whether that be in the Wairau (Fig. 12.12) or further North in the Kaipara where the District Council was being kept well informed of progress (Fig. 12.13).

Mosquito's bite felt

Anderton says control cost may be too much

Wellington: Taxpayer funds poured into the battle against the "Aussie mozzie" — the southern salt marsh mosquito species — have passed $51 million as Biosecurity NZ takes over the eradication effort from public health officials.

But the biosecurity involvement also signals the Government is considering the possibility of throwing in the towel.

"While it's a pest we can do something about, it does come at a significant cost," Biosecurity Minister Jim Anderton said on Monday night. "The Government has already spent $48 million on this programme, and it still keeps spreading."

"We're really keen to get rid of it, but at some point it's going to become a cost-benefit exercise," Mr Anderton said.

"Right now, there's not enough information available to make final decisions. That's what we've asked Biosecurity New Zealand to provide".

Though the latest $11 million for control measures brings to more than $51 million the money taxpayers have spent on it, in the 2004 budget the Government cut back its spending.

In 2004, it allocated extra spending of $6.288 million ($1.572 million in each of four years to 2009) for surveillance, down from the $1.602 million allocated in the 2004 financial year. But in the 2004 budget, it slashed the spending specifically on eradicating the mosquito by 75%, from $12.794 million to $3.01 million.

Biosecurity NZ took over the Health Ministry's bid to eradicate the pest in July, shortly after three new breeding sites were discovered in a remote area of the Coromandel Peninsula, just north of Colville.

The mosquito, *Ochlerotatus camptorhynchus*, was first found in Napier in December 1998, and has since been identified at Muriwai, Mahia, Porangahau, Kaipara and Mangawhai harbours, Whitford, Whangaporaoa and in the South Island, at Wairau, near Blenheim.

It is an aggressive biter with the potential to cause significant nuisance for people, livestock and birds, and health authorities fear that if the spread continues, the nation will be put at risk of an epidemic of Ross River virus with huge costs in terms of both the economy and lifestyle.

Medical research has shown that a "virgin soil epidemic" could be started by just one mosquito picking up the virus from an infected tourist or returning NZ traveller. There have been a dozen known cases of Ross River virus in New Zealand since 1980, in people though to have caught it overseas, including a Waikato patient last year.

If salt marsh mosquitoes in New Zealand picked up the virus, it could become permanently established in feral animals, such as possums, or farm livestock. — NZPA

Figure 12.10 *Otago Daily Times*, 30 August 2006.

Southern saltmarsh mosquito confirmed

The Ministry of Health has confirmed the southern saltmarsh mosquito was found in Shakespear Park, Whangaparaoa, last Tuesday.

The Ministry of Health's chief technical officer, Sally Gilbert, says specimens found at Shakespear Park were sent to Richard Russell, associate professor with the Department of Medicine at the University of Sydney and founding director of the Department of Medical Entomology.

Mr Russell has been studying mosquitoes for over 30 years and has worked for the World Health Organisation in 12 countries including Australia, Asia, the Pacific and South America, dealing extensively with the saltmarsh mosquito.

An adult southern saltmarsh mosquito.

"Yes, the results confirmed it was the southern saltmarsh mosquito," Mrs Gilbert says.

Five adult traps have been set in the area, but reports so far indicate there is no sign of adult mosquitoes.

Last week the New Zealand Biosecure response team treated suspicious areas at Shakespear Park with a slow-release agent which lasts up to 20 days but does not kill the larvae.

"The treatment doesn't kill the larvae but stops pupae emerging into adults," Mrs Gilbert says.

A Biosecure team has started to survey potential habitats from Long Bay north to Puhoi.

The team will be checking dry areas and more index sites where Mrs Gilbert says they expect to find more larvae.

Last week in Auckland City 30 exotic larvae were found on a ship in the main port, adding to the outbreak of exotic mosquitoes in Auckland and Rodney district.

The Ministry of Health is worried the adults may have hatched while the ship was still in port.

– Rebecca Milne

Figure 12.11 *Rodney Times*, 5 February 2004.

Mozzie busters bite back with gusto

In May, an outbreak of the southern saltmarsh mosquito was found in Marlborough. Five New Zealand BioSecure staff are now permanently posted in Blenheim to eradicate it. KATE TRINGHAM talks to locality manager Steve Crarer to find out more.

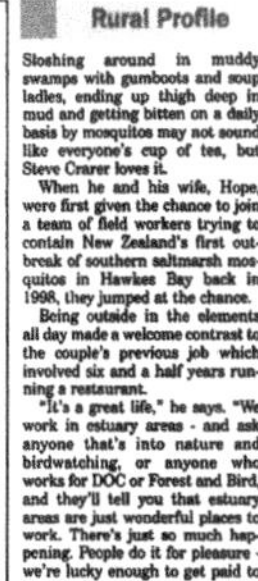

Rural Profile

Sloshing around in muddy swamps with gumboots and soup ladles, ending up thigh deep in mud and getting bitten on a daily basis by mosquitos may not sound like everyone's cup of tea, but Steve Crarer loves it.

When he and his wife, Hope, were first given the chance to join a team of field workers trying to contain New Zealand's first outbreak of southern saltmarsh mosquitos in Hawkes Bay back in 1998, they jumped at the chance.

Being outside in the elements all day made a welcome contrast to the couple's previous job which involved six and a half years running a restaurant.

"It's a great life," he says. "We work in estuary areas - and ask anyone that's into nature and birdwatching, or anyone who works for DOC or Forest and Bird, and they'll tell you that estuary areas are just wonderful places to work. There's just so much happening. People do it for pleasure - we're lucky enough to get paid to be there."

The company BioSecure, which runs the current eradication programme, was created in 2001 after it was discovered that desoite suc-

MOSQUITO BUSTERS: New Zealand BioSecure locality manager Steve Crarer, front, with his eradication team at the mouth of the Blind River. From left Rob Waihi, Hope Crarer, Adam Mason and Dan Edwards. Photo Scott Hammond 43151

Figure 12.12 *Marlborough Express,* 15 October 2004.

Eradication programme having desired effect

The Kaipara Harbour southern saltmarsh mosquito eradication programme is working well, with the last adult trapped in September, 2003, although small numbers of larvae are still appearing.

In a report to the Rodney District Council at a recent meeting, environmental health manager Ian Farrell said the eradication plan was being 'fully implemented' and 31 trap sites were monitored, including at Mangawhai and Whitford.

They were checked twice weekly for adults and larvae, he said. The last adult was trapped at Mangawhai in December, 2002 and at Whitford in November, 2002. If no adults or larvae were found for two years, the programme would be considered successful, Mr Farrell said.

Figure 12.13 *Rodney Times,* 5 February 2004.

Hunting down that very last mosquito

Some milestones: The Kaipara re-emergence > the Coromandel detection, a remarkable success for the National Surveillance Programme > Programme transfer to MAF > New contractors for the Eradication programme > Final days of treatment in the Wairau.

There had been a couple of minor hiccups that had to be dealt with, with the re-emergence at Kaipara Harbour being one (see Chapter 8).

Nevertheless, *Campto* was to have one last blast and this was on the Coromandel Peninsula.

However, this time we were very well prepared for it. The establishment of the National Surveillance Programme, (at the recommendation of the TAG) had provided a sensitive detection system for saltmarsh mosquitoes. The response was executed as a well-practised drill: delimiting survey, assessment and then treatment (Fig. 12.14).

Mosquito experts fly in to check Colville investations

Members of the national southern saltmarsh mosquito surveillance programme (NSP) flew into the Coromandel yesterday to determine the extent of the insect's infestation in Colville.

The programme detected three new breeding sites of the Australian insect last week at Colville; two on the west coast and one on the east.

The mosquito carries the Ross River virus (RRV), a viral infection endemic in Australia, and has affected people from Papua New Guinea, the Solomon Islands and some Pacific Islands.

It causes persistent muscle pain, flu symptoms and rashes sometimes for years. It also affects livestock and birds. No cases have been reported in New Zealand, but if the insect's spread is not kept in check health officials warn RRV could turn into an epidemic.

The mosquito was found in New Zealand in 1998 in Napier, and has spread to eight sites including Muriwai, Whangaparaoa, and the Kaipara and Mangawhai Harbours, despite a $40 million effort to eradicate the species.

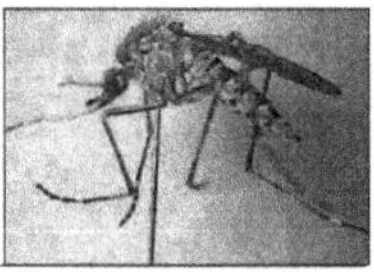

The southern saltmarsh mosquito.

NSP spokesman John Gardner says it is unclear how the insect managed to get to the Coromandel. This week five teams of surveyors, including a helicopter team, will comb the breeding sites for clues and take preventative measures to slow the progress of the bugs.

The mosquito is believed to have infested an area of about 10ha, and the teams will use the organic insecticide BTI.

The NSP is funded by the Ministry of Health and is designed to monitor saltmarsh habitat for exotic mosquitoes throughout New Zealand. "I guess the only good thing to come out of this is proof the detection methods used by the teams work," Mr Gardner says. "Now they will go out about 5km from each breeding site and look for evidence. We will speak to locals, landowners, iwi; anyone around the site who could offer help so we can form a plan to deal with the infestation."

Anyone who has seen the mosquitoes or experienced an unusual bite should ring 0800 MOSSIE (669 943) or the Ministry of Agriculture and Forestry exotic pest and diseases hotline 0800 809 966.

Figure 12.14 *Hauraki Herald,* 16 May 2006.

By 2008, the end was in sight. Eradication was completed already and the end of the Kaipara programme was well reported (Fig. 12.15).

The Wairau was to be the last eradication area to be signed off. In late 2010, a joint statement by the Minister for Biosecurity, Hon. David Carter and the Minister of Health, Hon. Tony Ryall was made, that acknowledged that this was the first time an eradication of this type of mosquito had been completed successfully.

It is contended that the development and implementation of a comprehensive communications strategy was a 'force enabler' for the eradication programme. A well-informed public became the 'cheerleaders' for eradication and were only too willing (in 99.99% of the

Bad biter hits the dust

THE ozzie mozzie is gone from Rodney.

After six years of treatment, the southern saltmarsh mosquito has been eliminated from the Kaipara Harbour area.

No sign of the potential flu-like virus carrier – either adult, larva or egg – has been found for two years, says MAF Biosecurity New Zealand.

That ends six years of treatment and surveillance involving ground and helicopter crews, and a public education programme to find the nasty day-time biter.

Traps were used to locate the mosquito, along with reports from those suspected to have been bitten, while a growth regulator, insecticide and a natural bacteria, bacillus thuringiensis – Bti for short, were used to combat it.

The discovery of the southern saltmarsh mosquito in Napier in 1998 led to a full-scale operation to search out and eradicate this Australian pest. Aside from a viscious bite, the mosquito was also a carrier of the Ross River virus – a severe flu-like illness, and posed a serious health risk.

The government approved a $30 million mosquito eradication programme for the Kaipara in July 2002, a step up from the original control and containment programme initially run for about a year.

Since 1998 this mosquito had been found in eight different areas of New Zealand, including Mangawhai where it was eradicated in 2004, the Whangaparaoa Peninsula where it was eliminated in April 2007, and also around the Kaipara Harbour.

The Kaipara Harbour is the largest natural harbour in the southern hemisphere, and the area around the harbour that had to be monitored and cleared of the mosquito amounted to about 2800 hectares.

In its country of origin Australia, and in the United States also, this pest is a real problem, but because the land masses are so large compared to New Zealand, eradication is not an option. Those countries operate control programmes only.

In New Zealand it has been possible to aim at complete eradication. Just two remaining areas are under surveillance – Wairau in Marlborough, and the Coromandel which is not far off being declared ozzie mozzie free as well, says a Biosecurity New Zealand spokesperson.

So far, it has cost about $72 million nationally during the 10 years of war on the pest.

National eradication of the southern saltmarsh mosquito is expected to be declared in June 2010.

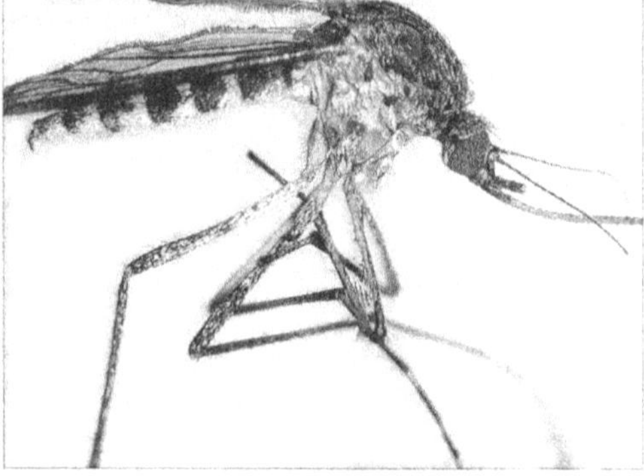

Ozzie mozzie: The southern saltmarsh mosquito is no longer in the Kaipara Harbour area.

Chopper drop: During the mosquito eradication operation helicopters were used to drop S-methoprene insecticide granules on land surrounding the Kaipara Harbour.

Figure 12.15 *Rodney Times,* 11 September 2008.

cases) to assist the operational teams with their tasks. The bureaucrats and communications gurus in government can take some of the kudos for delivering the communications at the strategic level, but the hard work done by the operational staff, who had to deal with the stakeholders face to face, was probably more significant in terms of gaining public confidence and support.

This work began with the Napier MRC in Napier, led by Steve Garner, who initiated a drive to get the local community/stakeholders on board. Napier was a proving ground for communications that was to be a valuable asset when the expansion of the programme to Tairawhiti occurred. Bryn Gradwell led operations at this site and he further developed and refined the communications methodology. NZ BioSecure was able to transfer this hard-won expertise to the Kaipara region. The communications strategy for this phase became the blueprint for subsequent eradication programmes in the Wairau and Coromandel. This communications plan provides an excellent template for the communications support required for any other public health issue.

Looking back, it would seem to be remarkable that the long-running programme managed to sustain the support of the communities where it operated. The media commentary was almost always supportive of the eradication activities and at no point in time did a 'counter eradication' constituency develop and coalesce into a vocal dissident faction. The *NZ Herald*, for example provided objective information to inform its readers of the programme (Fig. 12.16).

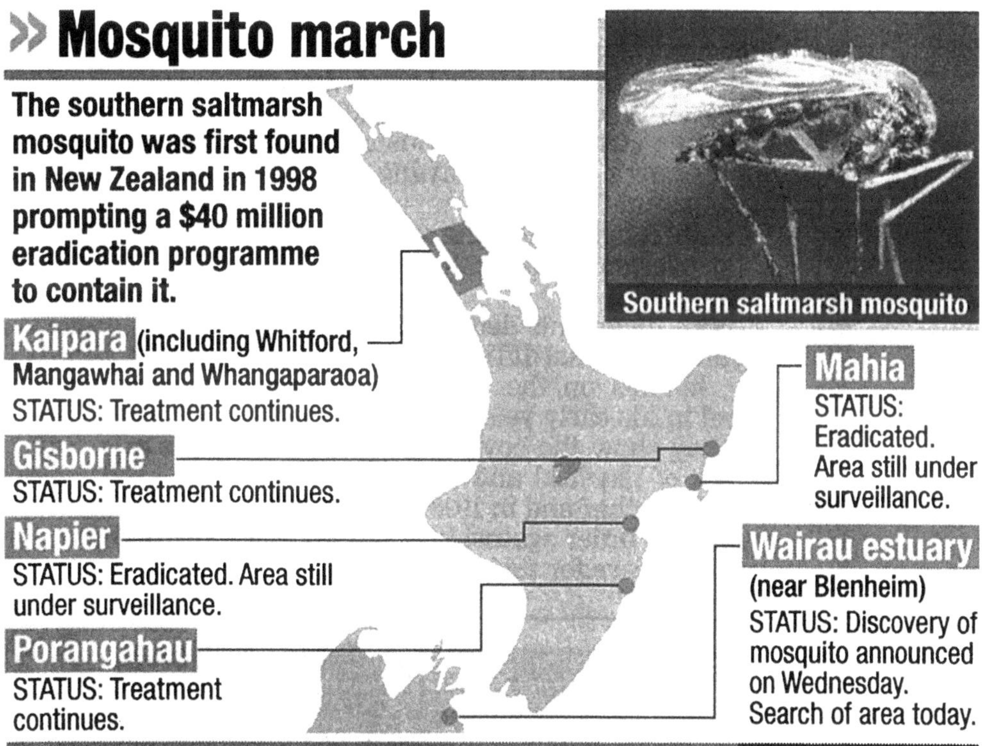

Figure 12.16 *NZ Herald,* 14 May 2004.

Acknowledgements

We are indebted to the various journalists and editors of the NZ media for their fair and thorough coverage of the incursion response, and in particular to Heugh Chappell (Radio NZ) for making his personal news clippings available, and to APN Group, Fairfax Media, the *Gisborne Herald* and NZPA for permission to use their articles in this publication.

References

Ministry of Health (2006) *Kaipara Eradication Plan*. Unpublished report by the Ministry of Health, Wellington, NZ.

Van Riel CBM (2007) *Essentials of Corporate Communication*. Routledge, Abingdon, UK.

WHO (2005) *WHO Outbreak Communications Guidelines*. World Health Organization, Geneva.

WHO (2006) *Framework to Guiding Public Health Policy in Areas of Scientific Uncertainty*. World Health Organization, Geneva.

WHO (2009) *Guidance for Pandemic Communications*. World Health Organization, Geneva.

13

Reflections on a successful eradication programme

Richard Russell, Henry Dowler, JR Gardner, Sally Gilbert and Matthew Stone

> *'SSM was declared eradicated from New Zealand by Minister for Biosecurity, David Carter and Minister of Health Tony Ryall on 1 July 2010.' (http://www.biosecurity.govt.nz/pests/southern-saltmarsh-mosquito).*

Although the *Campto* eradication program ran for more than 10 years (well beyond what might have been initially anticipated, because of its unusual pattern of spread) and cost approximately NZ$70 million (also well beyond what might have been initially anticipated, because of its seasonal recurrences and disparate appearances), it was finally successful and so represents an internationally significant achievement in the history of mosquito control.

The criteria that were used for declaring local or regional elimination, and finally national eradication, were the absence of any life forms (eggs, larvae, pupae, adults) found during active surveillance following at least three inundation events over 2 years. Following the declaration of national eradication, a National Surveillance Program has continued to keep a watching brief on the possibility that the mosquito might yet again arise somewhere or be again introduced.

Because of the nature of the biology of the species in surviving through adverse seasonal conditions, and the fact that the pathway/s by which the species originally entered NZ and disseminated remains uncertain, there should be concerns for future such, or similar, occurrences. However, the success of the programme was indeed an international triumph and, for those directly involved, and those who observed from near or far, its success allows for a consideration of what factors provided for its success, what lessons were learnt along the way and what is left as its legacy for NZ. To that end, this chapter is a compilation of personal views, from many of those most intimately involved, of the 'essential experiences' and the 'signal successes' of the *Campto* eradication programme.

Being prepared

NZ was not caught with its pants down. For some years before 1998, it had been recognised there was a risk of introduction of exotic mosquitoes of public health concern (NZ had a history of introductions and incursions) and there were efforts to establish a means to manage the risk. Along with its biosecurity partners, the MoH had spent quite a lot of time and effort before the incursion, preparing for an eradication response. The 1997 Kay Review informed the development of an eradication contingency plan and training for local public health staff on mosquito exclusion and containment, and surveillance and control techniques (Kay 1997). There had been significant investment in training public health officers in mosquito surveillance and response.

Although the eradication of a mosquito from NZ had never before been attempted, the MoH, and its public health unit officers, were experienced in responding to interceptions of exotic mosquitoes of public health significance. Work was also well underway on technical assessments of the suitability of insecticides that would need to be used in an eradication response. Even though saltmarsh mosquitoes such as *Campto* and *Ae. vigilax*, and freshwater surface pool species such as *Cx. annulirostris* or some *Anopheles* mosquitoes, were not seen as the highest risk (i.e. compared with container-breeding mosquitoes such as *Ae. aegypti* and *Ae. albopictus*), some potential habitat areas had already been identified and consideration given to climate and establishment risks.

This preparedness must have contributed substantially to ensuring a high-quality initial response to the Napier discovery of *Campto* in December 1998 and, ultimately, to the long-term success of the national eradication efforts that were founded on the Napier experience.

New, risk-based national surveillance arrangements had only been in place for a short time before the discovery of *Campto* in Napier. Although the presence of the mosquito became evident through public complaints, the identification and confirmation of the new species in NZ was certainly due to well trained, motivated local public health staff and the availability of a MoH-funded mosquito identification service.

Having good people

The most important aspect of the successful eradication programme was the people who participated in it. The work of journalists informing the public, and the support of the public, particularly those who were directly affected by the mosquito and eradication programme, cannot be underestimated. The commitment of a large number of individuals was essential to the success of the programme. Stakeholder relationships were crucial and there was generous and unstinting support from other agencies, local government, NGOs, *Iwi* (local tribes) and Maori, manufacturers and the public health sector (MoH, public health units and private providers).

Many people, both at the national level and in the field, worked unpaid hours late into the night and at weekends to deliver the eradication programme. Field staff kept their enthusiasm through adverse weather events and difficult terrain to undertake physically demanding work to a consistently high standard. Field staff maintained their commitment even when the numbers of *Campto* declined to nothing, and the success of eradication in a zone meant their employment was also drawing to an end.

The MoH had a crucial role in leading a response that primarily impacted on human health and directly affected people, using proven systems and processes such as communications, risk management and experience in dealing with communities. There was an ability to deploy public health unit staff to identify and check habitat in their regions, as well as to second health protection officers to support incursion responses in other areas. Roles and responsibilities were clear at national, regional and local levels, including roles and responsibilities across agencies.

Having good leadership

The eradication programme would not have been possible without strong and decisive leadership from government and within the government agencies. Leadership was evident at all levels – technical, political, managerial and in all aspects of service delivery. Without such leadership, other success factors would have been largely absent. However, the courage of Ministers and senior MoH officials to agree to fund what was to be a world-first achievement cannot be underestimated.

Gaining and retaining the confidence of government was crucial. The eradication programme was long-term and highly expensive. Ensuring Ministers were well briefed with progress reports, but also with urgent advice on any new incursions or emergent issues, meant they did not receive any surprises. Advice was clear and evidence-based, with options – and their potential consequences – described so Ministers could make well-informed decisions. The government was given confidence in the ability of the MoH and the MAF to develop and deliver the eradication because of some key features of the programme.

Using established systems and processes

There had been significant investment in training public health officers in mosquito surveillance and response. MoH officials had developed good networks and relationships with other agencies within NZ, and with international experts (primarily in Australia but also more widely).

The MoH and MAF also had existing generic systems for emergency response, communications, and risk management that could be adapted to a specific response. There were good legislative frameworks in place, including relatively recently promulgated statutes that covered biosecurity responses. The MoH had developed guidance, templates and standard operating procedures for using the legislative tools (e.g. declaring unwanted organisms, use of emergency provisions), and had statutory officers appointed. Officials had strong relationships with other agencies, including local government, to get their support for legislative requirements relating to their portfolios.

Accessing expert advice

The eradication programme used the best available scientific, technical and other advice available. Areas of uncertainty were identified throughout the programme and consideration given to whether scientific research could provide further information to underpin the programme within the forecast timeframes of the programme. Scientific research was balanced with operational and field experiences. Overseas experiences were considered and adapted for the NZ experience. The impacts of the eradication programme were considered, as well as the impacts of the pest. Expert advice was obtained on mosquito biology and ecology, treatment agent impacts and efficacy, surveillance sensitivity and specificity.

Having international help

The willing assistance provided by Australians mosquito experts was used to develop and implement the eradication programme. Confirmation of the initial identification came from Richard Russell in Sydney, and the core group of specialist technical advisors established immediately after by the MoH came from Australia, including Brian Kay and Darryl

McGinn who, with assistance from David Sullivan from the US, designed the programme and played a major part in getting the show up, plugging gaps and running it. Brian Kay and Darryl McGinn continued to provide advice via the MoH's TAG. Richard Russell and Scott Ritchie, who had reviewed the NZ national surveillance programmes in NZ in 2002 (Ritchie and Russell 2002), and Craig Williams, came from Australia to participate MAF's TAG. Overall, Australian scientists participated in, or carried out, scientific research to help the programme and provided *ad hoc* advice and information throughout the programme.

This willing advice and support came about not just because of professional (or commercial) interest. It was built on many positive personal relationships established and maintained over decades between entomologists, health officials, scientists and colleagues in both countries.

The partnership worked brilliantly. When Australian mosquito experts realised NZ was grappling with an Australian mosquito, offers of help and advice – much of it offered freely – flooded in to the MoH. When the Australian experts realised that the Kiwis were determined that, come hell or **high water** (quite appropriate given the life cycle of *Campto*), the 'Aussie mossie' was not going to be allowed to stay on NZ's shores (meaning eradication rather than the usual Australian control and containment programmes), their enthusiasm and interest were even more pronounced.

Developing rigorous analyses

Ensuring that rigorous health impact assessments and cost–benefit analyses were developed from the beginning meant that the basis for decision making was underpinned by robust and consistent advice. The development of the underlying assumptions, using international experts and best practice, ensured these tools were able to be adapted, updated and applied throughout the programme. The assumptions within these tools included, for example, how to deal with intangible costs and benefits, and the epidemiology of potential epidemics.

Having legislation that works

There were good legislative frameworks in place, including relatively recently promulgated statutes that covered biosecurity responses. The *Biosecurity Act 1993* proved to be a very effective statute under which the eradication programmes were run. Powers afforded by the legislation were substantial and commensurate with the risks being managed. Although a great deal of operational freedom was enabled by the Act, there were also appropriate checks and balances that helped to guard against unwise decision making or actions. MAF had developed policies, guidance, templates and standard operating procedures for using the legislative tools (e.g. declaring unwanted organisms, use of emergency provisions) and made these freely available to the MoH. Officials had strong relationships with other agencies, including local government, to get their support for legislative requirements relating to their portfolios.

Planning effectively

The eradication plans were both strategic and operational. The strategic eradication plan described a consistent framework for all eradication activities. The strategic plan included reassessment rules and criteria for completing the treatment phase of an eradication programme, reducing the frequency or intensity of surveillance after treatments have been completed, and declaring eradication. The strategic plan was then implemented at a local level using detailed operational plans, to give effect to the strategic plan but recognising local conditions and challenges. There was flexibility built into the plans, such as using *Bti* to treat areas used for organic farming and other operations, and the ability to scale up from local to regional to national responses. The eradication model was comprehensive and included treatment of habitat, surveillance (delimiting and eradication), habitat modification and mitigation, national surveillance, and monitoring non-target effects.

The establishment of a nationally delivered surveillance system, centrally coordinated and consistent with established international best practice, fundamentally provided for the success of *Campto* eradication in NZ. It provided a 'centre of excellence', through the establishment of NZ BioSecure, whereby there was a consistent approach to surveillance delivered at high quality with attention to detail engendered from thorough training from recognised experts. It provided an additional expert and experienced capacity that could be deployed to respond to suspected or confirmed incursions. Many of those field staff who began their training when the first incursion was reported in 1998 remain active today with the National Surveillance Programme, almost 15 years later. The success of now having dedicated expertise in NZ was no better demonstrated than when technical staff from the National Surveillance Programme discovered *Campto* at Coromandel in 2006.

Managing risk

The MoH had well-established systems to manage risks that arose during the programme, with strong in-house support from financial, legal, audit and communications experts. Problems and emergent issues, whether emergency funding, lost specimens or new incursions, were able to be dealt with rapidly, comprehensively and appropriately.

In the course of setting up and implementing the initial Napier eradication response, it was clear that the NZ Government had faith in the team they had deployed to do the job. Empowering people on the front line to make difficult operational decisions and to consequently expend significant resources, while accepting that many decisions would be made with imperfect information, was a sign of true leadership and an understanding of what was at stake if the field operations were impeded by bureaucratic 'road blocks'.

Communicating effectively

When the requirement to mount a response to a public health event was identified, the need for, and the content of, a communication strategy was one of the earliest considerations. The strategy was reviewed and revised as the process continued. There was a continuous effort to better understand people's concerns about the risks and how to communicate in a way that was most effective.

The idea of exterminating incredibly annoying and potentially dangerous mosquitoes was not at all a hard thing to sell to the public. There were also almost no adverse economic impacts arising from eradication activities. However, the public, politicians and several government departments certainly still needed to be convinced that the eradication actions were appropriate and safe and that the high operational costs were justified.

On the public front, having communication specialists imbedded in the eradication programme from the outset proved to be invaluable. On many occasions they were able to anticipate public concerns and were crucial in determining how best to prepare for, and respond to, such concerns. The MoH had a culture of informed consent – that is, giving a priority to getting **all** the information out to the public (an example was including information about the lost specimens in the media statement released that announced *Campto* had been found in the Gisborne region). The MoH 'informed patient' model (i.e. assuming the public were intelligent and would make their own minds up) needed full and frank information (i.e. not the 1960s model of 'doctor knows best' or 'trust us, it's good for you').

There was a highly proactive communications strategy, including via schools, early childhood centres, medical centres, councils and public notices – which was ongoing, with regular updates being provided. Proactive advice was given to residents about treatments being made nearby, with very careful use of language, such as we 'treated' areas, we didn't 'spray' them. *Bti* was awarded organic status, thus could be used in sensitive areas (e.g. organic orchards, vineyards and the saltworks) and so local commercial operations were not going to be compromised. Treatments occurred in non-residential areas only and we provided full information of when and where treatments were taking place – information to let people make choices about how they managed any perceived or potential risks to themselves from the mosquitoes or exposures to treatment agents, so could stay away if they felt it necessary.

Very full information was made available, including handouts on the various treatment agents used and all scientific reports were placed on the website, with reviews of health and environmental assessments of the products to keep up to date on the latest evidence of any potential risks. Also, there were careful and full responses to any *Official Information Act* requests, queries or concerns from the public, media, environmental groups, and so on. There was good delineation of roles and responsibilities between national and local programme delivery, and information was available locally (i.e. the programme was run locally and local people were on hand to talk to any concerned residents).

What was an outstanding success in all the local eradication programmes was the attention paid by planning and operational staff to communication and consultation with the stakeholders. Almost without exception, these operated without adverse commentary or publicity. Involving as many stakeholders and those considered affected persons at the earliest opportunity, paid dividends. They all became partners in one form or another, assisting the delivery of the factual information rather than establishing controversy and argument.

Being able to draw on strong, positive personal relationships with key media leaders was undoubtedly also a major success factor in terms of 'selling' the eradication idea. Ensuring those media leaders understood what was really at stake if exotic mosquitoes were allowed to establish permanently in NZ resulted in a great deal of positive coverage. Media statements were supported with Q&A and other background information for journalists to use if they wished to add more detail.

Overall, the challenge was how to deliver a communications strategy at both the strategic and operational level. At the strategic level, this was done so that the NZ public could become well informed about:

- the target species
- the potential health risks associated with the target species
- the high nuisance values associated with the mosquito
- the safety of the eradication products that were to be used.

At the operational level, it was all about:

- winning the confidence and cooperation of local land occupiers so that broad-scale delivery of the eradication product could be conducted with the willing compliance of the local community.

The golden rule for good communications

Say what you know, say what you **don't** know, and say what you are doing about it.

Maintaining effective field operations

Typical mosquito larval control programs are reactive: adults are detected, likely larval habitat is sampled and, if larvae are found, the habitat is treated. Instead, the concentration was on detecting larval populations and, once larvae were found in a site, the whole site was routinely repeatedly treated with residual S-methoprene granules for a year or more. Thus, the treatment strategy assumed that there were going to be larvae in there and thus they would have to be treated. This allowed the program to eliminate very small populations that would have escaped a sampling program.

However, despite confidence that had developed after Kaipara Harbour had been conquered, it is salient to remember that ProLink XR-G had just been introduced into the marketplace and there was little experience of its performance and decay under various environmental factors (see Chapter 6). Similarly, one needs to reflect on the inadequacies of the original R22 helicopter (see Chapter 5) and inexperience of Helicopters Hawkes Bay in treating mosquito larvae with what must have seemed to be ridiculously small amounts of product. With considerable ingenuity from both pilot and public health officers, helicopters and application equipment were upgraded to morph into a very sophisticated outfit. Thus, the job was done, and the chronology and duration of the various local eradications can be seen in Table 13.1.

Table 13.1. The chronology and duration of the *Campto* eradication programmes in NZ, 1999–2010

	S-methoprene treatment started	Last adult detected	Last larvae detected	Treatment ceased	Eradication completed
Napier	Aug-99	Apr-00	Aug-00	Apr-01	Jul-02
Mahia	Nov-00	Apr-01	Aug-01	Aug-02	Aug-03
Tairawhiti	Nov-00	Sep-02	Sep-02	Jun-03	Sep-04
Porangahau	Nov-00	Jun-02	Aug-02	Apr-03	Sep-04
Mangawhai	Oct-02	Dec-02	Nil	Apr-03	Dec-04
Whitford	Oct-02	Apr-02	Nov-02	Mar-04	Nov-04
Kaipara (north)	Oct-02	Sep-03	Feb-04	Jun-04	Jun-06
Kaipara (south)	Oct-02	Apr-06	Jul-06	Apr-08	Aug-08
Whangaparaoa	Jan-04	Nil	May-05	Apr-06	Apr-07
Wairau/ Grassmere	Feb-05	Oct-06	Jun-08	Jun-09	Jun-10
Coromandel	Sep-06	Nov-06	Apr-07	Dec-08	May-09

Maintaining confidence

It was a tough decision in 1999 to attempt eradication of *Campto*. It was also a tough decision, perhaps even tougher, to continue with the programme following the re-evaluation in 2007. Long campaigns, characterised by periodic setbacks, undermine confidence – among the public, delivery agencies and their workers, programme owners and funders. Those involved in the eradication programme had to justify it on an ongoing basis and set up the systems to ensure operational and strategic intelligence was continually providing information on the likelihood of successful eradication. The development and rigorous implementation of reassessment rules was pivotal to retaining the confidence of government.

Satisfying the funders

Everybody who invests wants to know the likely return. Every government or Minister that is asked to provide funding for an incursion response will want to know the likelihood of successfully eradicating the organism at the heart of the programme. It's a natural question to ask and a very difficult one for scientists to answer. In 1999, Brian Kay gulped when asked this question by the NZ Government. 'Ninety per cent' he said, knowing damn well that answers such as 50:50 would not encourage anyone to spend money.

Aside from the usual caveats, any scientist will make when faced with such a question, there will be variability (different people providing different answers) and uncertainty (the same person might couch their answer as a range) in the answers received. Who should one believe in such circumstances? How can a decision maker that does not have the scientific expertise to engage in the issue trust the advice they are receiving. However, there are processes that can be used to tackle this challenge in a robust way: bringing together a range of experts, providing them with the best intelligence available so that they were making informed judgements and breaking down the problem into component chunks so their advice on various matters were isolated and focussed.

Using intelligence to drive responses

Intelligence is essential to running a good response (Stone 2008). The *Campto* response usefully demonstrated the difference between operational and strategic intelligence. In the *Campto* response context, however, operational intelligence was about designing risk-based surveillance and using the information our surveillance systems give us to direct our control operations. It was about using all the information available on control operations to ensure they were planned and delivered most effectively and efficiently. It was also about thinking through spread mechanisms, designing movement controls that considered the major risks and looping back to ensure spread risk is a key input to risk-based surveillance. In this sense, operational intelligence answers how to optimise the likelihood of success.

The science that supports operational intelligence typically comes from both the readily available scientific literature and from the applied settings of response operations. In the eradication programme, there were deliberate connections between surveillance, control operations and movement control to ensure the historical data from the programme were analysed to inform future operations.

However, control programmes lasting longer than 1, 3 or 5 years, and with big budgets to match, needed a new approach to intelligence gathering. Science takes time and costs money, but every biosecurity incursion response starts with a great deal of uncertainty.

There are uncertainties about the organism, in particular its biology and how that interacts with our surveillance, spread and control mechanisms, and how the organism will behave in a new setting.

The change in responsibility for the programme from the MoH to the MAF in 2007 provided another opportunity to step back and assess the intelligence systems operating in the response, and include additional international and national experts in the assessments. The investment in a science programme, both to make best use of the operational data from the history of the response to direct future operations and to procure science to answer key questions that were driving uncertainty, was acknowledged by the TAG that operated after 2007 as being critical to their confidence in the programme.

Overall, the course of the eradication programme has reinforced the importance of understanding the many aspects of pest biology, environmental factors, pest and host distribution, and spread mechanisms that govern the establishment and dispersal of exotic species. It has been the combination of this information that has provided a picture of what happened in the past and possible future scenarios for which there can be preparation and planning. This information relies heavily on the best available science, and tools such as population genetics and spread models help to reduce uncertainty for decision makers.

Providing an ongoing capability

An important and valuable outcome of the program has been the building of a substantial national capacity for mosquito surveillance and management (coming from a very small and limited base). The team from the NZ BioSecure Entomology Laboratory sought and obtained their professional training in Australia, USA and in NZ because of their involvement in the *Campto* eradication programme and border health control initiatives. There is now a dedicated team of entomologists with more than a decade of work with NZ indigenous mosquitoes and exotics entering NZ, and a specialist laboratory now able to service the resultant national surveillance programme and the MoH's border surveillance. This 'home-grown' expertise did not exist before the *Campto* invasion.

Preparing for the future

Although we don't know the pathway by which *Campto* entered NZ at Napier (if, indeed, Napier was the place of first entry), the fact that it hasn't been repeated in more than 10 years (since the Napier programme ceased), would indicate that it was a novel or rare pathway or event at that particular time. However, given that the pathways by which *Campto* both entered and became distributed within NZ remain unknown and, considering the increasing international transport of cargo and people, the country arguably remains vulnerable to further introductions of this and other significant ground-pool species (e.g. *Ae. vigilax* and *Cx. annulirostris*, and particularly from Australia). Additionally, the historical and ongoing importations of container-breeding species (e.g. *Ae. aegypti* and *Ae. albopictus*, and particularly from Melanesia and Polynesia), indicate that the environment of NZ is susceptible to the establishment of a range of species and the local transmission of disease-causing pathogens.

Internationally, the Asian 'tiger' mosquito *Ae. albopictus*, a vector for dengue viruses and chikungunya virus, has been seemingly introduced 'almost everywhere' (including North and South America, Africa, Europe and the Pacific) in recent decades, while the north Asian *Ae. japonicus* has been introduced to North America and Europe, and *Cx.*

gelidus and *Ae. vexans* have been introduced to Australia from south-east Asia during the same timeframe.

Apart from the international transport of mosquito species, there are records of mosquito-borne pathogens being transported (with infected humans or other animal reservoirs, or vector mosquitoes) and causing outbreaks of disease beyond their traditional boundaries, such as Ross River virus from Australia to a range of Pacific islands (Melanesia and Polynesia, 1979–80) and more recently again in Fiji (2003–04), West Nile virus from Europe/Middle East to USA in 1999 (with establishment), a new Chikungunya virus strain from East Africa to India to Europe (2006–07), Zika virus from Africa/Asia to Micronesia (Yap Island in 2007), Usutu virus from Africa to Austria (infecting birds) in 2001 and Italy (infecting humans) in 2009.

NZ vulnerability

Both mosquitoes (pest and/or vector species) and mosquito-borne human pathogens (particularly arboviruses) could arrive through air- and seaports receiving international passengers and/or cargo. NZ ports receive vessels from neighbouring regions with mosquitoes of public health significance and these regions also provide sources of mosquito-borne human pathogens that can be disseminated internationally with their mosquito vectors or with human carriers.

The list of mosquito species imported to (but not established in) NZ from various regions (including Asia, Australia, the Pacific and the Americas) in recent decades includes *Aedes aegypti, Ae. albopictus, Ae. alternans, Ae. cooki, Ae. japonicus, Ae. polynesiensis, Ae. sierrensis, Ae. togoi, Ae. vexans, Ae. vigilax, Ae. vittiger, Ae. (Stegomyia) ? species, Anopheles albimanus, Culex annulirostris, Cx. australicus, Cx. fuscocephala, Cx. gelidus, Cx. ocossa, Cx. pallens, Cx. sitiens, Cx. tritaeniorhynchus, Mansonia titillans, Toxorhynchites speciosus, Tripteroides bambusa, Tp. tasmaniensis, Uranotaenia novobscura* and *Verrallina funerea.*

NZ susceptibility

Whether all the mosquito species intercepted at the NZ border (or others yet to be detected) could become established in NZ and cause pest or disease issues is unlikely, but (based on the climatic characteristics of the native distribution and international spread) it is highly likely that some at least could be introduced successfully to particular regions of NZ.

Historically, NZ has been free of endemic arbovirus activity threatening humans, although Whataroa virus is established in bird populations and transmitted by local mosquitoes (see Chapter 1). However, this fortunate situation has been considered to be threatened by invasive mosquitoes, tourism and climate change, and recent research (Kramer *et al.* 2011) has indicated that some NZ local indigenous mosquito species (including *Ae. antipodeus, Cx. pervigilans* and *Opifex fuscus*) and some local introduced species (including *Ae. australis, Ae. notoscriptus* and *Cx. quinquefasciatus*), can be highly competent vectors of certain arboviruses (particularly alphaviruses) and may pose a threat to human health if one of these viruses was to be introduced at a time when the mosquito populations were abundant and in close contact with humans, and the climatic conditions were favourable for virus transmission.

There has been a concern about the impact of climate change (Howden-Chapman *et al.* 2010). In terms of vector-borne diseases, parts of the North Island may become more suitable for *Ae. aegypti* and *Ae. albopictus*, which are competent vectors of dengue viruses, but the risk of dengue in NZ may remain below the temperature threshold for local transmission, even beyond 2050 according to some scenarios (Hales *et al.* 2002). Although some mosquito vectors for Ross River virus are already established in NZ, others such as *Ae. vigilax* and *Cx. annulirostris*, which could be introduced from Australia, may find more favourable conditions to establish in NZ with a warmer climate. From experience in Australia, where warm and wet conditions are associated with outbreaks of Ross River virus, there may be increased potential for outbreaks of Ross River virus in a warmer NZ.

NZ's capacity to respond

As mentioned above, an important and valuable outcome of the program has been the building of a substantial national capacity for mosquito surveillance and management. There is now widespread experience in broad-scale mosquito surveillance and management, with a specialist laboratory to service the resultant National Surveillance Programme managed by the MAF and the sea- and airport surveillance, which remains the responsibility of the MoH. While it might be that one could say, with respect to another 'exotic invasion', that it still is not **'if'** but **'when'**, the next time will see NZ well prepared to respond and deal with the issues.

Conclusion

The successful elimination of *Campto* from NZ is notable in several ways. Saltmarsh mosquitoes are especially prolific, with huge populations migrating and infesting coastal communities in many countries. It is truly astounding to think that a species that breeds in the millions within saltmarsh pools and mangrove pools could actually be eradicated! This is the only successful elimination of a saltmarsh or flood water *Aedes* mosquito anywhere in the world.

Acknowledgements

Apart from the co-authors, the following provided thoughts or words for this chapter: Mark Bullians, Mark Disbury, Ian Gear, Bryn Gradwell, Brian Kay, Scott Ritchie and David Yard.

References

Hales S, de Wet NJM, Woodward A (2002) Potential effect of population and climate changes on global distribution of dengue fever: an empirical model. *Lancet* **360**, 830–834. doi:10.1016/S0140-6736(02)09964-6

Howden-Chapman P, Chapman R, Hales S, Britton E, Wilson N (2010) Climate change and human health: impact and adaptation issues for New Zealand. In *Climate Change Adaptation in New Zealand: Future Scenarios and some Sectoral Perspectives.* (Eds RAC Nottage, DS Wratt, JF Bornman and K Jones) pp. 112 – 121. New Zealand Climate Change Centre, Wellington, NZ.

Kay BH (1997) 'Review of New Zealand Programme for exclusion and surveillance of exotic mosquitoes of public health significance'. Unpublished report to the Ministry of Health, Wellington, NZ.

Kramer LD, Chin P, Cane RP, Kauffman EB, Mackereth G (2011) Vector competence of New Zealand mosquitoes for selected arboviruses. *The American Journal of Tropical Medicine and Hygiene* **85**, 182–189. doi:10.4269/ajtmh.2011.11-0078

Ritchie SA, Russell RC (2002) 'A Review of the New Zealand Mosquito Surveillance Programme'. Report for the Ministry of Health, Wellington, NZ.

Stone M (2008) Decision support analyses during the evaluation of the Southern Saltmarsh Mosquito (*Aedes camptorhynchus*) eradication programme. In *Proceedings of the Combined Meeting of the Food Safety, Animal Welfare and Biosecurity and Epidemiology and Animal Health Management Branches of the NZ Veterinary* Association. 25–28 June 2008, Wellington NZ. pp. 65–78. Publication No.273 VetLearn. Massey University, Palmerston North, NZ.

www.ingramcontent.com/pod-product-compliance
Lightning Source LLC
LaVergne TN
LVHW060629110826
845147LV00014B/880

* 9 7 8 1 4 8 6 3 0 0 5 7 0 *